Constru Practice

Also of interest

Strategic Management in Construction
Second Edition
David Langford and Steven Male

Construction Reports 1944–98
Edited by David Langford and Mike Murray

Best Value in Construction
Edited by John Kelly, Roy Morledge & Sara Wilkinson
Published in conjunction with the RICS Research Foundation

Benchmarking in Construction
Steven McCabe

Value Management of Construction Projects
John Kelly, Steven Male & Drummond Graham

Research Methods in Construction
Richard Fellows and Anita Liu

Construction Management in Practice

Second edition

Richard Fellows
David Langford
Robert Newcombe
Sydney Urry

Blackwell
Science

Editorial Offices:
Osney Mead, Oxford OX2 0EL
25 John Street, London WC1N 2BS
23 Ainslie Place, Edinburgh EH3 6AJ
350 Main Street, Malden
MA 02148 5018, USA
54 University Street, Carlton
Victoria 3053, Australia
10, rue Casimir Delavigne
75006 Paris, France

Other Editorial Offices:

Blackwell Wissenschafts-Verlag GmbH
Kurfürstendamm 57
10707 Berlin, Germany

Blackwell Science KK
MG Kodenmacho Building
7–10 Kodenmacho Nihombashi
Chuo-ku, Tokyo 104, Japan

Iowa State University Press
A Blackwell Science Company
2121 S. State Avenue
Ames, Iowa 50014-8300, USA

First Edition published 1983 by Construction Press
Reprinted by Longman Scientific & Technical 1988, 1990, 1991, 1993
Reprinted by Addison Wesley Longman Limited 1998
Second Edition published 2002 by Blackwell Science Ltd

Set in 10/12½ pt Palatino
by DP Photosetting, Aylesbury, Bucks
Printed and bound in Great Britain by
MPG Books, Bodmin, Cornwall

DISTRIBUTORS

Marston Book Services Ltd
PO Box 269
Abingdon
Oxon OX14 4YN
(*Orders:* Tel: 01235 465500
Fax: 01235 465555)

USA
Blackwell Science, Inc.
Commerce Place
350 Main Street
Malden, MA 02148 5018
(*Orders:* Tel: 800 759 6102
781 388 8250
Fax: 781 388 8255)

Canada
Login Brothers Book Company
324 Saulteaux Crescent
Winnipeg, Manitoba R3J 3T2
(*Orders:* Tel: 204 837-2987
Fax: 204 837-3116)

Australia
Blackwell Science Pty Ltd
54 University Street
Carlton, Victoria 3053
(*Orders:* Tel: 03 9347 0300
Fax: 03 9347 5001)

A catalogue record for this title is available from the British Library

ISBN 0-632-06402-1

Library of Congress
Cataloging-in-Publication Data is available

Contents

Preface

The first edition of *Construction Management in Practice* appeared in 1982. Since then there have been significant changes in the construction industry and its practices (outlined in chapter 1) and the new edition reflects these changes.

Chapters 2 and 3 on strategic management have been completely rewritten to reflect recent developments in strategic thinking and the way in which construction firms are now thinking more strategically.

Trades unions have always been undermined in the construction industry by the self-employment system, which has increased in the last 20 years; this was exacerbated by the introduction of legislation to limit the power of unions during the Thatcher era. The chapter on industrial relations, whilst still dealing with the mechanics of collective bargaining, focuses more on management roles and empowerment of the work force.

As is pointed out in chapter 1, accident statistics for the construction industry have improved but are still far too high. Chapter 5 revisits the problem, again emphasising the current magnitude of the problem and reviewing the more stringent legislation which has been put in place since 1980. In particular, it emphasises the changing attitudes to health and safety and the need for the construction industry to adopt safer working practices.

Two chapters in the first edition on manpower planning and personnel management have been consolidated into a single chapter on managing people. The human resource management (HRM) function is still undervalued in most construction firms and the need to audit current and future needs for staff is often avoided by citing the need in the industry for firms to retain flexibility. The result is the well-known skill shortages, which the industry perpetually suffers. Until HRM issues are included in the strategic thinking and planning of construction firms, this problem will continue. This leads to the second aspect of HRM dealt with in this chapter – the need to attract and offer career development opportunities to young people. Staff and operative levels are discussed, as is the recent use of learning networks to create learning organisations in the construction industry.

Chapters 7, 8 and 9 have been revised to reflect changes in financial management and accounting practices since 1982.

Chapter 10 has been rewritten to place greater emphasis on the *use* of quantitative methods rather than the techniques themselves. The quantitative analysis of construction operations, which is the subject of chapter 11, has changed little and simply needed updating.

The purpose of the second edition remains the same – to alert students and young practitioners to the practical implications of managing in a dynamic construction industry environment. The standard is that of university degree level courses and the examinations of the professions allied to the industry.

We have aimed to present these concepts as succinctly as possible and demonstrate relevance to the day-to-day problems of the industry. To this end we have included many illustrative examples in the text and review questions at the end of each chapter.

Despite careful checking, some errors may remain and any criticism or correction will be gratefully acknowledged.

RFF Hong Kong
DAL Glasgow
RN Reading

Authors' note

When the first edition of this book was written all the writers worked at Brunel University in West London. Since its publication the personal circumstances of all the authors has changed. The most painful to record is the death of Syd Urry in June 1999. This book is dedicated to him. His life and works were a great inspiration to all of the remaining authors.

At the time of his death Syd had completed his revisions to chapter 10 and, most poignantly, had written the first paragraph of the revisions to chapter 11. His work was completed by the other authors and we hope that we have done justice to Syd's memory by our attempt at the revisions that Syd had planned.

The other authors have experienced changes since the first edition. They moved *en masse* from Brunel to the University of Bath in 1987 where they worked for three years. Bob Newcombe was then appointed Director of Executive Development in the Department of Construction Management and Engineering at the University of Reading in 1990. Richard Fellows moved to take up residence in Hong Kong and Dave Langford holds a chair in the Department of Civil Engineering at Strathclyde University.

Despite these changes in location the authors meet regularly and are actively engaged in sharing ideas about the business of construction.

Acknowledgements

This book could not have been prepared without the help of many people and the authors owe much to the work of others whose names appear in the references or bibliography.

We are particularly grateful to the team at Blackwell Science, especially Julia Burden, who stepped in at a point of crisis.

Finally, our thanks to Audrey Urry who, during a period of great sadness, conscientiously searched Syd's files and sent the manuscripts to the authors.

Abbreviations

ACAS	Advisory, Conciliation and Arbitration Services
ACT	Advance corporation text
AE	Annual equivalent
AEEU	Amalgamated Engineering and Electrical Union
AEU	Amalgamated Engineering Union
AGM	Annual general meeting
AIM	Alternative Investment Market
APA	Amount per annum
APTC	Administrative, professional, technical and clerical
ASF	Annual sinking fund
ASP&D	Amalgamated Society of Painters and Decorators
BAS	Building Advisory Service
BCIS	Building Cost Information Service (of the RICS)
BDA	Brick Development Association
BEC	Building Employer Confederation
BOO	Build-Own-Operate
BOOT	Build-Own-Operate-Transfer
BOT	Build-Operate-Transfer
BPR	Business process engineering
BQ	Bill of quantities
BRE	Building Research Establishment
BWF	British Woodworking Federation
CBI	Confederation of British Industry
CCA	Cement and Concrete Association, Current cost accounting
CCT	Compulsory competitive tendering
CEC	Construction Employers' Confederation
CECA	Civil Engineering Contractors' Association
CECCB	Civil Engineering Construction Conciliation Board
CIBSE	Chartered Institute of Building Service Engineers
CIC	Chartered Institute of Construction
CIJC	Construction Industry Joint Council
CIOB	Chartered Institute of Building
CIRIA	Construction Industry Research and Information Association
CITB	Construction Industry Training Board

CMBA	Central Master Builders Association
CMI	Construction Market Intelligence
CPD	Continuing professional development
CPM	Critical path method
CRE	Commission for Racial Equality
CSCS	Construction skills certification scheme
DCF	Discounted cash flow
DETR	Department of the Environment, Transport and the Regions
DFBO	Design-Build-Finance-Operate
DLO	Direct labour organizations
DoE	Department of Employment
DTI	Department of Trade and Industry
EC	European Community
ECITB	Engineering Construction Industry Training Board
EEF	Engineering Employers' Federation
EGCI	Export Group for the Constructional Industries
EIU	Economist Intelligence Unit
EOQ	Economic order quantity
EP	English Partnerships
EPSRC	Engineering and Physical Sciences Research Council
ERI	Effective rate of interest
FBSC	Federation of Building Specialist Contractors
FE	Further education
FRS(s)	Financial Reporting Standard(s)
GAAP	Generally accepted accounting practice
GDP	Gross domestic product
GDV	Gross development value
GMB	General, Municipal and Boilermakers' Union
HBF	House Builders' Federation
HCA	Historic cost accounting
HRM	Human resource management
HSC	Health and Safety Commission
HSE	Health and Safety Executive
ICE	Institution of Civil Engineers
ICFC	Industrial and Commercial Finance Corporation
IRR	Internal rate of return
ISE	International Stock Exchange
JCT	Joint Contracts Tribunal

LDDC	London Docklands Development Corporation
LOSC	Labour-only subcontracting
MAC	Mastic Asphalt Council
MAUT	Multi-attribute utility theory
MCG	Major contractors' group
ME	Monetary expectation
NAMB	National Association of Master Builders
NCF	National Contractors' Federation
NEB	National Enterprise Board
NEDO	National Economic Development Office
NFB	National Federation of Builders
NFBTE	National Federation of Building Trades Employers
NFSC	National Federation of Specialist Contractors
NJCBI	National Joint Council for the Building Industry
NPV	Net present value
NTO	National Training Organization
NVQ	National Vocational Qualification
ODPs	Office development permits
OECD	Organization for Economic Cooperation and Development
PCA	Plant Contractors' Association
PE	Price-earnings ratio
PERT	Programme evaluation and review technique
PFI	Private finance initiative
PPC	Probable profit contribution
PQS	Private quantity surveyor
PV	Present value
PW	Present worth
RAMP	Risk analysis and management of projects
RI	Rate of interest
RIBA	Royal Institute of British Architects
RICS	Royal Institution of Chartered Surveyors
RSA	Regional selective assistance
RT	Rate of tax
SBAC	Scottish Building Apprenticeship Council
SBEF	Scottish Building Employers' Federation
SF	Sinking fund
SME	Small–medium enterprise
SMM	Standard Method of Measurement of Building Works, Seventh Edition

SSAP	Statement of standard accounting practice
SVQ	Scottish Vocational Qualifications
SWOT	Strengths, weaknesses, opportunities and threats
TEC	Training and Enterprise Council
TGWU	Transport and General Workers' Union
TML	Trans Manche Link
TMOs	Temporary Multi-organizations
TQM	Total quality management
TRADA	Timber Research and Development Association
TSA	Training Services Agency
TUC	Trades Union Congress
UCATT	Union of Construction Allied Trades and Technicians
UDC	Urban Development Corporation
UMA	Union membership agreement
USM	Unlisted Securities Market
VDU	Visual display unit
VFM	Value for money
VM	Value management
WRA	Working Rule Agreement
YP	Years purchase

1 Construction Management in Practice

The construction industry offers an exciting, dynamic environment in which to work. The stream of unique projects, the on-site working environment, and the ingenuity and innovation required to solve one-off problems demand a creative approach to work which is seldom found elsewhere. To which can be added the satisfaction of making tangible changes to the micro-environment which the building or structure occupies. Buildings are monuments to the people who built them, for good or ill.

Since the first edition of *Construction Management in Practice* appeared in 1982, there have been significant changes in the construction industry and its practices. Markets are now more global in character. There has been a significant decline in public sector construction activity and an increase in the importance of the private client - especially the private corporate client. Technological change, driven by the rise in the use of information technology, has increased site and off-site productivity and enabled pre-project modelling of designs within the computer. Procurement patterns have moved away from traditional design-tender-build methods towards design and build and construction management and partnering methods. The rise of the private finance initiative (PFI) and private public partnerships with build-operate-transfer contracts has forced more emphasis on life-cycle costing of buildings and structures. The Department of Environment, Transport and the Regions (DETR) (Egan) Report (1998) introduced the concept of key performance indicators for the construction industry and triggered the recent *Respect for People (2000)* initiative. Managing the supply chain has become a mantra for the industry and lean construction is currently a buzzword.

Some things have not changed. Safety on sites is still a major problem with far too many operatives killed or maimed (see Chapter 5). Fragmentation and confrontation are still endemic in the industry's culture and are slow to change in spite of new methods of contract creating co-operative opportunities. Too many high-profile projects still come in late and vastly over budget. There is still an image problem for the construction industry which has led to declining entrance levels at craft and degree levels in colleges and universities. The gender balance is still predominately male with only 12 per cent of women employed in the industry.

These changes provide the impetus for the second edition of *Construction Management in Practice* and also the context.

Construction management in this book means management in the context of the construction industry, *not* the procurement method. Although the characteristics of the construction industry and the nature of managing construction are common to most countries, this book focuses on the context of the UK construction industry.

As stated earlier, whilst much has changed in the UK construction industry in the last two decades, much has stayed the same. This is pointed out in an important report entitled *A bridge to the future*, which, in 1998, reviewed the past, present and likely future of the UK construction industry. Whilst there are clear trends to modernise the industry, particularly in the wake of the Latham (1994) and Egan (1998) reports, the fundamental characteristics of the industry of the early eighties, identified in the first edition, remain but with subtle changes. Other characteristics have also emerged in the intervening period. In the first chapter therefore, the following topics will be discussed:

1.1 The characteristics of the construction industry
1.2 Construction management in practice
1.3 Strategic management
1.4 Managing people
1.5 Managing money
1.6 Decision-making techniques

1.1 Characteristics of the construction industry

The size of the construction industry in the UK is impressive, both in terms of output and employment. Construction output in 1999 was £6.5 billion (7 per cent of GDP), of which 54 per cent was new work and 46 per cent repair and maintenance. The industry employs around 1.5 million people.

Any study of industry statistics reveals a large industry comprising mainly small firms. Large firms of 1200+ employees undertake a mere 13 per cent of new work and repairs and maintenance. Small firms (1–7 employees) and SMEs (small-medium enterprises, 8–114) undertake a further 61 per cent of work leaving medium sized firms (115–1199) with the remaining 26 per cent. Types of firm include general contractors, who undertake a range of building and civil engineering work, specialist trade contractors, specialist management consulting firms undertaking project management, and consultants – architects, quantity surveyors and engineers. In addition, there are firms offering a combined design and build service and some offering a 'one-stop shop' to clients. Peripheral services, such as materials and component supply and plant hire, remain separate but are becoming more integrated into the industry's activities as the trend towards greater prefabrication leads to increasing off-site activities. This fragmentation of the supply chain contributes to a complex industrial structure.

Whilst the alleged unique nature of the industry is now challenged by inter-industry benchmarking studies, the combination of characteristics creates a management challenge which few other industries face. Some of the more important industry characteristics which influence construction management in practice are outlined below.

1.1.1 Size of firms

As already indicated, construction is a large industry of small firms. The Construction Market Intelligence Division (CMI) of the DETR published *Construction Statistics Annual 2000 Edition* which shows that the percentage of small, medium and large firms in October 1999 was small firms (1–7) 93.2 per cent (n154, 337), medium (8–114) 6.4 per cent (n10, 640) and large (115–1200+) 0.35 per cent (n 584).

The trend towards mergers and takeovers prevalent in the 1980s has not fundamentally changed the size profile of the industry – no single firm or group of firms has a monopoly. This size profile has obvious implications for construction management practice, which will be discussed in later chapters.

1.1.2 Construction projects

The industry is a project-based industry. Firms undertake a range of discrete projects of relatively long duration, constructed outside and geographically dispersed and fixed. The majority of such projects are still custom designed to a client's requirements, and designed and built for a price established through the competitive tendering system, which still operates extensively in the industry. In the 1990s other methods of delivery – design and build, construction management, partnering, etc. – have made inroads into the domination of competitive tendering which occurred during the 70s and 80s when public sector clients commissioned a significant proportion of construction work. Individually, projects frequently constitute a significant proportion of a firm's workload with serious consequences if things go wrong. Project decisions become strategic decisions, an issue we shall discuss in the next two chapters.

1.1.3 Workforce

CMI Manpower statistics for the third quarter of 1999 show that of the 1.41 million employed in the industry, 37% were self-employed. This percentage fluctuates between 25 per cent and 40 per cent depending on the level of demand on the industry. Managerial and professional staff accounted for 16 per cent of the workforce.

Operatives are predominately young, male and self-employed, according to CMI statistics. This is a response to fluctuating workloads and employment legislation and has always frustrated the unionisation of labour, a point discussed in chapter 4. Building production managers and staff have traditionally come from a trades background but the trend is towards staff from technician and degree courses. The professions – architects, engineers and quantity surveyors – have, in the last 125 years, developed a sophisticated system of registration and training, administered by their respective institutions. A relatively recent development has been the abandonment of mandatory fee scales.

1.1.4 Ease of entry to the industry

Whilst the professional institutions (RIBA, RICS, ICE, CIBSE, etc.) have an effective form of registration and control over their members, there have in the past been few constraints on setting up a building contracting business. The Chartered Institute of Building has recently set up the Chartered Builder Scheme but this only affects registered companies. Previous voluntary schemes have largely failed to attract membership.

The system of interim payments during construction projects, coupled with extensive credit concessions for materials purchasing and highly developed plant hiring facilities, mean that small firms entering sectors of the repair and maintenance market have minimal capital requirements. The ease of entry into this market has encouraged an influx of hopeful entrepreneurs. Sadly, their demise has often been equally easy, though much more painful for their clients, creditors and staff. Television exposure of these so-called 'cowboy builders' has prompted the DETR to set up a research programme, resulting in a proposal for the registration of small builders.

For larger projects, entry is restricted to companies of sufficient status who have a track record of successfully completing these types of project.

1.1.5 Separation of design and production

The separation of design and production in the building industry, and the consequent difficulties that can arise during construction projects, has been the subject of a large number of industry reports from Emmerson (1962) to Egan (1998). The report by Flanagan *et al.* (1998) points out that 'the industry has been shaped by the needs of the public sector' and that 'design has traditionally been separated from production by class and training'. With the decline of public sector contracts there has been a trend towards using other, more integrated procurement approaches such as design and build, construction management and partnering. However, the dominant procurement option is still the traditional design-tender-build process, which perpetuates the sequential separation of design from production.

1.1.6 *The nature of demand*

The demand for construction projects is essentially what economists call 'derived' demand. It is derived from the need for buildings in which to live, to manufacture or store goods, or in which to operate various services. Building is thus strongly related to the state of health of the general economy and to the level of interest rates and business activity in particular. The fact that buildings are capital items makes them natural targets for expenditure cuts by both government and the private sector. This has led to the characteristic fluctuations in demand, which are familiar to construction firms, and to the more permanent downturn which occurred in the 1980s and early 1990s.

1.1.7 *The Government's role*

There has been a significant shift from public funding of construction projects to public/private initiatives of which PFI is a prime example.

The role of adjusting interest rates was given to the Bank of England following the 1997 Labour party General Election victory. The Bank stimulates or stifles demand by the judicious use of interest rates. At the time of writing there is concern about rising house prices and talk of raising interest rates to dampen demand and control price rises.

1.1.8 *Other characteristics*

There are a number of other characteristics of the construction industry in the UK which have emerged in the last twenty years. The rise of partnering, increasing globalisation of construction activities, joint ventures and strategic alliances, mergers and takeovers of household names, an attack on cost and waste in the industry using techniques such as value management, business process re-engineering and lean construction and the inevitable increase in the use of IT and the Internet are slowly changing the nature of the industry.

1.2 Construction management in practice

No single book can claim to cover management in an industry as large and complex as the construction industry. It is therefore necessary to be selective. This book is focused on the strategic level in construction businesses; other excellent books deal with managing projects (Walker

1998) and managing human resources in the industry (Langford *et al.* 1995).

Three key areas where strategic thinking and actions can create competitive advantage for construction firms are:

- strategic management (Chapters 2 and 3)
- managing people (Chapters 4, 5 and 6)
- managing money (Chapters 7, 8 and 9)

A fourth key aspect is the process by which decisions are made in these areas. So a fourth key area is:

- Decision-making techniques (Chapters 10 and 11)

1.3 Strategic management

Chapter 2 aims to explain the role of corporate strategic management within the construction industry. It traces the evolution of strategic management thinking through succeeding schools of thought from the 1960s to the present day. The chapter thus presents a comprehensive overview of strategic management theory and practice.

Chapter 3 is specifically concerned with strategic management systems or the ways in which strategy is developed within organisations. Three pure systems, strategic planning, strategic vision and strategic learning, are shown to be points on a continuum of strategic decision-making.

1.4 Managing people

Construction is a labour intensive industry where the phrase, 'people are our greatest asset' is often quoted. Three key aspects of managing people in this industry are dealt with in the next three chapters.

Industrial relations legislation and trades union power have always been undermined in the construction industry by the widespread use of labour only subcontracting. Chapter 4 argues that the formal industrial relations systems have had to be adapted to the unique construction environment but still have a contribution to make in protecting workers, establishing minimum pay levels and representing the industry at government level.

The crucial aspect of health and safety of construction people is addressed in Chapter 5. It is argued that safety issues do not just start once site activities commence but that designers have a duty to design for safe production. An understanding of the health and safety regulations is seen as a minimum requirement, which must be translated into safety policies, attitudes and activities by senior management.

Obtaining and developing people for construction firms is the subject of Chapter 6. The mapping and matching of a company's human resources to its strategic plans is a vital role for senior management. A formal system for inducting, developing and training people at all levels is a key part of a company's human relation's strategy. The creation of a genuine learning organisation is still an aspiration for most construction firms but a process is presented whereby it may be achieved.

1.5 Managing money

The next three chapters deal with a topic which would come top of most contractors strategy list – how to obtain and manage finance.

Chapter 7 deals with the financing of business units. As stated earlier the construction industry is a large project-based industry of small firms; this has implications for the types of business units which exist and for the sources of capital which operate. Short-term capital requirements are particularly important to contractors. The capital structure of the firm is clearly a strategic decision which senior management must determine.

Monitoring and managing finance is the subject of Chapter 8. Often survival is the motivation for the close monitoring of costs and revenues seen in the construction industry. Clients, consultants and contractors have a joint interest in ensuring that project costs are monitored and controlled and this is dealt with in some depth. A more strategic issue is investment appraisal for projects and a number of techniques is reviewed.

Chapter 9 is concerned with financial performance. In particular, decisions about bidding and risk are key strategic decisions in an industry in which the majority of projects are still let in competition for both professional and contracting firms. Financial reporting is a legal requirement for companies and the understanding and interpreting of financial reports and accounts is discussed.

1.6 Decision-making techniques

The previous chapters lead naturally into a review of decision-making techniques in Chapters 10 and 11. The emphasis is on quantitative aids to decision-making which require only elementary mathematics; numerical examples show how these techniques can be applied to problems of optimisation. In particular, they are concerned with achieving economies of time and cost under conditions of limited resources. These chapters give quantitative support to the chapters on strategic management and, particularly, to the chapters on financial management.

Such is the scope of the book. Each chapter starts with an overview of the content of the chapter and concludes with a summary of the main ideas together with relevant questions for consideration and suggestions for further reading.

References and bibliography

Construction Management Forum (1991) *Report and Guidance*, Centre for Strategic Studies in Construction, University of Reading.

Egan, J. (1998) *Rethinking Construction* DETR, HMSO.

Emmerson Report (1962) *Survey of Problems before the Construction Industries*, HMSO.

Flanagan, R. *et al* (1998) *A bridge to the future*, Thomas Telford.

Langford, D. *et al* (1995) *Human Resources Management in Construction*, Longman.

Latham, M. (1994) *Building the Team*, HMSO.

Walker, A. (1998) *Project Management in Construction*, Third Edition, Blackwell Science.

2 Introduction to Strategic Management

2.1 Concepts and definitions

Ansoff (1984), one of the foremost writers and consultants on strategic management, defines strategic management as follows:

> Strategic management is a systematic approach to a major and increasingly important responsibility of general management: to position and relate the firm to its environment in a way which will ensure its continued success and make it secure from surprises.

As we will see shortly, this definition is too limited in its premises. A 'systematic approach' may well not be used. The 'responsibility of general management' may now be shared with other employees and stakeholders in the business, and making an organization 'secure from surprises' is now seen as idealistic.

The definition by Bowman and Asch (1987) is:

> Strategic management is the process of making and implementing strategic decisions ... it is about the process of strategic change.

Again this definition is too narrow in focusing solely on the process of making and implementing strategic decisions although it does introduce the important principle that it is a process which brings about strategic change.

Even Johnson and Scholes' (1999) definition has a strong 'planning flavour' about it:

> Strategic Management includes strategic analysis, in which the strategist seeks to understand the strategic position of the organization, strategic choice which is to do with the formulation of possible courses of action, their evaluation and the choice between them, and strategy implementation, which is concerned with both planning how the strategy can be put into effect, and managing the changes required.

All these definitions assume that a logical and systematic planning approach is to be adopted, whereas as we shall see in Chapter 3, there are alternative ways of undertaking strategic management.

The definition that will be adopted for this chapter is:

> Strategic management is a system for producing strategies within an organizational infrastructure responding to an environmental context.

Strategic management is sometimes referred to as corporate strategy, business policy, strategic planning, etc., which all deal with the same general area of study.

Having defined strategic management let us turn to the other key term - *strategy*. In a provocative exposition, Henry Mintzberg (1994) points out that 'the word strategy has long been used implicitly in different ways even if it has traditionally been defined in only one'. He then presents five definitions of strategy - as plan, pattern, position, perspective and ploy.

Strategy is a *plan* when it is perceived as some sort of consciously intended course of action. By this definition, strategies have two essential characteristics: they are made in advance of the actions to which they apply and they are developed consciously and purposefully. The idea of strategy as a plan was the original approach to strategic management and still dominates the literature on the subject as we shall see in Chapter 3.

The definition of strategy as a *pattern* - specifically, a pattern in a stream of actions - implies consistency in behaviour whether or not intended. Under this definition the strategy of an organization can be read by observing the strategy that emerges from its past and present actions. Whereas planning represents a deliberate, *intended strategy*, a pattern of actions imply an *emergent strategy*.

A third definition is that strategy is a *position* - specifically a means of identifying where an organization locates itself in its 'environment', which for a business firm means its market. Strategy becomes a niche in the market or environment where the firm concentrates its efforts. For example, a residential, speculative housing developer may choose to focus within the housing market on superior individual executive houses or mass-produced, lower-priced properties.

The fourth definition, that strategy is *perspective*, implies that strategy is a perspective shared by members of an organization, through their intentions or by their actions. The term currently used to describe this approach is *culture* or the shared values of the members of the organization. There are overtones of an ideology or vision to which all members of the organization subscribe.

The final definition of strategy as a *ploy* has political implications and is seen as a specific manoeuvre intended to outwit an opponent or competitor. There is an element of bluff about this approach as the real strategy may be the threat of a course of action rather than the action itself. For example, a property developer may bid for a site it does not intend to buy in order to push up the price for a competitor to the level where its development will be uneconomic.

2.2 The evolution of strategic management

In the three decades since strategic management emerged as a distinct discipline, it has evolved through a number of schools of thought. These schools of thought in roughly chronological order are as follows:

- the strategic planning school
- the strategy - structure school
- the power - culture school
- the competitive advantage school
- the incrementalists school
- the synthesis school

We shall summarize the main approaches and ideas of each school at this point but references to original sources of these ideas will be given to enable a fuller exploration.

2.2.1 *The strategic planning school*

Igor Ansoff is often credited as the founder of this school and indeed the doyen of strategic management. His book, *Corporate Strategy*, originally published in 1965, was the first to offer a comprehensive theoretical framework for strategic planning. It was novel in introducing a number of key concepts:

1. Three classes of decisions were identified - *strategic*, *administrative* and *operational* - each with their own focus and scope.
2. A comprehensive model of the strategic planning process is presented in the book in the form of a series of flow charts. A specific sequence of steps is recommended starting with the recognition of a 'gap' between the desired future performance of the organization and the performance that will be achieved following its present strategies, the setting of objectives and targets, the controlled and conscious formulation of strategies to meet the objectives and the mobilization and control of organizational resources to implement the strategies. The comprehensiveness of Ansoff's model can be judged by the fact that his *summary* flow chart contains 57 boxes!
3. Ansoff was one of the first writers to identify generic strategies, in this case *market penetration*, i.e. gaining a larger market share, *expansion strategies* into new products or markets and *diversification* into new markets and new products simultaneously, which became known as *conglomeration* strategy.
4. The concept of *synergy* was introduced in this context for the first time. Ansoff originally defined synergy as the 2 + 2 = 5 effect or the whole is greater than the sum of the parts. Synergy can be achieved in mar-

keting, operations, distribution, management, etc. This concept is particularly important when companies are considering a merger or acquisition where the joint effects of synergy may achieve economies of scale and operation.

A number of other researchers and writers including Andrews (1987), Argenti (1980), Hussey (1975), and Steiner (1969) refined and developed Ansoff's ideas.

In summary, the planning literature presents a highly formalized procedure, decomposed into an elaborated sequence of steps supported by techniques and checklists, and executed in a mechanical, almost military, fashion.

The basic premises are that the formulation and implementation of strategy can be clearly separated, that it is the role of the chief executive to formulate strategy and control its implementation and that generic strategies will emerge fully developed from the process.

Mintzberg (1994), in a book entitled *The Rise and Fall of Strategic Planning*, identifies four fundamental fallacies on which the strategic planning approach is based:

1. The fallacy of *formalization*: the assumption that a formal system of analysis, however detailed, can produce strategies whose main characteristic is that they require synthesis rather than analysis.
2. The fallacy of *detachment*: the view that formalization of strategy can be separated from implementation, or strategy from tactics, or thinking detached from action. Mintzberg's (1973) research on the nature of managerial work revealed that the separation of planning and action is a myth.
3. The fallacy of *quantification*: the belief that strategy is driven by hard data. Mintzberg's (1973) research and other research by Peters (1987), Stewart (1967), Kotter (1982) and Kanter (1983), has revealed that managers at all levels rely significantly on soft data, offering them instant feedback, in order to do their jobs.
4. The fallacy of *predetermination*: the assumption that the future will be stable or an extrapolation of the past and is therefore predictable. This belief allows strategists the luxury of formulating and implementing long-term strategies.

Mintzberg points out that these four assumptions were built on very little research evidence; in fact, such research as there is tends to show these assumptions to be total fallacies.

The rise in popularity of strategic planning in the late 60s and early 70s was halted by its failure to predict the 1973 oil crisis. The fallacy of predetermination was clearly and cruelly exposed and the blithe assurance and credibility of the strategic planners has not been the same since.

2.2.2 The strategy-structure school

The strategy-structure school, developed in parallel with the strategic planning school, hypothesized that if business strategy is the process of formulating and implementing a strategic plan and strategic management is the dynamic element which steers and motivates business action, then the organization structure of the firm is the framework within which both strategy and strategic management occur. Business strategy and strategic management are both meaningless outside the context of a business organization; it follows, therefore, that strategy and structure must be interrelated. This link between strategy and structure, well-known by astute businessmen, was verified by an extensive research programme conducted under the auspices of Harvard University (Channon 1973) and Manchester Business School (Channon 1978). The impetus for this programme came from an earlier theory which was derived by Chandler (1962) from his study of the growth of four American corporations – DuPont, General Motors, Standard Oil Company, and Sears Roebuck. The essence of his theory was that 'structure follows strategy' or 'that a company's strategy in time determines its structure'. The positive relationship found to exist between strategy and structure was then empirically tested and refined by Chandler using data from the 70 largest US corporations in nine industries. His key finding was that if an organization's structure did not match its strategy then financial performance declined.

The study by Channon (1978) of the strategic, structural, and financial histories of the largest 100 service industry companies for the period 1950–1974 is particularly relevant as it includes seven construction and property corporations – Richard Costain, Taylor Woodrow, John Laing, George Wimpey, Trafalgar House Investments, Wood Hall Trust, and London and Northern Securities. *Strategy* was defined as the extent of diversification and international activity together with the acquisition policy adopted. *Structure* was classified in terms of the formal organization structure (i.e. functional, holding company, multi-divisional, critical function) and the corporate leadership mode (i.e. entrepreneurial, family and professional management). *Financial performance* was measured using conventional accounting ratios for growth and profitability.

The findings of the research confirmed Chandler's thesis that structure follows strategy, for as the companies diversified from offering a single product or service to offering a range of related or unrelated services or products so the pressure for organizational change proved irresistible if efficiency was to be maintained, with functional forms giving way to the more appropriate holding company and multi-divisional structures.

This strategy-structure school has been criticized for the limited definition and scope of the variables considered. Strategy is defined solely in terms of expansion and diversification, structure solely in terms of the formal organization structure and performance solely in financial terms.

Peters and Waterman (1982) attempted to expand the strategy-structure perspective with their 7-S framework. The five Ss added to strategy and structure are systems, style, staff, skills, and shared values, see Figure 2.1.

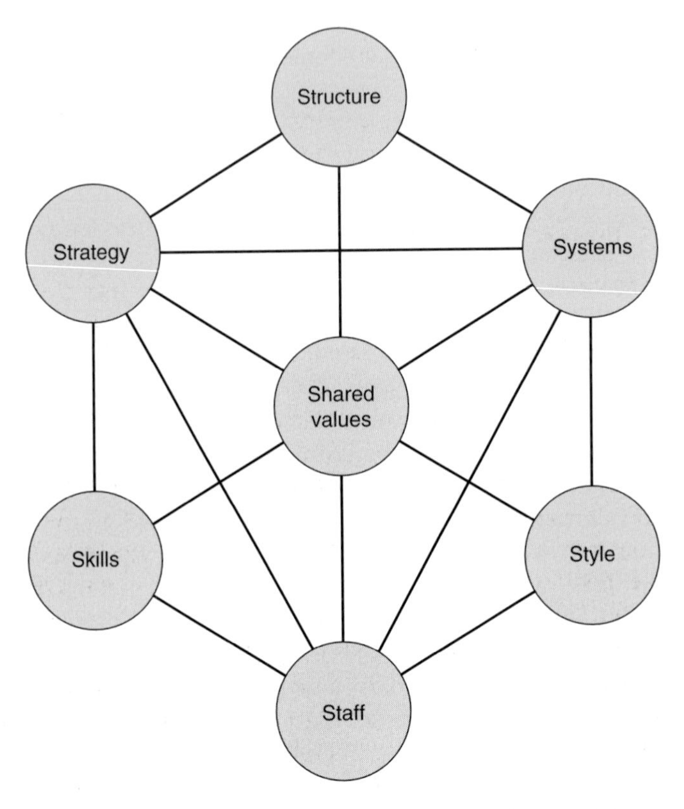

Fig. 2.1 The 7-S model. (Reproduced with permission of McKinsey and Company.)

The seven Ss are defined as:

- *Strategy*: plan or course of action leading to the allocation of a firm's scarce resources, over time, to reach identified goals.
- *Structure*: characterization of the organization chart (i.e. functional, decentralized, etc.)
- *Systems*: proceduralized reports and routinized processes such as meeting formats.
- *Staff*: 'demographic' description of important personnel categories within the firm (i.e. engineers, entrepreneurs, MBAs, etc.) 'Staff' is not meant in line-staff terms.
- *Style*: characterization of how key managers behave in achieving the organization's goals; also the cultural style of the organization.

- *Skills*: distinctive capabilities, key personnel or the firm as a whole.
- *Shared values*: the significant meanings or guiding concepts that an organization imbues in its members.

The first three components are seen as the hard assets of the organization with the next three being seen as the complementary soft assets; both are mediated through the shared values of the people involved in the organization. In Peters and Waterman's view, strategic management equals all seven Ss!

The 7-S framework provides a bridge between the strategy-structure school and the power-culture school.

2.2.3 *The power-culture school*

The view of the organization as a coalition of powerful individuals and interest groups stems from the pioneering research of Cyert and March (1963). Their view of the organization as a 'shifting multi-goal coalition' was novel at the time it was proposed but subsequent research by Mintzberg (1983), Johnson and Scholes (1999) and others (Bacharach and Lawler 1980, Bass 1990, Cohen and Bradford 1990, Greiner and Schein 1988) has confirmed that the power bases of the main actors and indeed the actors themselves 'shift' over time and that organizations operate with a system of multiple and often conflicting objectives. The way in which the potential conflicts are handled through a system of sequential attention to conflicting objectives and the use of 'side payments' to 'buy off' opposition gives a very strong political flavour to the strategy-making process. Strategy thus evolves through a bargaining process between the key actors or interest groups and dramatic changes in strategy may emerge as the power of the participants rises and wanes over time.

The use of power in the *political* sense is well defined by Mintzberg (1989):

> What do we mean by politics in organizations? An organization may be described as functioning on the basis of a number of systems of influence: *authority, ideology, expertise, politics*. The first three can be considered legitimate in some sense. Authority is based on legally sanctioned power, ideology on widely accepted beliefs, expertise on power that is officially certified. The system of politics in contrast, reflects power that is technically illegitimate (or probably more accurately, alegitimate), in the means it uses and sometimes also in the ends it promotes. In other words, political power in the organization (unlike government) is not formally authorized, widely accepted or officially certified. The result is that the political activity is usually divisive and conflictive, pitting individuals or groups against the more legitimate systems of influence and, when those systems are weak, against each other.

Does this sound familiar? It is in this political arena that strategy as *ploy* comes into its own.

Culture in organizations

Traditionally strategy has been viewed as the response of an organization to its external environment. Recent research, however, has revealed that faced with a similar environment, organizations respond differently; these differences are accounted for by the influence of the organization's internal culture on strategic decision-making.

Organizational culture is the deeper level of basic assumptions and beliefs that are shared by members of an organization which operate unconsciously and define in a basic 'taken for granted' fashion an organization's view of itself and its environment. This set of basic assumptions and beliefs can stem from a number of sources such as functional, e.g. marketing or finance; professional, e.g. architect or engineer; national, e.g. British or Japanese; industrial, such as construction or aerospace; and organizational, represented by the shared values of the members of the organization.

Two of these cultural frames of reference have a particularly strong influence on strategy:

- the industry frame of reference, which is termed an *industry recipe*
- the organizational frame of reference, which is termed the *organizational ideology* or *paradigm*

The *industry recipe* is particularly strong in the construction industry, the flavour of which is captured in the perceived view of the traditional design-tender-build method of procurement, in which design is separated from construction and the system characterized by fragmentation, friction and mistrust. Newer forms of procurement, such as construction management and partnering, are being used as vehicles to change the industry recipe, albeit slowly.

This industry recipe contaminates the internal culture of organizations involved in construction activities, from clients to contractors to professional practices.

The *organizational ideology* or *paradigm* of construction firms tends to mirror the industry recipe and is coupled with lack of security and low levels of loyalty engendered by the temporary nature of projects and employment. Again, some organizations are trying to break this mould by operating more democratic and participative organizations which encourage loyalty and commitment.

The extent to which strategy is 'top down', made by the chief executive, or 'bottom up', evolving anywhere in the organization, is to a large extent influenced by the culture of the organization.

As far as strategy is concerned, culture acts as a screen, screening out strategies which are not compatible with the organizational paradigm or frame of reference of the managers. It can also be a device for distributing strategic decision-making throughout the organization.

2.2.4 *The competitive advantage school*

Michael Porter (1980, 1985) has put the classical economic theories of market forces into a framework for analysing the nature of competitive advantage in a market and the position and power of a company in that market.

Defining the market

The first step in analysing industry structure is to define the market (or, as Porter calls it, the industry). A market is an interaction between a grouping of customers who have similar requirements for particular goods or services on the one hand, and a strategic group of competitors competing to meet these requirements on the other hand. The boundary around the market is defined in terms of similarities in what is demanded and in terms of the closeness of the competition.

For example, when analysing the construction industry, managers would be ill-advised to regard construction as one market. This is because the 'construction' market contains some very different groupings of customers and competitors. It would be more accurate to talk about a market for private houses because there we can find a reasonably clear group of customers – families who share a similar need for a home – and a relatively constant strategic group of large house builders who account for most of the houses built. It might even be necessary to break this market down further, for example, into a small-house market for first-time buyers, a market for medium-sized homes and a market for luxury homes. In addition, there is a quite separate market for large structures: roads, office blocks and hospitals, for example. Here again we find customers looking for much the same thing: organizations with the ability to manage large and complex construction programmes which take long periods of time to complete. We also find that there is a relatively constant group of about twenty or so companies in the UK that compete to fulfil this need. Then there is a market for small structures and repair work. In this market customers are not looking for a management service but for those who can do the work directly. The competitor group is a shifting population of small companies. Each of these markets is very different from the other and therefore different strategies may be appropriate in those different markets. There are no hard and fast rules about drawing the boundaries around a market, but how that boundary is drawn will have a major effect on the analysis.

Analysing the structure of the market

Having defined the market, the next step in the analysis is to identify the structure of that market. We do this by analysing the five competitive forces that shape the prices firms can charge, the costs they have to bear, and the investment they must undertake to compete. In other words, the competitive forces determine the relative market power of competitors, the kind of competition they can engage in, the factors that give some of them a competitive advantage, and the relative attractiveness of that market compared to others. These competitive forces are shown in Figure 2.2.

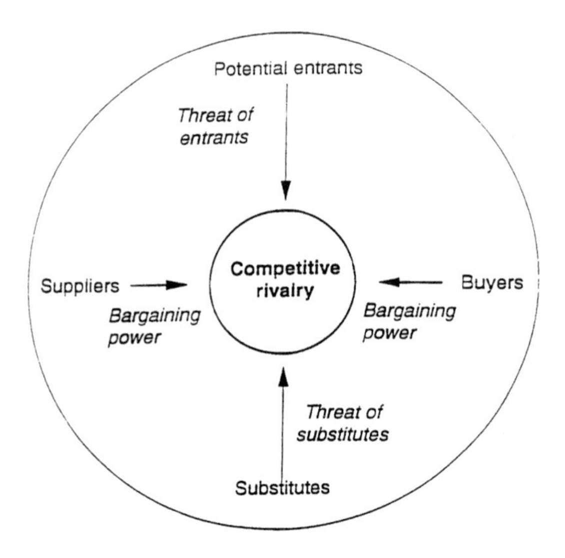

Fig. 2.2 Competitive forces. After Porter (1980).

1. *The threat of new entrants.* New entrants would obviously add production capacity and seek to fill that capacity by eating into the market share of existing firms. This would lead to lower prices and hence lower profits. Those already in the market will therefore seek to use or build barriers against the entry of others and if they succeed in this they will be able to hold prices above the minimum level necessary to keep firms in the business. Entry barriers make excess profit possible and so make the market more attractive to firms. One of the most common entry barriers is the need to invest large sums in the production facilities necessary to produce the product, for example, in the automobile industry. This is not true of the construction markets, but in some of them we still find entry barriers. For example, in the market for large structures it is rare to see new entrants because it takes time to acquire skills of managing large construction projects. On the other

hand, in the market for small domestic extensions and repairs, the skills are easier to acquire and we therefore find firms coming and going. Where the entry barriers are low, the threat of new entrants will force firms to compete on cost or try to build a market niche and focus their strategies on that niche.

2. *The threat of substitute goods for services.* Another threat that will force competitors to keep their prices low will be that of some substitute for their product or service. While in the construction industry the threat of substitute products is limited, for example, a bridge cannot be substituted for a school, the threat of substitute services has become more acute with the introduction of a wider variety of procurement approaches. Clients can select design and build or construction management as an alternative to the traditional method of procurement, therefore these represent substitute services.
3. *The bargaining power of buyers.* Customers or clients have bargaining power if there are few of them or if they are buying a product or service which will facilitate what they themselves are producing and what they buy represents a high proportion of their costs or critical elements of their quality. The importance of what they are buying will make them more demanding customers. A buyer is also powerful when purchasing a significant proportion of firm's output. The existence of a group of powerful buyers means that a firm has to keep its prices low. All these conditions pertain to the construction industry. Clients are relatively few in number and are buying a product, e.g. a building, which represents a large investment, the quality of which is critical to the client's production. Individual projects form a significant part of a construction firm's output. Where clients form a long-term relationship with particular construction firms, they are often able to exercise a control over the strategy and pricing of that firm.
4. *The bargaining power of suppliers.* Suppliers are powerful in relation to a particular firm when they are few in number, if they are supplying a product or service which is crucial in cost or quality terms to the business of the particular buyer, or if they are not supplying a major part of their own output to that particular customer. Powerful suppliers will mean that a firm's costs will be higher than they otherwise would have been, making profits lower and the market less attractive. The bargaining power of suppliers and subcontractors or specialists trades contractors in the construction industry is typically low because there are many alternative firms prepared to provide the products or services. The exception is where a supplier or subcontractor is nominated by the client to provide a particular product or service or where the unique nature of the product or service makes substitution impossible.
5. *The rivalry among existing competitors.* If rivalry is intense, marketing costs are higher or prices lower than they would otherwise have been.

Once again the market is less attractive. Traditionally the construction industry has been recognized as an industry where competition is fierce because of competitive tendering.

Once the competitive forces have been identified, a judgement has to be made on which are the relatively most important because they could be working in different directions. Thus, a firm may have weak buyers but powerful suppliers, or there may be little in the way of entry threats but rivalry among those already in the market may be intense. Analysing the forces and making judgements on their relative importance should lead managers to see how they can deal with and influence the forces.

Identifying competitive advantage

The analysis indicates the source of competitive advantage. For example, the private housing market in the South-East of England has weak buyers because there are many families purchasing homes and there is very little threat from substitutes – people mostly live in houses rather than tents or caravans. It is the suppliers of land to build the houses who are in a powerful position because land is so scarce in the South-East. Once a house-building company has built up a land bank, that company is in a powerful position because its land bank represents an entry barrier to others. The source of competitive advantage lies primarily in the ability to acquire land. In a less densely populated area, however, the ability to acquire land will not be the source of competitive advantage, instead it may be the ability to arrange finance for buyers or the ability to build houses at low cost. When we analyse the structure of a market, we are identifying the position a firm should occupy and the posture it should adopt to generate acceptable performance. The purpose of identifying a market structure, and the sources of competitive advantages it creates, is to draw conclusions about the strategy that will match or fit the environment. When managers select strategies that are adapted to the environment in this way, then, according to the technically rational, analytical approach, they will be successful. Porter expresses this relationship between the form of competitive advantage and the matching strategy in a matrix of generic strategies. Porter identifies two types of *competitive advantage*: firms compete either on the basis of lower *cost* than their rivals, or on the basis of *differentiation*, where they provide some unique and superior value to the buyer, allowing them to charge higher prices. Firms can use these competitive advantages in either a broad or a narrow way – their *competitive scope*. Advantage and scope together create four kinds of strategy: *cost leadership*, *differentiation*, *cost focus*, or *differentiation focus*, which we will discuss in more detail in Chapter 3.

Kenichi Ohmae (1982), in his book *The Mind of the Strategist*, agrees with

Porter that the secret of strategic success is competitive advantage but from a Japanese perspective identifies three possible responses:

- customer-based strategies
- corporate-based strategies
- competitor-based strategies

Two more recent contributors to the competitive advantage school are Gary Hamel and CK Prahalad (1994) with their best-selling book *Competing for the Future*. They introduce concepts such as:

- competing for industry foresight
- strategic intent
- strategic architecture
- strategy as stretch
- strategy as leverage
- competing to shape the future
- core competences

They argue that traditional strategists try to position their organizations as cleverly as possible in existing markets whereas companies should try to reinvent their whole industry: *competing for industry foresight* by following a vision, what Hamel and Prahalad (1994) call *strategic intent*. Strategic intent is defined as having ambitions that are out of all proportion to the organization's resources and capabilities. This is the antithesis of traditional SWOT analysis which seeks to match organizational capabilities to market opportunities.

In *competing to shape the future* companies will need to adopt *strategic architecture*: a sense of what benefits they want to deliver to their customers and what delivery mechanisms they will use to deliver those benefits. Three concepts are crucial to strategic architecture.

The first is to view strategy as *stretch* rather than *fit*, as in traditional strategic planning. Stretch is achieved by proposing a very ambitious vision for the future of the organization (*strategic intent*) without telling people how to get there. In other words, stretching employees without stressing them unreasonably.

The second is *leveraging resources*, which means that a 'David' can compete with a 'Goliath' by outmanoeuvring a competitor with superior resources. This is achieved by 'getting to the future first' so that physical resources become less and less important as a source of competitive advantage. This means strategically out-thinking opponents.

The third key concept is *core competences*: viewing organizations as a collection of skills rather than as a collection of products/services and business units. It involves identifying what you are good at, venturing only into areas where your core skills can be applied to create new products and services, and buying only companies that can add substantially to the organization's portfolio of skills.

To quote Hamel and Prahalad:

> The need to think differently about strategy cannot be divorced from the need to think differently about organizations. Mobilizing employees at all levels around a strategic intent, leveraging resources across organizational boundaries, finding and exploiting 'white space' opportunities, redeploying core competences, consistently amazing customers, exploring new competitive space through expeditionary marketing, and building up banner brands all require new ways of thinking about the organization. Just as the current language and practice of strategy is not up to the challenge of competition for the future, neither is the current language and practice of organizational change.

This further reinforces the link between strategy and structure outlined earlier.

2.2.5 The incrementalists school

Quinn's (1980) research into the decision-making process of a number of companies revealed that most strategic decisions are made outside the formal planning system. Effective managers accept the high level of uncertainty and ambiguity they have to face and do not plan everything. They preserve the flexibility of the organization to deal with the unforeseen as it happens. The key points that Quinn made about the strategic decision-making mode are as follows:

1. Effective managers do not manage strategically in a piecemeal manner. They have a clear view of what they are trying to achieve, where they are trying to take the business. The destination is thus intended.
2. But the route to that destination, the strategy itself, is not intended from the start in any comprehensive way. Effective managers know that the environment in which they have to operate is uncertain and ambiguous, they therefore retain flexibility by holding the method of reaching the goal open.
3. The strategy itself then emerges from the interaction between different groupings of people in the organization, different groupings with different amounts of power, different requirements for and access to information, different time-spans and parochial interests. These different pressures are orchestrated by senior managers. The top is always reassessing, integrating and organizing.
4. The strategy *emerges* or evolves in small *incremental*, opportunistic steps. But such evolution is not piecemeal or haphazard because of

the agreed purpose and the role of top management in reassessing what is happening. It is this that provides the logic in the incremental action.
5. The result is an organization that is feeling its way to a known goal, opportunistically learning as it goes.

In Quinn's model of the strategy process, the organization is driven by central intention with respect to the goal, but there is no prior intention as to how that goal is to be achieved; the route to the goal is discovered through a logical process of taking one small step at a time. In *logical incrementalism*, overall strategy emerges from step-by-step, trial and error actions occurring in a number of different places in the organization. For example, some may be making an acquisition while others are restructuring the reporting structure. These separate initiatives are promoted by champions, each attacking a class of strategic issues. The top executives manage the process, orchestrating and sustaining some logic in it. It is this that makes it a purposeful, proactive technique. Urgent, interim, piecemeal decisions shape the organization's future, but they do so in an orderly, logical way.

Since in this model it is assumed to be possible to decide in advance where the organization is going, it too, like the strategic planning approach, can apply only in predictable conditions or conditions of contained change. The degree of order and logic it describes is not possible in truly open-ended situations.

To cope with truly open-ended situations, Lindblom (1959) also describes the process of strategic decision-making as incremental but to him it is a form of 'muddling through'. His observations are derived from decision-making in public sector organizations but they have implications for private sector organizations too. Because in complex situations it is not possible to identify all the objectives of different groups of people affected by an issue, policies are chosen directly. Instead of working from a statement of desired ends to the means required to achieve it (Quinn's approach) managers chose the ends and the means simultaneously.

This means that we cannot judge a policy according to how well it achieves a given end, instead we judge a policy according to whether it is itself desirable or not. A good policy is thus simply one that gets widespread support. It is then carried out in incremental stages, preserving flexibility to change it as conditions change. In this approach, dramatically new policies are not considered. New policies have to be close to existing ones and limited comparisons are made. This makes it unnecessary to undertake fundamental inquiries. The procedure also involves ignoring important possible consequences of policies, but serious lasting mistakes can be avoided because the changes are being made in small steps.

2.2.6 *The synthesis school*

The current view of strategic management is a synthesis or integration of the respective schools of thought into a holistic view of strategic management. Two approaches to synthesis within strategic management are the study of *patterns of strategic change* over time and the identification of different *strategy development routes.*

Patterns of strategic change

Henry Mintzberg's historical studies of organizations over many decades show that global or *transformational* change did take place. However, it was infrequent. More typically, organizations changed *incrementally,* during which times strategies formed gradually; there were periods of *continuity,* in which established strategy remained unchanged and also periods of *flux,* in which strategies did change but in no very clear direction. Figure 2.3 illustrates these patterns.

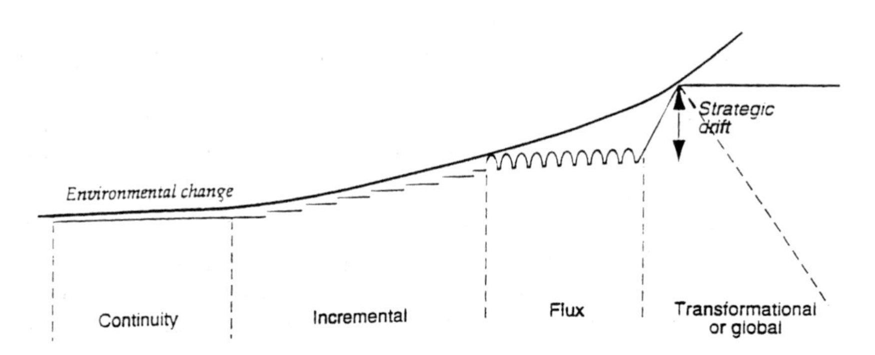

Fig. 2.3 Patterns of strategic change. After Johnson and Scholes (1993).

During the time-slice shown in the model the first phase is one of consolidation which is characterized by *continuity* in an established strategy; *incremental* changes in strategy may then occur as described by Quinn but within an overall direction set by top management; a lack of clear direction, perhaps brought about by a change of corporate leadership or an open-ended change situation may lead to a period of *flux* where strategies change but in no very clear direction – what Lindblom describes as 'muddling through'. Depending on the length of this period of flux and the rate of change in the environment the organization may drift drastically out of alignment with its environment, necessitating global or *transformational* changes in strategy to bring it back into line with its environment.

Strategy development routes

Figure 2.4 summarizes the various ways in which strategy may be realized.

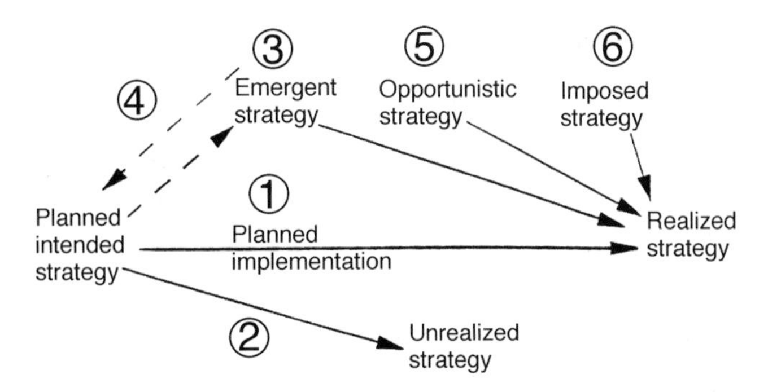

Fig. 2.4 Strategy development routes. After Johnson and Scholes (1993).

Route 1 is the *strategic planning* route where a planned, intended strategy is realized through planned implementation. Research suggests that less than 10 per cent of realized strategies are achieved through this route.

Route 2, the *unrealized strategy*, is the fate of planned, intended strategies which are found to be inappropriate or unachievable. If the research cited above is true, then 90 per cent of planned, intended strategies will follow this route.

Route 3 is the *emergent strategy* which will emerge from what Mintzberg defines as a *pattern* of decisions and actions taken in an incremental way by people in the organization; this could be logical incremetalism if it evolves within an intended strategic trajectory or simply flux or muddling through if there is no over-arching goal.

Route 4 suggests that the emergent strategy may become the planned intended strategy of the organization over a period of time, as discussed earlier.

Route 5, the *opportunistic strategy*, may be based on pure serendipity or, as Edward de Bono (1988) suggests, in one of his best books entitled *Opportunities*, may be harvested systematically by an organization through a programme of Opportunity Search which is described in the book.

Route 6 allows for the fact that strategy may be *imposed* on an organization, perhaps by a parent organization or a government/client organization upon whom the business depends for survival.

This model describes the range of routes through which organizations may achieve their *realized strategy* – the strategy which is ultimately adopted.

Summary

We have traced the evolution of strategic management through six schools of thought, which have become increasingly sophisticated in their treatment of the subject.

The *synthesis school* integrates the previous approaches into models which illustrate the complexity of strategic management.

Questions

1. Use Porter's five forces model to analyse the market or industry faced by your organization.
2. From the analysis above, which of Porter's generic strategies has your organization adopted and why?
3. Use the model of strategy development routes to trace the way in which strategies have evolved within your own organization.

References and bibliography

Andrews, K.R. (1987) *The Concept of Corporate Strategy* (Third Edition), Richard D., Irwin Inc.

Ansoff, H.I. (1965) *Corporate Strategy*, McGraw-Hill.

Ansoff, H.I. (1984) *Implanting Strategic Management*, Prentice Hall International.

Argenti, J. (1980) *Practical Corporate Planning*, George Allen & Unwin.

Bacharach, S.B. and Lawler, E.L. (1980) *Power and Politics in Organizations*, Jossey-Bass.

Bass, B. (1990) *Bass and Stodgill's Handbook of Leadership* (Third Edition), The Free Press.

Bowman, C. and Asch, D. (1987) *Strategic Management*, Macmillan.

Chandler, A.D. (1962) *Strategy and Structure*, MIT Press.

Channon, D.F. (1973) *The Strategy and Structure of British Enterprise*, Macmillan.

Channon, D.F. (1978) *The Service Industries, Strategy, Structure and Financial Performance*, Macmillan.

Cohen, A.R. and Bradford, D.L. (1990) *Influence Without Authority*, Wiley.

Cyert, R.M. and March, J.G. (1963) *A Behavioral Theory of the Firm*, Prentice Hall.

De Bono, Edward (1988) *Opportunities*, Penguin.

Greiner, E. and Schein, V.E. (1988) *Power and Organization Development*, Addison-Wesley.

Hamel, G. and Prahalad, C.K. (1994) *Competing for the Future*, Harvard Business School Press.

Hussey, D. (1975) *Corporate Planning: Theory and Practice*, Pergamon Press.

Johnson, G. and Scholes, K. (1999) *Exploring Corporate Strategy*, Prentice Hall International (UK) Ltd.

Kanter, R.M. (1983) *The Change Masters*, Simon & Schuster.

Kotter, J.P. (1982) *The General Managers*, Free Press.

Langford, D. and Male, S. (2001) *Strategic Management in Construction* (Second edition), Blackwell Science.

Lindblom, C.E. (1959) 'The Science of Muddling Through', *Public Administration Review*, Volume 19, *American Society for Public Administration*, Spring 1959 (pp. 79–88).

Mintzberg, H. (1973) *The Nature of Managerial Work*, Harper & Row.

Mintzberg, H. (1978) 'Patterns of strategy formation', *Management Science*, Volume 24, Number 9 (pp 934–48).

Mintzberg, H. (1983) *Power In and Around Organizations*, Prentice Hall.

Mintzberg, H. (1994) *The Rise and Fall of Strategic Planning*, Prentice Hall International (UK) Ltd.

Ohmae, K. (1982) *The Mind of the Strategist*, McGraw-Hill.

Peters, T.H. and Waterman, R.H. (1982) *In Search of Excellence*, Harper & Row.

Peters, T. (1987) *Thriving on Chaos*, Macmillan.

Porter, M.E. (1980) *Competitive Strategy: Techniques for Analysing Industries and Competitors*, Free Press.

Porter, M.E. (1985) *Competitive Advantage: Creating and Sustaining Superior Performance*, Free Press.

Quinn, J.B. (1980) *Strategies for Change – Logical Incrementalism*, Richard D. Irwin.

Steiner, G.A. (1969) *Top Management Planning*, Macmillan.

Stewart, R. (1967) *Managers and Their Jobs*, Macmillan.

3 Strategy Systems

The strategy system or process is at the heart of strategic management because it seeks to answer the question of how organizations make important decisions and link them together to form strategies. As we have seen in our reviews of the schools of thought, the assumption of the early writers was that strategic planning was the single and best way of taking strategic decisions.

However, Mintzberg's definition of strategy as not only plan but pattern, position, perspective and ploy presents a broader perception of the strategy process. This broader view is captured in Figure 2.4 which showed a variety of sources of strategy and approaches to realizing strategy.

Current thinking is that the strategy-making process or system is a continuum with *strategic planning*, a top-down approach, at one end and *strategic learning*, a bottom-up approach, at the other. Between these two extremes there may be any number of strategy-making processes but we shall focus on just one called *strategic vision*.

In the strategic planning mode, formal analysis is used to plan explicit, integrated strategies for the future. In the strategic vision mode, a strong leader establishes a broad vision for the organization within which the details of strategic direction are worked out. Under the strategic learning mode, the organization adapts in small, disjointed steps to a difficult environment.

We shall now describe these three strategy systems or processes separately before considering how they may be combined into a hybrid strategy process for an organization.

3.1 Strategic planning

We have already outlined the assumptions underlying the strategic planning approach to formulating and implementing strategy:

- *formalization* of the process by analysis and systems which can be programmed
- *detachment* of thinking from action and strategy from operations or tactics
- *quantification* as a basis for the strategy process through the generation of hard data

- *predetermination* of the future based on the assumption of a stable and predictable environment

These assumptions are exemplified by Ackoff (1970) who isolates the three chief characteristics of the strategic planning mode:

1. Planning is something we do in advance of taking action; that is it is *anticipatory decision-making*.
2. Planning is required when the future state that we desire involves a set of interdependent decisions; that is, a *system of decisions*.
3. Planning is a process that is directed towards producing one or more *future states* which are desired and which are not expected to occur unless something is done.

As Mintzberg (1973) comments:

> Formal planning demands rationality in the economists sense of the term – the systematic attainments of goals stated in precise, quantitative terms. The key actor in the process is the analyst, who uses his scientific techniques to develop formal, comprehensive plans.

Mintzberg further points out that the strategic planning mode exhibits three essential characteristics:

1. In the planning mode the analyst plays a major role in strategy-making.
2. The planning mode focuses on systematic analysis, particular in the assessment of the cost and benefits of competing proposals.
3. The planning mode is characterized above all by the integration of decisions and strategies.

An organization plans in the belief that decisions made together in one systematic process will be less likely to conflict and more likely to complement each other than if they were made independently. Thus, strategic planning is a process whereby an organization's strategy is designed essentially at one point in time in a comprehensive process (all major decisions made are interrelated). Because of this, planning forces the organization to think of global strategies and to develop an explicit sense of strategic direction (Mintzberg 1973).

The planning mode is oriented to systematic, comprehensive analysis and is used in the belief that formal analysis can provide an understanding of the environment sufficient to influence it.

The strategic planning process consists of a sequential series of steps which are taken in a very mechanistic fashion. The number of steps involved ranges from Ansoff's large number of steps and accompanying routines to Johnson and Scholes three-step process of strategic analysis, strategic choice and strategy implementation. Probably the clearest and most practical exposition of the strategic planning process is developed by Argenti (1990) in his set of manuals for strategic planners under the title of *Argenti System of Corporate Planning*.

Argenti's opening assertion is that in all planning systems one has to go through a process, or sequence of stages, in order to produce a plan. In the Argenti system there are ten stages as outlined below:

Stage 1 Preparation: to start the process and to select a planning assistant.
Stage 2 Objectives and targets: to define the aims and ambitions of the organization.
Stage 3 Forecasts and gaps: to calculate the size of the strategic task.
Stage 4 Strengths and weaknesses: to decide what the organization is especially good at and bad at.
Stage 5 Threats and opportunities: to identify what major changes the future may hold for the organization.
Stage 6 Alternative strategies: to list all the alternative strategies available to the organization.
Stage 7 Selecting the strategies: to decide which set of strategies is the most suitable.
Stage 8 Evaluation: to determine whether this set is good enough.
Stage 9 Action plans: to draw up detailed plans and budgets.
Stage 10 Monitoring: continuously study progress over the years.

Argenti adds, 'these ten stages are the same for all types and sizes of organization – all planning teams should move through the stages exactly as they are presented in their manuals.'

Stages One to Eight are usually termed the formulation stage and Stages Nine and Ten the implementation stage.

Stage One envisages the selection of a strategic planning team from within the organization together with the appointment of a strategic planner to manage the process.

At Stage Two objectives are quantified in the form of targets and compared at Stage Three with forecasts of future performance based on the extrapolation of existing strategies. Argenti explains, 'a target is what you *want* to achieve; a forecast is what you *expect* to achieve. The difference between targets and forecasts identifies the *gaps* which any proposed strategies must fill.'

3.1.1 The SWOT analysis (Strengths, Weaknesses, Opportunities, Threats)

It is widely recognized that the heart of the whole strategic planning system is Stages Four and Five, often called the SWOT analysis, which identifies the *strengths and weaknesses* of the organization for comparison with the *opportunities and threats* in the future environment. It is at this point in the process that many of the writers on strategic planning, including Ansoff, Steiner, and Porter, produce comprehensive checklists to enable this process to be carried out systematically. Argenti produces his own checklist for assessing the strengths and weaknesses of the organization as follows:

- finance and legal
- facilities
- purchasing
- products and services
- research and development
- marketing
- distribution
- employees
- management
- position in the industry

Each of these is defined with examples and participants in the planning process are asked to rate their own organization against these criteria.

Participants are then asked to attempt to rank the strengths and weaknesses in order of strategic significance. A simple A, B, C classification is suggested where:

- A means that the item is of critical strategic importance and absolutely must be accounted for in the strategies of the company. Without this item in the strategies they would be valueless.
- B means that the item is of strategic importance and should be in the corporate strategies if possible; if not it must be attended to elsewhere by the planning team.
- C means that the item is not of corporate significance but may be of departmental or sectional importance and should be dealt with by a relevant senior executive.

Argenti concludes that this stage (Stage Four) is probably the most important for most organizations.

A similar checklist is proposed by Argenti to analyse the threats and opportunities facing the organization:

- political changes
- economic changes
- social changes
- technological changes
- changes in your own industry
- other changes

Again definitions and examples are provided and participants are expected to insert entries in the checklist against each criteria for their own organization.

As with strengths and weaknesses, participants are again asked to rank these threats and opportunities in order of significance for their organization. The ranking system suggested for opportunities and threats classifies each item by *probability* and by *impact*. That is, they are ranked according to the chances of the threat or opportunity actually taking place and by the impact it would have on the organization if it did. Trends or events are classified on these two dimensions using a *High, Medium,* or *Low* classification (*H, M, L*). Using this procedure generates the following categories:

1. *HH* (high high): this is obviously an enormously important threat or opportunity because it is both very likely to occur and will have a great impact on the organization. Plainly it is something of such outstanding importance that the strategies which are devised in Stage Seven must, *absolutely must*, take it into account. Without this item the strategies would be useless.
2. *HM* or *MH*: also very important because it either highly probably will occur and will have a considerable impact on the organization, or it is quite likely to occur and would have a major impact if it did. The strategies must take account of any threats and opportunities to which they have given one or two Hs.
3. *LH*: this is a trend or event which it is thought is unlikely to occur but it is believed would have a most severe impact if it did. So while this may not have to be included in the main strategies, something must be done about it - perhaps a 'disaster plan' or 'contingency plan' needs to be devised.
4. *HL, LM, ML, MM*: if these can be included in the main strategies then they should be; but these are probably not significantly important to need further very serious consideration in the strategic planning exercise and might be better dealt with on an *ad hoc* basis by the organization's other senior executives.
5. *LL*: not worthy of further consideration in the strategic planning exercise.

3.1.2 Total strategic summary

At Stage Six, all this information is collected together into a chart called the Total Strategic Summary. This chart displays the targets and forecast in graphical format, extracts the major strengths and weakness, opportunities and threats from the SWOT analysis and summarizes them in the form of what Argenti calls a *cruciform chart*. An example of a cruciform chart for a UK building products manufacturer is shown in Figure 3.1.

Strengths		**Weaknesses**	
Marketing	A	High overheads	A
Merchants	B	New product development	A
		Dying products	B
		International	B
Opportunities		**Threats**	
International	HM	Declining home market	HH
New product home	HM	Over capacity	HH
		Foreign competition	HM

Fig. 3.1 Cruciform chart.

The chart shows that the company is strong in marketing and has access to a wide range of builders merchants. Its weaknesses are its high overheads, the lack of new products, the growing obsolescence of its existing product range, and its lack of a presence in overseas markets. The firm recognizes opportunities in the international market for its existing products and also the market for new products in the UK. However, this has to be set in the context of a shrinking home market with many competitors and the possibility of entry into the UK by foreign competitors.

It is intended that the cruciform chart summarizes and displays the results of the SWOT analysis in such a way that the strategic situation of the organization can be seen at a glance. The chart can now be 'read' to identify the key issues that proposed strategies must address. The reading of the cruciform chart will reveal the balance of strengths and weaknesses and opportunities and threats in terms of both number and weighting. If, for example, strengths outweigh weaknesses, then the organization may be justified in adopting a bold and aggressive set of strategies; conversely if weaknesses outweigh strengths, as in the example above, then a more defensive set of strategies may be called for.

Further analysis may be conducted by noting whether the same factor appears more than once in a cruciform chart which will identify it as a factor at the central core of the organization's strategy. In the example shown, 'international' occurs in two quadrants, with the related factor of foreign 'competition' also reinforcing the need for an overseas strategy.

Equally, the relationships between factors in various quadrants may reveal key issues and possible strategies. The weakness of the company in terms of new products and the opportunity for new products in the home market coupled with the firm's excellence in marketing are an obvious direction for a strategy of new product development. What other cluster of factors indicate new directions for strategy?

The SWOT analysis conducted using the cruciform chart will identify a number of key issues which any strategies must address. The conclusion of this alternative strategies stage of the process is to generate a number of possible strategies for each of the key issues.

As Mintzberg (1994) points out, the process by which strategies are created is not addressed in the strategic planning literature. It is assumed that the system will automatically and magically produce the required strategies to deal with the key issues. This is an issue we shall come back to shortly in talking about the evidence for the strategic planning process.

Selecting strategies

Stage Seven of the process requires the planning team to select the strategies. The strategies must meet all the following requirements:

- They must give the organization a very good chance of hitting its targets or even exceeding them.
- At the same time they must protect the organization from the risks that might drag its performance below target levels.
- They must make full use of all the organization's most impressive strengths, certainly all those ranked with an 'A'.
- They must correct or neutralize all major weaknesses, certainly those ranked 'A'. They must not build on these weaknesses, nor even those ranked 'B'. A strategy built on major weaknesses could be disastrous.
- They must either eliminate or reduce the impact of all the threats that were ranked as 'HH' and those with any 'H' in their ranking and perhaps also those ranked with a 'M'.
- They must exploit any opportunity ranked 'HH'.
- They must conform to the 'flavour' of the strategies, e.g. bold, defensive, high risk, etc.
- They must address the key issues that were identified at the previous stage.

Argenti contends that there will be one permutation of strategies that will come closer than any other to meeting this complex specification. He recommends consideration of the inevitable generic strategies under the headings of 'Relevant Strategies – strategies that appear to be highly appropriate to the way the world is moving at the end of this century' and 'Hazardous Strategies – strategies which may be highly effective in skilled hands ... but may be dangerous in other circumstances, especially when the balance of the cruciform chart is unfavourable'.

The final stage of the formulation phase (Stage Eight) of the strategic planning process is to evaluate the strategies. It is at this stage that the analysts and 'number crunchers' come into their own. Another checklist called the *Evaluation Report* is produced in which all the facts, figures, and detailed costings and calculations are contained under the following headings:

- *The salient points*: for example, if a North of England versus a South of England geographic decentralization is being considered then facts like the size of the market in each area, the availability of premises or sites, availability and salary levels of employees, etc. would be considered.
- *Resources required*: these would include capital, management skills, other skills, premises, facilities, etc.
- *Major assumptions*: this would include both pessimistic and optimistic assumptions.
- *The calculations*: this would include detailed financial projections of turnover, profits and sales for the planning period. It is suggested that both pessimistic and optimistic forecasts are made.
- *What might go wrong*: this would identify the trends and events that might occur to which strategies might be particularly sensitive and a calculation made of how they would affect the attainment of targets.
- *What might be disastrous*: this would highlight the events or combinations of events that might be disastrous for the organization and how the organization will be protected from these events.
- *Previous strategies*: any record of this strategy being previously attempted by the organization or others and its results.
- *Hazardous and relevant strategies*: identifying strategies which are particularly hazardous and especially relevant to the organization's strategic position.
- *Strengths, weaknesses, opportunities and threats*: check that the strategies account for all the 'As' and the 'Hs', and perhaps some of the 'Bs' and 'Ms'.
- *Conclusions*: a statement of the planning team's confidence that the strategies will enable the organization to achieve its targets.

This stage completes the formulation phase of the strategic planning process; the remaining two stages are concerned with implementation of the selected strategies.

Implementation

Stage Nine entails deciding which parts of the strategic plan require *action plans* or *project plans*, deciding who will be the project manager for each, deciding who will draw up the plans, what form these should take, by what date they should be available and who will approve them. A cascade of plans may then be formulated for the different functions and levels of the business, for example, marketing plans, production plans, project plans, etc.

As soon as a sufficient number of projects have been defined these will need to be summarized in the form of a *business plan*. This will ensure that the individual plans dovetail together, that they do not all call for cash at the same time, that they are not incompatible, and so on. A business plan will be required in any case if the organization has to go to the market to obtain finance and resources.

The final stage in the whole process involves the continuous *monitoring* of the strategies and action/project plans. The purpose of Stage Ten is to give the organization the opportunity at regular intervals (three-monthly is suggested) both to control the strategic progress of the organization and to review the whole strategic direction that has been selected. This will involve the normal control loop of measuring performance against targets set in the strategies, noting unacceptable variances from the targets, and taking corrective action in the form of either adjusting the performance to meet the strategies or changing the strategies.

We have discussed in some detail the strategic planning process in order to compare its mechanistic nature with the alternative systems that have been identified.

A critique of strategic planning

As strategic planning dominates the strategic management literature, what evidence is there of its implementation and effectiveness? The answer to this question, according to Mintzberg (1994) and others, is very little. The McGill University research programme which tracked the strategies of a number of companies over many decades came to the conclusion that there was very little evidence to support the mechanistic strategic planning process just described. Rather the research shows that the strategy process is sophisticated and subtle, involving the most complex of subconscious cognitive and social processes. Strategists rely on soft, non-quantitative detail which requires them to be connected to the detailed operation of the organization, not detached as the strategic planners suggest. Formulation and implementation are not separate activities but merge into a fluid process of strategic learning. The context within which strategists work is dynamic, unpredictable and unknowable – not predetermined as the strategic planning process assumes. Strategies emerge, rather than being deliberately enacted, involving all kinds of people throughout the organization rather than the process remaining the prerogative of the chief executive and the top team. Strategies are not formally planned but emerge informally, often within the envelope of an entrepreneur's vision for the organization. This is the logical incrementalism described by Quinn (1980).

Therefore, strategy *formation* (a term coined by Mintzberg to describe the seamless process of formulation and implementation he had observed) is a process of strategic learning which proceeds in 'fits and starts', relies heavily on serendipity and the recognition of unexpected patterns in evolving events. The basis for strategic learning is a deep understanding by the strategists of the industry and context within which they are operating which enables them to recognize and exploit discontinuities in the environment before competitors. The result is novel strategies which often defy allocation to the generic strategic categories so important to the strategic planning theorists. This requires that the strategists should be what Peters and Waterman (1982) called 'hands on – value driven', not detached from the daily detail of organizational life but connected to the daily detail in order to abstract the strategic messages which are the key to strategy-making. To obtain this soft information, strategists need to be in touch with customers, clients, suppliers, competitors, government, projects and so on. This is not to conclude that hard data derived from analysis is not a useful aid to decision-making but it needs to be combined with soft data and the tacit knowledge which only line managers have.

Mintzberg suggests that it is the line managers using a combination of hard and soft data, derived from their tacit and intimate knowledge of the organization's operations and the industry context who have the necessary intuitive skills to develop novel strategies. This process, unlike the strategic planning process, requires insight, i.e. a penetrating understanding of complex issues and creative synthesis, aspects which the formalization process of strategic planning discourages.

Having said this there are some situations in which strategic planning has had some success. Where organizations are large enough to afford expensive analysis, have clear operating goals, and operate in a predictable and stable environment, then strategic planning may be possible. The large business or public sector organization which has a monopoly or near monopoly over its markets could operate a strategic planning system.

In the absence of such relatively rare conditions, different approaches to strategic decision-making must be used. Two popular approaches discovered from research by Mintzberg and Quinn which reflect a more informal approach to strategic decision-making, are *strategic vision* and *strategic learning*. In both cases the dichotomy of separation of formulation from implementation of strategy which forms the central tenet of the strategic planning system is abandoned. In the *strategic vision* approach, the formulator implements the strategy; in the *strategic learning* approach, the implementator formulates the strategy. The strategic vision approach is dependent on a single creative strategist; the strategic learning system depends on a variety of actors being capable of experimenting and then integrating their incremental strategies.

3.2 Strategic vision

Strategic vision has traditionally been called the entrepreneurial mode of strategy-making and is well known both in the economics and management literature. There is probably as much written on the role of the entrepreneur in modern business as there is on strategic planning. However, in contrast to the strategic planning literature, which is largely written by academics and which has a strong theoretical base, the entrepreneurial literature tends to be written by practitioners in a very anecdotal and prescriptive fashion. Business biographies, from Alfred P Sloan's (1965) *My Years with General Motors* to John Harvey Jones' (1994) *Getting it Together* are early and recent examples of the genre. Management consultants have also found the concept of the entrepreneur an attractive basis for discussion from Drucker (1970) to Peters (1987). The entrepreneur is the folk hero in much of the management literature, in both America and Europe. The idea of a single individual with a burning vision creating and driving an innovative and vibrant organization is an attractive role model in societies where individualism is a dominant concept.

The visionary or entrepreneurial mode of strategy – making has four chief characteristics:

1. Strategy-making is dominated by the active search for new opportunities. The visionary approach focuses the organization on opportunities rather than problems.
2. Power is centralized in the hands of the chief executive. In the strategic vision mode, power rests with one individual capable of committing the organization to bold courses of action. He rules by fiat, relying on personal power and sometimes on charisma. While there may be no formal strategic plan for the organization, typically strategy is guided by the entrepreneur's own vision of direction for his organization.
3. Strategy-making is characterized by dramatic leaps forward in the face of uncertainty. Strategy moves forward by the taking of large, bold decisions; the chief executive seeks out and thrives in conditions of uncertainty where the organization can make dramatic gains.
4. Growth is the dominant goal of the entrepreneurial organization. Entrepreneurs have been shown to be driven by the need for achievement, and the clearest recognition of achievement is the growth of the organization which they run. Channon's (1978) research showed that entrepreneurial leaders achieved much higher levels of corporate growth than either family or professionally led businesses.

The strategic vision approach is a more flexible way to deal with an uncertain world. Vision sets the broad direction of strategy while leaving the details to be worked out in an incremental way. Thus, referring back to Mintzberg's set of definitions for strategy, the vision may be expressed as a broad *perspective*, which may be *deliberate*, but specific *positions* can *emerge*. When the unexpected occurs, provided the vision is sufficiently robust, the organization can adapt, it can learn. Substantial but limited change can thus be accommodated while retaining the central vision.

Peters (1987) stresses the need for organizations to have a vision. The leader should develop and live an enabling and empowering vision. Effective leadership is marked by a core philosophy and a vision of how the enterprise wishes to make its mark. The vision must be specific enough to act as a 'tie breaker' between organizational units (e.g. quality versus volume) and general enough to allow the taking of bold initiatives in today's ever-changing environment. Peters argues that the chief executive should take every opportunity to stress and exemplify the vision. He suggests that the vision must replace the rule book and the policy manual and presumably the strategic plan.

Peters goes on to identify the characteristics of a vision:

- Effective visions are inspiring.
- Effective visions are clear and challenging and concerned with excellence.
- Effective visions make sense in the market place by allowing flexibility within consistency.
- Effective visions must be stable but constantly challenged and changed at the margin – the paradox of maintaining a stable vision in a changing world through incremental adjustments to the vision.
- Effective visions are beacons and controls when all is changing – the vision as a *beacon* depends on leaders at all levels exhibiting behaviour which is consistent with the vision at all times. As a *control* it is a replacement for the rules book and the procedures manual.
- Effective visions are aimed at empowering people – particularly they allow people space to make strategic experiments within the visions' parameters.
- Effective visions are focused on the future but aware of the past – the most effective vision allows us to draw upon enduring themes which give confidence in shaping the future of the organization.
- Effective visions are lived in the details – this stresses the point we made earlier about the nature of strategic management and the idea of the formulator also being the implementer.

The visionary mode requires that strategic vision comes from one powerful individual. The environment must be malleable, the organization oriented towards growth and the strategy able to shift boldly at the whim of the entrepreneur. These conditions are typically found in small and/or young organizations although more and more larger organizations are trying to encourage 'intrapreneurship' within the organization. Organizations in deep crisis often import a visionary leader to drag the organization on to a new strategic trajectory. Strategic vision is most likely to be found in business organizations but also in public sector organizations where strong leadership is allowed. Construction contractors and professional practices are often driven by the vision of a dynamic individual, particularly during the early years and in periods of rapid growth or change, e.g. Sir John Laing and Sir Ove Arup.

3.3 Strategic learning

In situations of turbulent change, when even strategic vision cannot cope, then the organization may have to revert to pure learning approach – to experiment with strategies in an incremental way, adapting and adopting those which seem to work. This is purely an emergent strategy of trial and error and trial again where the implementer becomes the formulator of strategy. The process becomes one of strategy formation rather than strategy formulation, by individuals anywhere in the organization. The *realized strategy* is the *emergent strategy*.

Many construction companies evolve strategy in this emergent way when projects are obtained in an unpredictable and *ad hoc* way, so that the organization strategy becomes the pattern of projects which it is successful in obtaining. There is a strong flavour of Lindblom's 'science of muddling through'.

Strategic learning organizations have four major characteristics:

1. Clear objectives do not exist as in the strategic vision organization. This is a response to the complexities of a rapidly changing environment but may also reflect the division of power among members of a powerful coalition.
2. The strategy-making process is characterized by the 'reactive' solution to existing problems rather than the 'proactive' search for new opportunities. The difficulties of working in a turbulent environment together with the lack of clear goals and overarching vision often preclude a proactive approach.
3. Strategy-making proceeds in incremental, serial steps – because its environment is complex, the strategic learning organization finds that feedback is a crucial ingredient in strategy-making. The organization moves forward in incremental steps, which evolve in a serial fashion so that feedback can be received and the course adjusted as it moves along.
4. Disjointed decisions are characteristics of strategic learning. The objective of integrated, compatible and coordinated decisions, which is a feature of the strategic planning mode, is not possible under strategic learning. It is simply easier and less expensive to make decisions in a disjointed fashion so that each is treated independently and little attention is paid to the problems of coordination. Strategy-making is fragmented, but at least the strategist remains flexible, free to adapt to the needs of the moment.

A fuller description of the process is given by Mintzberg (1994) in what he calls the 'grass roots' model of strategy formation:

1. *Strategies grow initially like weeds in a garden,* they are not cultivated like tomatoes in a greenhouse. The process of strategy formation should not be overmanaged: it is more important to let patterns emerge than to force an artificial consistency among strategies prematurely.
2. *These strategies can take root in all kinds of places,* virtually anywhere people have the capacity to learn and the resources to support that capacity. It may be that an individual or a unit in touch with a particular opportunity creates its own pattern. Strategies may emerge through serendipity or by people experimenting with ideas until they converge on something that works. Equally, the independent actions of a number of people may converge into a strategic theme gradually and spontaneously. The external environment may impose a particular pattern of strategy on to an unsuspecting organization. Organizations cannot always plan when their strategies will emerge, let alone plan the strategies themselves.
3. *Emergent strategies become organizational when they become collective,* that is when the patterns shape the behaviour of the organization as a whole. Emergent strategies can sometimes displace the existing deliberate ones when they are reflected in the realized strategy of the organization.
4. *The processes by which emergent strategies become the intended strategies of the organizations may be conscious and managed or unconscious and ad hoc.* Strategies may unconsciously evolve or be consciously intended within the minds of the formal and informal strategists within the organization. Equally, emergent strategies may be managed or simply spread by collective action.
5. *New strategies, which may be emerging continuously, tend to pervade the organization during periods of flux and transformation* (see Figure 2.3)
6. *To manage this process is not to preconceive strategies but to recognize their emergence and intervene when appropriate.* This is the opposite to the strategic planning theorist's view of the role of the strategist. The strategic manager's role in the strategic learning context is to create the climate within which a wide variety of strategies can emerge and to be prepared to invest in promising initiatives. It also entails allowing people to make mistakes on the premise that the successful strategies will outweigh the unsuccessful strategies and give the organization a creative competitive edge.

The strategic learning approach may seem to be extreme but, in a sense, it is no more extreme than the strategic planning approach. The two systems define the end points of a continuum along which real world strategy-making behaviour must lie. Sometimes organizations may tend towards the more deliberate end of the continuum, where clear thought has to precede action, because the future seems reasonably predictable; in this case formulation may proceed implementation. But in times of complex and difficult change when new strategies have to be worked out in a process of learning, the organization must lean towards the emergent end of the spectrum. The dichotomy between formulation and implementation is abandoned and the formulators implement in a more centralized way, as under the strategic vision approach, or the implementers formulate in a more decentralized grass roots way, as in strategic learning. In both the case of strategic vision and strategic learning thinking is reconnected directly to acting.

The strategic learning approach seems to be associated with collegiate-type organizations, e.g. universities, hospitals, etc., but in any organization where a participative approach to management exists and empowerment is enacted throughout the organization the strategic learning approach is likely to blossom. Construction management firms and professional partnerships are likely to have this approach to strategy formation.

3.4 A hybrid approach

It must be recognized that the three strategic planning systems discussed are pure types and may coexist in a real world organization. It may be, for example, that the marketing department of an organization operates in a visionary mode while the production or project division of the organization operates in a planning mode. Equally, in a diversified organization the centre may use a planning approach whereas the divisions or business units of the organization may rely more on a strategic learning approach.

In some situations no strategy-making system will be required at all, or at least very little, for example, monopolistic organizations in stable environments. Equally, in these situations planning may be necessary and possible.

Strategic planning is the least flexible of the strategic systems discussed and will be inappropriate where unexpected events occur or the dominant coalition in the organization cannot agree on specific action. The best managed and most successful of large organizations will exhibit a hybrid approach to its strategy process.

Summary

In this chapter we have reviewed three approaches to making strategic decisions.

The *strategic planning* approach has been described in some detail as a mechanistic system which relies on a systematic and analytical process. It is a 'top-down' approach. The system has been shown to be appropriate for stable and predictable conditions but is unable to cope with a rapidly changing environment. A number of other criticisms of the system have also been raised.

Entrepreneurs often adopt the *strategic vision* approach which allows flexibility of strategic decision-making within the parameters of the vision. This is a combination of 'top-down' and 'bottom-up' approaches.

The *strategic learning* approach is necessary when conditions of extreme turbulence mean that the inability to predict the future dictates that incremental 'trial and error' steps are the only way to proceed. This is what Mintzberg calls strategy *formation*. It is a 'bottom-up' system.

These three systems are pure types and real organizations are likely to exhibit a combination of these approaches.

Questions

1. Conduct a SWOT analysis of your organization and produce a cruciform chart as shown in Fig 3.1.
2. Examine your organization for evidence of the *strategic vision* or *strategic learning* approaches to strategy formation, either currently or in its recent past.

References and bibliography

Ackoff, R.L. (1970) *A Concept of Corporate Planning*, Wiley.

Argenti, J. (1990) *The Argenti System of Corporate Planning*, Argenti Systems Ltd.

Channon, D.F. (1978) *The Service Industries: Strategy, Structure and Financial Performance*, Macmillan.

Drucker, P.F. (1980) 'Entrepreneurship in Business Enterprise', *Journal of Business Policy*, Autumn (pp. 3–12).

Jones, John Harvey (1994) *Getting It Together*, Ulverscroft Large Print Editions.

Mintzberg, H. (1973) *The Nature of Managerial Work*, Harper & Row.

Mintzberg, H. (1994) *The Rise and Fall of Strategic Planning*, Prentice Hall International (UK) Ltd.

Mintzberg, H. and Quinn, J.B. (1992) *The Strategy Process* (Second Edition), Prentice Hall.

Peters, T.H. and Waterman, R.H. (1982) *In Search of Excellence*, Harper & Row.

Peters, T. (1987) *Thriving on Chaos*, Pan Books.

Quinn, J.B. (1980) *Strategies for Change: Logical Incrementalism*, Richard D. Irwin.

Sloan, A.P. (1965) *My Years with General Motors*, Sidgwick & Jackson.

4 Industrial Relations

In most developed countries the topic of industrial relations is a matter that has to be managed in a structured way. Certainly the industrial relations problems which beset construction managers are not as torrid as were experienced in the 1960s and 1970s. However the collective agreements which bond together employers and trade unions still have some force and the collective agreements reflect the well-established pattern of bargaining within the industry and the relative weakness of the trade unions allied to the construction industry.

4.1 The history and development of industrial relations

Although relationships between employers and employees can be traced back to the medieval guilds, the modern period of industrial relations commenced at the beginning of the nineteenth century when the building industry underwent a profound change.

Prior to the Industrial Revolution, the client undertook to organize the building project with master craftsmen of each principal trade being asked to undertake specific aspects of the work, not unlike contemporary arrangements in construction management procurement systems. *Plus ça change!* With the nature and magnitude of the type of buildings required to provide the infrastructure for an emerging capitalist nation, the role of the master builder was enhanced because they now tendered for the complete building.

This change had its implications for established rules governing the conduct of industrial relations; tradesmen resented the development of the master builder because this impinged upon their opportunities to become master craftsmen themselves. As Cole (1953) observed, 'The general contractor was apt to be intolerant of the traditional rules and customs of the various trades.'

In the main, bargaining concerning wages was localized, and, with the development of the piecework system of payment, was often individually rather than collectively bargained. The new master builders employed workers at the rate of wages pertaining to a town or locality and expected them to work under the customary conditions; hence, 'custom' has become a basic component of understanding between employees and employers. The piecework aspect to this was a challenge to their ability to organize effectively. Marx (1867) summed up the new system of payment when he described the piecework system as 'the form of wages most suited to the capitalist method of production'.

Despite the legal barriers imposed by the Combinations Acts of 1799 and 1800, building workers and building employers formed organizations. In the first instance workers formed combinations under the guise of friendly societies and employers took advantage of the jurisdiction of friendly local magistrates to collect together to discuss common problems.

The Combination Acts allowed sentences on any working man, and invariably it was men, of three months in jail or two months' hard labour if he was found guilty of combining with others for the object of increasing wages or decreasing hours. Sentence was by two magistrates, not a judge. Appeal against sentence was not allowed unless £20 surety was given; this represented approximately twenty weeks' wages for a bricklayer at the time. A fine of £10 was levied upon anyone helping with the expenses for anyone convicted under the Acts. The Acts also forbade employers' combinations, but there is little evidence of enforcement against employers.

The friendly societies of building workers existed until repeal of the Combination Acts in 1824, after which unions of building workers developed rapidly: they had a strong base from which to work. Hilton (1968) accurately suggested that:

> ...Certainly the building unions which made themselves effective in the immediate post repeal [of the Combination Acts] period, did not act as if they were newly formed, naïve and inexperienced organizations. Instead they bore that quality of militancy and independence against which all the statutes over the last 500 years had complained.

The trade unions in building grew rapidly, and as soon as eight years after the repeal the first national union was formed, which attempted to weld together the multitude of local associations. In 1832 the Operative Builders Union was formed, whose aim was to 'advance and equalize the prize of labouring every craft' (Postgate 1923). In 1833 its membership was 49,000. However, employers' associations retaliated with the 'Document' which attempted to gain pledges from employees that they would not join the Operative Builders Union. The 'Document' asked employees to sign the following statement (Postgate 1923):

> We the undersigned do hereby declare that we are in no way connected with the General Union of Building Trades and we do not and will not contribute to the support of such members of this said union, as we are or may be out of work in consequence of belonging to such unions.

Clearly, to seek such a pledge would embitter industrial relations and several localized strikes ensued. Clarke (1992) records widespread strikes in 1834 in London, Birmingham and Manchester over workers' refusal to sign the 'Document'.

As the Industrial Revolution intensified, the massive civil engineering projects generated large contracting organizations, which further strengthened the employers at the expense of the trade unions. The 1850s found trade union organization – the Central Master Builders Association – which, in 1859, organized a 'lock-out' of 24,000 London-based builders after a union had called for a nine-hour working day. This lock-out demonstrated that trade unions also required national organization to coordinate recruitment and policy. However, while there was general agreement to move towards a central trade union authority, there were disagreements over the nature and structure of such a body. Two views could be detected: the *traditionalists* and the *amalgamationists*. The traditionalists saw the builder's union based upon local democracy with a decentralized administration and decisions concerning strike action to be taken at the work face. The amalgamationists saw the union movement best served by caution, restraint, conciliation, with a centralized administration who would vet any strike action considered by workers.

The amalgamationists' argument succeeded for two reasons. First, the employers were seeking to establish standardized working rules and the amalgamationists' preference for negotiation as opposed to industrial action matched this initiative. Second, the government favoured the more moderate policy of this group and had influenced the climate of opinion which favoured their rise to supremacy.

By the late 1870s the case for national institutes for the conduct of industrial relations had been established and the master builders widened their organization to form the National Association of Master Builders of Great Britain, with 64 employers and 19 local associations. By 1892 it had grown to 1,300 members (NFBTE 1978) and in 1899 it formally became the National Federation of Building Trades Employers (NFBTE).

Despite these national organizations, industrial relations still retained its local character, with a plethora of district rules and conventions.

It was not until 1904 that a nationwide basis for industry-wide bargaining was established. Yet the scope of such bargaining was limited until 1920 when the National Wages and Conditions Council was formed with a view to negotiating a national working rule agreement for the building industry. This agreement 'covered rates of wages, extra payments, overtime, payment for night gangs and travelling and lodging allowances to be fixed nationally with provision for regional and local variations. This progress marked the birth of the NJCBI' (NFBTE 1978).

The National Wages and Conditions Council was superseded by the National Joint Council for the Building Industry (NJCBI) in 1926. The unions proposed to this body that wage rates be stabilized, with a five-day, forty-hour week and the principles of tool money and a guaranteed working week established. Most of these demands were not established as working conditions until after World War II, when the unions' strength was enhanced by a shortage in the supply of tradesmen. The buoyant economic conditions of the 1950s had allowed employees to make steady gains in wages and conditions. Throughout the 1960s union membership began to decline, despite the reduction of the working week from 46.5 hours to 44 hours in 1960, a further fall to 42 hours in 1961 and, by 1965, stabilized at 40 hours per week. It fell to 39 hours in 1982 and by 1998 it stood at 37.5 hours.

Union membership in construction continued to fall and it has been mentioned that the relative weakness has been one reason for the relatively stable industrial relations in building. Burgess (1975) identifies other reasons for this pattern. He suggests that the pattern of relationships 'lacks well-defined phases', which is partly due to the structure of the industry, in particular the relative absence of technological change and its complex relationship with cyclical fluctuations in the economy. The comparative industrial peace in the building industry has, no doubt, been partly due to the effectiveness of the negotiating machinery that has evolved, but generally it has been due to the weakness of the trade unions and 'the continuing craft sectionalism if not exclusiveness which divided workers among themselves' (Burgess 1975). This sectionalism has historically been reinforced by the growth in the numbers engaged in labour-only subcontractors. It was not until 1989 that UCATT retracted from a position which forbade labour-only workers membership.

The employers' associations' own history is almost as long as the trade unions' and the Building Employers Confederation is one of the oldest with a history of over 100 years.

From these roots the industrial relations machinery has evolved to incorporate a multitude of parties attached to the negotiating machinery in construction, and at this point it would be useful to examine their roles and responsibilities more fully.

4.2 The role of employers' associations in construction

The employers' associations have a number of functions. These may be seen as support for the industry in terms of the fellows' institution.

4.2.1 Economic questions

Employers have long recognized that by combining they were more able to withstand demands for better wages and conditions from trade unions.

It has been a long-established trade union negotiation tactic to 'leapfrog' in terms of wages and conditions, isolating one employer at a time. The employers' associations represent the employers' interests in a collective manner by undertaking negotiation with trade unions on a national basis over the questions of minimum acceptable standards of wages and conditions of employment. But this general role is supplemented by giving advice and assistance to individual member firms when dealing with their particular labour problems.

4.2.2 *Advisory*

Employers' associations provide members with an information service which is related to companies' trade or commercial functions. Areas included in this service would be the impact of legislation upon the building industry, a wage monitoring service, research and development progress reports.

In a sense, this particular service reflects the federal structure of the building employers' association, where much of the lubrication for industrial relations is provided by local associations with the central body acting as a coordinator.

4.2.3 *Regulatory*

Employers' associations regulate and administer agreements they have reached on behalf of their members and generally provide facilities for the settlement of disputes between unions and individual managements. In this respect, the employers' associations attempt to stabilize relationships between the parties to the industrial relations machinery.

4.2.4 *Representation*

Employers' associations, in common with trade unions, seek to have their point of view made known to the decision-makers. In particular, they will make representations to government, especially when seeking to amend specific legislation, but at a more general level, employers' associations will attempt to become party to economic planning in respect of the building industry.

4.2.5 *Technical and commercial service*

The structure of the construction industry, with the largest number of firms being concentrated into the small to medium-sized organizations (see Figure 4.1), means that an employers' association will often be asked to provide a technical and commercial advisory service. Issues which are likely to be foremost here are assistance in negotiating contractual conditions, legal advice, cost and estimating advice along with miscellaneous commercial advice on matters for which a small to medium-sized contractor would not have in-house expertise.

4.2.6 *Political influence*

The employees' association will seek to make representation to government to influence its policies in favour of construction industry interests.

4.3 The structure of employers' associations

There are several employers' associations allied to the construction industry. The largest of these is the Construction Confederation which has over 5,000 members and represents 8 per cent of the 80,000 firms identified as being within the industry. The employers' association in the construction industry is dominated by the Construction Confederation formed in 1997. As its name implies, it is an amalgamation of a number of distinct organizations. The constituent organizations cover a wide range of construction industry interests. The component groups are explained below.

4.3.1 *The Major Contractors' Group (MCG)*

This group represents 22 of the top contractors in the UK. Membership is limited to those with a turnover in excess of £200 million. In total the MCG carry out £20 billion of construction work - about 40 per cent of the total and much higher if one takes into account that such contractors are unlikely to engage in the repairs and maintenance market. The MCG is made up of the chairs or Chief Executives of the companies in membership.

The group works with the Government and other decision-makers on key issues affecting the industry such as the Private Finance Initiative (PFI) or the specific legislation such as the Construction Act (1997).

Allied to the MCG is the Export Group for the Constructional Industries (EGCI). This agency seeks to promote British construction abroad - this is a sizeable part of the major contractors' portfolio of work since it is a £5 billion export each year. In Europe it is linked to the European International Contractors.

4.3.2 *National Contractors' Federation (NCF)*

This group represents the second tier of construction contracting. Membership is limited to those companies with a turnover of £50 million. It has 25 members and serves a similar function to the MCG.

4.3.3 *National Federation of Builders (NFB)*

The NFB is the former Federation of Master Builders which dates back to 1872. It is the largest group in the confederation with 3,200 members and represents the interests of contractors with turnovers of up to £50 million.

Unlike the previous two contractors it has a regional and local organization. It provides a strong service base for the smaller contractor. These services would include credit references, insurance packages, debt collection and a guarantee scheme. Other facilities include group purchasing on fleet vehicles, etc. In order that the interests of the medium-sized builders are represented as well as the small builder, the NFB is split between a general builders' group and a smaller builders' group. The two groups are represented at all levels in the Federation.

4.3.4 *The Federation of Building Specialist Contractors (FBSC)*

This organization provides services to 800 member companies who tend to be specialist trade contractors. The affiliated specialist organizations are:

- Stone Federation of Great Britain
- National Association of Scaffolding Contractors
- Federation of Plastering and Drywall Contractors
- Painting and Decorating Federation
- Suspended Access Equipment Manufacturers Association
- National GRP Federation

4.3.5 *Civil Engineering Contractors' Association (CECA)*

This was formerly the Federation of Civil Engineering Contractors which became the CECA in 1996. It represents the civil engineering industry. It has 200 firms in membership and has developed seven regions who keep in touch with civil engineering clients such as the Government and local authorities. It is strong in lobbying the Government concerning infrastructure investment and repair and maintenance of the national infrastructure.

4.3.6 *House Builders' Federation (HBF)*

This is a very diverse federation since it accommodates the volume house builders through to those who build a few houses in a specific locality. It is organized into eight geographical regions with 50 companies having membership. Its role is two-fold, as a pressure group on the Government and an advisory service to members on land use, planning permission, social housing, public enquiries and marketing.

4.3.7 The British Woodworking Federation (BWF)

The organization dates back to 1904 and today represents the collective interests of companies manufacturing timber components from architraves to timber-framed houses. It functions as an advisory service to its 315 members. Much of the service provided to the members deals with offering advice on technical matters and advising in the development of UK and European Standards and Codes of Practice.

Since its incorporation into the Construction Confederation it has set up a quality assurance scheme which accredits certain products. For example, it has in place Product Accreditation Schemes for fire doors, timber windows and architectural joinery.

4.3.8 Scottish Building Employers' Federation (SBEF)

Uniquely, the SBEF has a strong local structure with 30 local associations. The function of the SBEF is the same as that for the National Federation of Builders for Scotland. The Scottish parliament is likely to strengthen the link that the SBEF has with the Scottish Executive. It represents Scottish construction firms on the Scottish Construction Industry Group.

Special representation is undertaken in respect of the position of Scottish Contract Law.

4.3.9 The Construction Confederation structure

The structure of the Construction Confederation is shown in Figure 4.1. As can be seen, it has four principal areas of activity.

It is hoped that the Construction Confederation harmonizes industrial relations in construction but its confederated state matches the industrial relations practice in industry. The employers do not seek strict control over the activities of their members. Even at the wages' negotiation table, the employers seek to establish a 'basic wage' below which construction workers should not fall and individual employers and trade unions are free to negotiate supplements to these basic terms and conditions agreed at national level. The local autonomy has been a response to the growing size of companies who have had the authority and confidence to negotiate separately with the trade unions. With the growing professionalism of management has come the ability to provide the advisory service which previously the employers' associations provided. Economic conditions have also weakened the employers' associations and contributed to their decline.

In the early 1980s the separate employers' federations claimed membership of over 20,000. The recession of the late 1980s and early 1990s suppressed membership. Furthermore, skill shortages in times of boom or oppressively harsh contract conditions in times when work is scarce have destabilized the authority of the formal system of industrial relations.

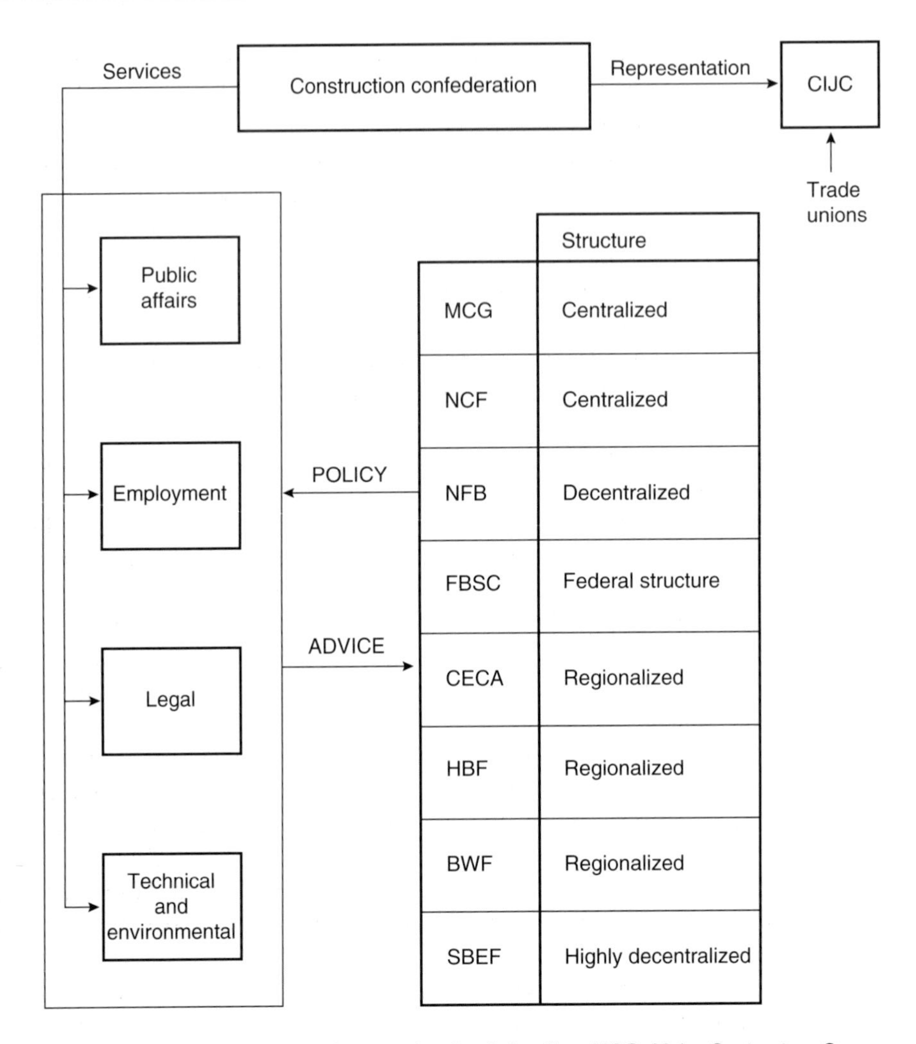

Fig. 4.1 The structure of the Construction Confederation. MCG, Major Contractors Group; NCF, National Contractors Federation; FBSC, Federation of Building Specialist Contractors; CECA – Civil Engineering Contractors Association; HBF, House Builders' Federation; BWF, British Woodworking Federation; SBEF, Scottish Building Employers Federation; CIJC, Construction Industry Joint Council; NFB, National Federation of Builders.

Despite this decentralization of bargaining, the Construction Confederation and its constituent members have played an invaluable role in determining the character of the industrial relations system in construction.

4.4 The role of the trade unions

The role of the trade unions in Britain has changed considerably over the last 20 years. The unions have suffered losses of membership, confidence and influence. The Labour Government (elected in 1997 and again in 2001) seems to be as enthusiastic about keeping the trade unions out of the political limelight as were their Conservative predecessors.

Construction managers may not always welcome the presence of trade unions on a construction site and a prevailing view existed that managers needed to preserve the right to manage. However, the industrial strife which characterized the 1970s (1972 saw the first and only national building strike) was a displaced struggle for power between unions and employers. The miners' strikes of 1982 and more particularly 1984 changed the contours of the balance of power between these competing groups. The recession of the late 1980s gave rise to further losses of membership. The 1990s, with their preferences for unfettered markets, shaped the profile of industrial relations. In such times euphemisms such as 'downsizing' and 'outsourcing' were being used to describe employment practices; inevitably conflict arose between managers and employees. The construction trade unions face difficulties as a consequence of this 'downsizing' for it often means that the framework for the union organization is skeletal. As members are recruited on one site, they may move to another unorganized place of work and drop out. As Drucker and White (1996) note, many construction workers are recruited time and time again.

The competition for recruits to the union is also fierce and this may undermine a general drive to boost numbers of union members. There still exists, despite calls for amalgamations, several trade unions and the structural fragility weakens the union cause. Few administrative, professional, technical and clerical (APTC) staff are recruited and this is surprising since it is often middle managers who are the most vulnerable in respect of job insecurity.

Differences of view are inevitable in these circumstances and it is necessary to recognize valid differences of opinion. Construction managers can benefit from improved communications which flow from strong relationships with trade unions. The partnering culture could be extended from business relationships to that between the employer and trade union.

Against this background the construction trade unions have a single role and that is to represent the members' interests, be it collectively or individually.

At a collective level, the trade unions are responsible for negotiating wages and conditions for their members. Wages negotiation is undertaken nationally inside the regulatory framework for negotiating minimum terms and conditions applicable in the construction industry. This framework is discussed in Section 4.5. Clearly the 'conditions' part of this negotiation can take place at the national level but the trade unions will be concerned to implement the agreement on sites and here the role is one of policing the standards agreed at the Construction Industry Joint Council (CIJC).

Also at the collective level will be the trade unions' involvement in safety management. Here true alignment of interests between construction managers and trade unions may be seen to be evident. The role of trade unions is also referred to in the chapter on safety management.

At the level of the individual, the trade unions are usually the first call for members who are seeking redress for legal claims for compensation in respect of injury or unfair dismissal. Of course this representational role may go as far as the courts but is more likely to stop short and the trade unions may act as a shock absorber in respect of individual grievances, disputes and disciplinary cases.

During the late 1980s the union, in an attempt to staunch dramatic losses in membership due to the recession, agreed to represent self-employed members and has made efforts at improving the conditions for the self-employed and supporting them in negotiations over piece rates and support services on site to be provided by major contractors. The ambiguous tax position of the 800,000 labour-only subcontractors was the subject of an Inland Revenue investigation and the trade unions were engaged in attempting to persuade contractors to take on subcontractors as directly employed staff. This is now fully implemented and Inland Revenue rules have encouraged employers to convert the self-employed into being directly employed.

4.4.1 The trade unions involved in the industry

Despite the long history of trade unionism in construction, the numbers of trade unionists in the building industry are small in comparison to the labour force as a whole. The reasons for this are manifold. The craft sectionalism and stiff competition for members has been mentioned, but other reasons can also be identified. Firstly, the turbulent business environment is prominent – the cycle of high levels of unemployment in the industry to times of skill shortage is seemingly ever shorter. Secondly, the imposition of Compulsory Competitive Tendering (CCT) for local government service coupled with the strong concentrations of union members in direct labour organizations (DLO) in local authorities meant that building workers were working in a market as opposed to a service environment. Indeed the decline in the importance of direct labour organizations has meant that the unions have had to recruit more extensively from the private sector. Figure 4.2 shows the pattern of changes in private sector and DLO employment. Additionally the hegemony of management in the 1980s deterred many trade unionists from being active in organizing and leading unions at the site. Finally, the industry is dependent upon self-employed labour which, despite the best efforts of the unions to recruit them, remain largely stubbornly hostile to organized labour.

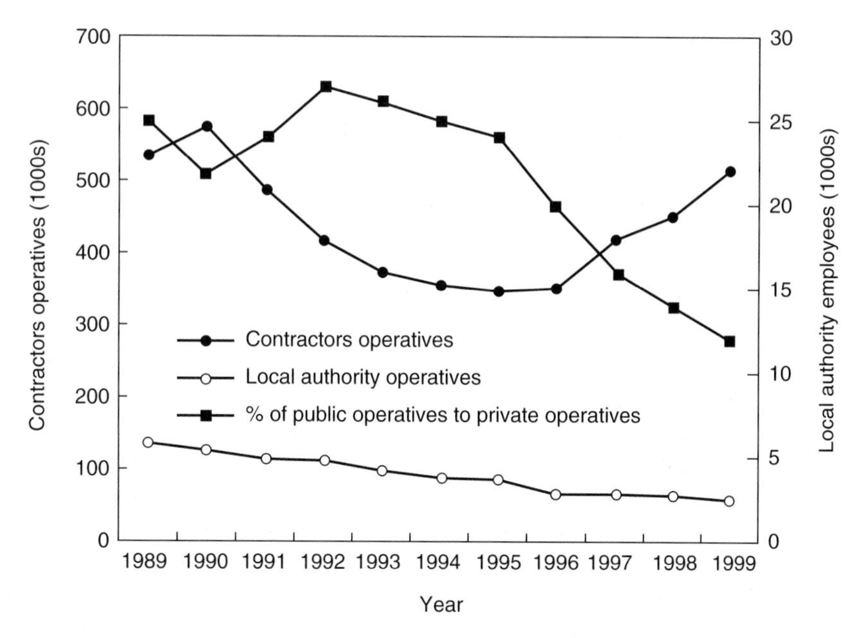

Fig. 4.2 Changes in public and private sector employment.

Notwithstanding these problems the trade unions have a presence. The unions involved in the industry are:

- Union of Construction Allied Trades and Technicians
- Transport and General Workers' Union
- Amalgamated Engineering and Electrical Union
- General, Municipal and Boilermakers Union

UCATT is the foremost of the building trade unions and is a product of the union amalgamations of 1972. At this time several of the old craft unions came together to form UCATT. The need for trade union mergers had been set out some four years earlier in the Royal Commission on Trade Union and Employers' Associations 1965–1968 (Donovan 1968). The report pointed out that there is scope for many more mergers between unions. In particular, it seems to us that problems caused by a multiplicity of unions organizing in individual factories would be considerably eased in a number of important industries if certain groups of craft unions could be encouraged to amalgamate. This is particularly true of engineering and construction.

Furthermore, George Woodcock, TUC General Secretary from 1960 to 1970, had encouraged the union movement to seek mergers wherever a common interest bound them together. Given the background of declining membership of the traditional trade crafts, amalgamation became a necessity rather than an ideological commitment. In 1972, agreement was reached between three principal building unions of the day, namely, the Amalgamated Society of Woodworkers (ASW), the Amalgamated Society of Painters and Decorators (ASP&D), and the Amalgamated Union of Building Trades Workers (AUBTW).

As with most trade unions, the local branch is the basic unit of the organization. UCATT has approximately 650 branches with the size of the branch varying according to location. Branches of UCATT meet on a monthly or fortnightly basis. UCATT has frequently been associated with mergers, the aversion to amalgamating with other unions has been riven by political differences between senior figures in the union movement but at the end of 1995 the union had just over 100,000 members. Significantly only 50,000 were employed in the private sector of the industry. The structure of UCATT is shown in Figure 4.3.

The other major force within the construction industry is the Transport and General Workers' Union (TGWU). This union, because of its general nature, is organized by trade groups, with building and civil engineering having its own trade group organization. This consists of national officers and regional organizers who are elected by members in branches allied to the construction industry. The structure is illustrated in Figure 4.4.

As can be seen, the grass-roots membership is organized into regional sections and, in a general union such as the TGWU, workers in a particular industry are organized into industry-based branches so that a region will have building trade branches situated around a geographical area or a particular site. These branches elect one delegate to serve on the Regional Trade Group, which meets on a quarterly basis and has a full-time official to serve it. Within the TGWU there are 14 different trade groups, each of which elects a delegate to serve on the Regional Committee which overviews the work of the region. At the same time, the Building Trade Group will elect delegates to the National Committee of Building Trade Groups. This body holds meetings on a quarterly basis and its prime function is to review National Joint Council business. The National Committee of Building Trade Groups also elects one delegate to serve on the Executive Council of the union. The Executive Council is served by the General Secretary of the union, who acts as its chief officer. This person is elected by secret ballot by the whole membership of the union.

The union holds a biannual conference to which delegates are elected from regions on a trade group basis. At these conferences, policy is established and the General Executive Council carries out the union's policy between conferences. The union claims some 23,000 members in the building trade group.

Amalgamated Engineering and Electrical Union

The Amalgamated Engineering and Electrical Union (AEEU) is a consolidated union which was produced in 1992 by mergers between the

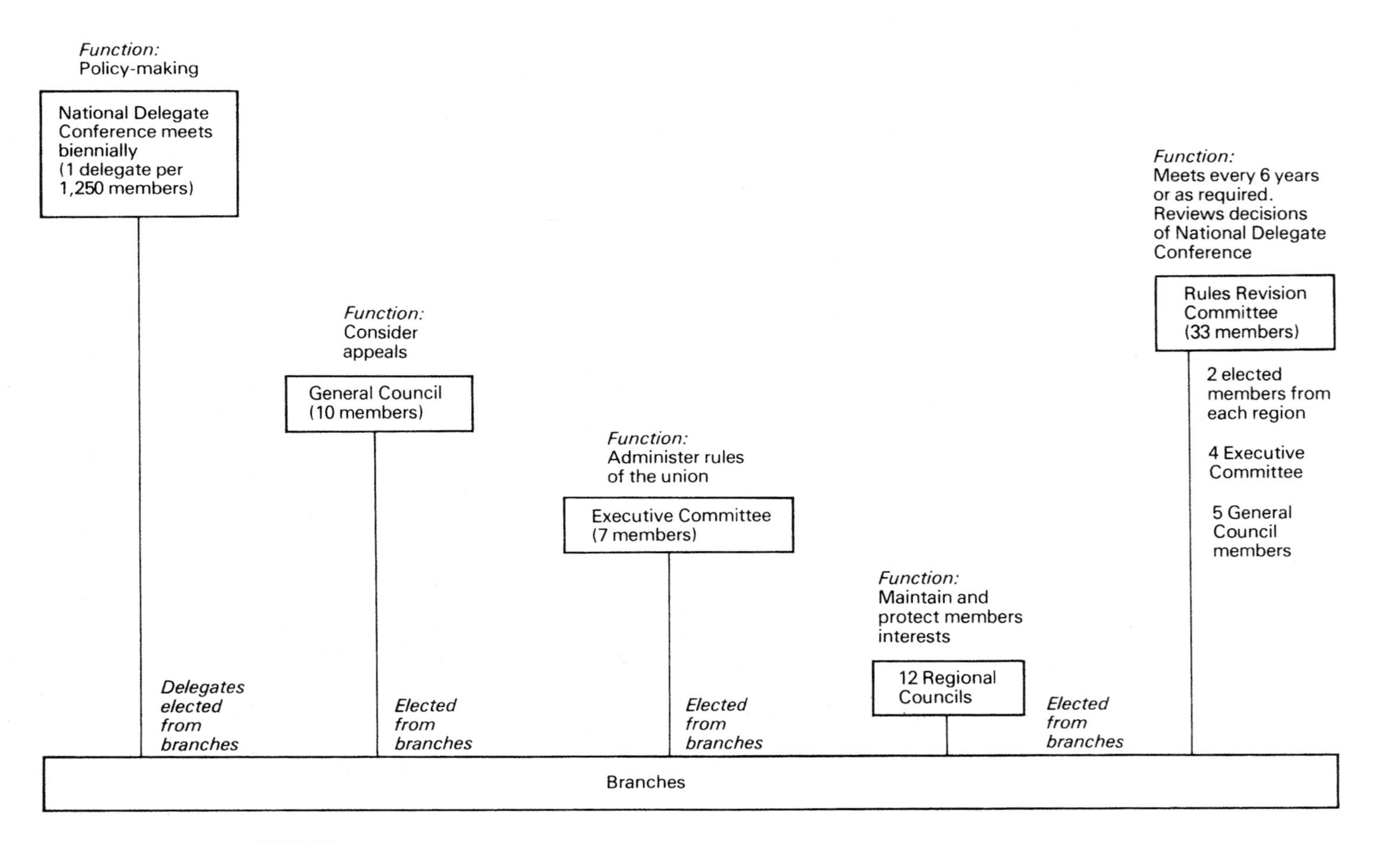

Fig. 4.3 The structure of UCATT.

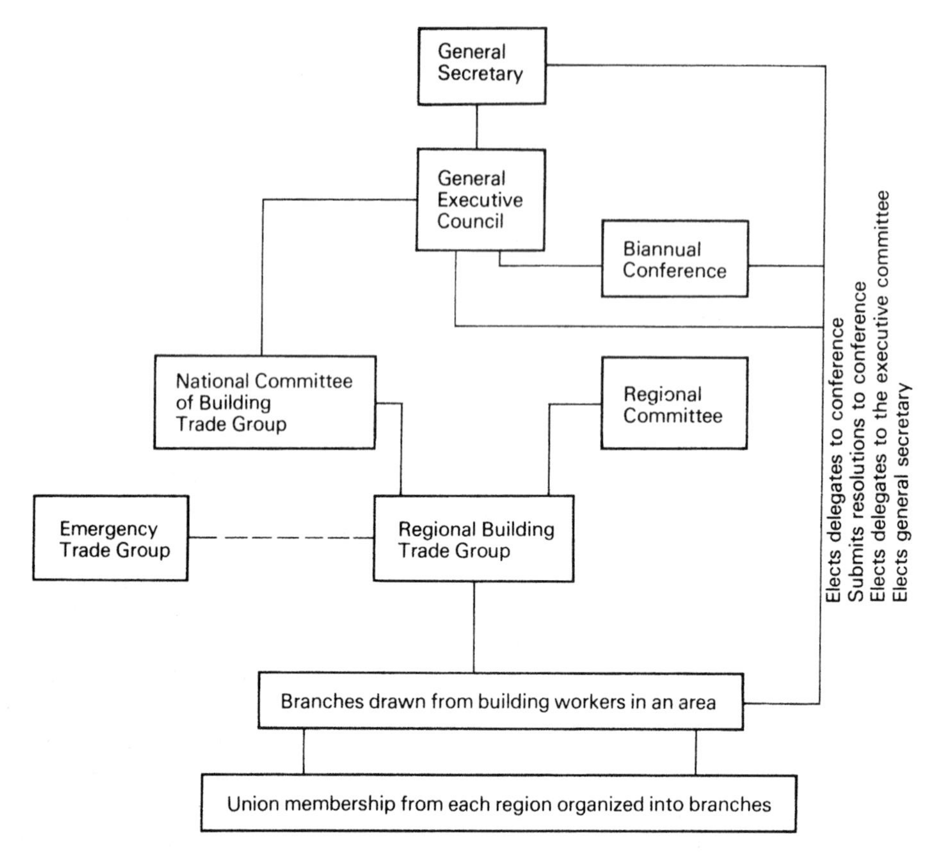

Functions of each committee:

Building Trade Group	– Each region if broken down into areas and one delegate is elected from each area to serve on the Regional Trade Group. Each Regional Trade Group has a full-time secretary. Meets quarterly.
Emergency Trade Group	– Three persons drawn from trade group + chairman + secretary. Discuss emergency matters.
Regional Committee	– All eleven trade groups elect delegates to serve on the regional committee. Administers the region.
National Committee of Trade Groups	– Quarterly meetings, discusses NJC matters. Delegates elected from regional trade groups.
Union Executive Committee	– One member elected from each national trade group. Delegates elected directly from the regional membership.
Biannual Conference	– The sovereign body of the union. Delegates elected from the region on a trade group basis. Monitors the work of the General Executive Council.
General Secretary	– The chief officer of the union – elected by ballot of the whole membership of the union.

Fig. 4.4 The structure of TGWU.

Amalgamated Engineering Union (AEU) (whose membership was mainly factory-based, skilled engineering trades, but contained a few thousand structural steelworkers) and the Electrical, Electronic, Plumbers, Telecommunications Trade Union (EEPTTU). In the context of construction, the AEEU represents the interests of the electricians and plumbers. The balance of membership leans heavily towards factory workers rather than construction personnel and the agency inside the union which deals with these trades is the Construction/Building group which is not recognized by the Construction Industry Joint Council for purposes of negotiation.

The General, Municipal and Boilermakers Union (GMB)

The origin of this union was as a general union which recruited strongly in the local authority sector. Consequently, it organizes many of the construction workers who are employed by local authorities. Its construction membership is around 70,000 with the largest majority being employed by furniture manufacturers, timber and woodworking shops and those mounting exhibitions.

While its role is less influential than UCATT or the TGWU, it is a party to the industry-wide agreements. In many ways, the GMB is well placed to recruit members in projects where private finance initiative (PFI) projects are undertaken. They seek to come to arrangements with promoters of such projects to ensure that construction workers and those who operate the facilities are in the GMB. The union is also very European-minded and in 1994 it formed a construction consortium with four European unions and a construction consortium which was bidding to run work in the London to the Channel Tunnel rail link.

4.4.2 The role of the shop stewards

Trade unions are now less likely to be seen as the fomenter of strikes and industrial unrest and their role is more likely to be resolvers of individual difficulties. Within this process the shop steward is likely to be an important person.

In the words of Clack (1967), shop stewards are 'the shock absorbers in the industrial relations machinery'. However the shop stewards will also be the 'policeman on the beat' in ensuring that the terms and conditions of the national agreements are upheld at the site. One of the critical tasks of the shop stewards, often in conjunction with the full-time officials, is to recruit members and through a system of 'check-off' ensure that the employers deduct the union subscription from the employees' wages. The union then collects the subscription in one tranche from the employer thus saving the shop steward the cumbersome duty of collecting subscriptions each week.

Other functions will be to ensure that the health and safety regulations are kept and here the shop steward can greatly assist management by adding his own voice to ensure that safety rules and regulations are obeyed. It may be easier for a worker to be seen to be responding to a colleague who has been elected than a manager who has not!

Each of the unions has a handbook which spells out the duties of the shop steward. Once elected by trade union members on site, the shop steward is officially accredited by the trade union and the union then applies to the employer for official recognition, which should not be unreasonably withheld. On large sites where more than one trade union is likely to have a presence, the unions may choose to elect a 'convenor'. The facilities provided to this official will obviously vary from site to site, but in certain circumstances could include facilities for interviewing new employees, accommodation for meetings, access to a telephone and even office equipment.

Additionally, employers are legally required to allow stewards time off for trade union duties without loss of pay; the cost of such provisions could be more than outweighed by the advantages of a healthy industrial relations climate. Moreover, this type of cooperation is broadly encouraged by the governments' Advisory, Conciliation and Arbitration Services in the code of practice concerning the facilities to be made available to union representatives.

More often than not managers use shop stewards as quasi-management, acting as a go-between in communicating management decisions. This is a misunderstanding of the role of the steward. Management must use its own line management to inform the workforce of its decisions; however, such decisions are more likely to find acceptance by the stewards if the stewards have been involved during some stage of the decision-making process.

4.5 Management responsibilities in industrial relations

The construction industry has a bifocal view of the management of industrial relations. At one level the employees have given their support to the sustenance of the industrial relations machinery in terms of ensuring that the formal collectively bargained agreements survive. The evolution of the Construction Industry Joint Council in 1998 is evidence of this. Similarly the trade unions have sought to extend the coverage of issues dealt with by the collective bargaining machinery.

However, as was highlighted in the section on the industrial relations machinery, there is a large distance between the formality of the national agreements and industrial relations management on site. Drucker and White (1996) describe this as 'proceduralism'. The gap between agreement and practice is shaped by the numerical and organizational weaknesses of the trade unions and while the employers' associations support the principles of national agreement, contractors forming the Major Contractors' Group seldom implement the national agreements to the letter. This is not to say that employers seek to worsen the national terms and conditions, frequently take-home pay is topped up to match local market conditions for labour.

This combination of a formal system of industrial relations being supported by an informal system means that the management responsibilities of industrial relations can be complex. In order to identify these, the section has been broken into four points:

- labour-only subcontractors and industrial relations
- the legal framework for the conduct of industrial relations
- managing industrial conflicts
- empowerment and industrial participation

4.5.1 Labour-only subcontracting (LOSC) and industrial relations

In the early days of subcontracting on a labour-only basis, Langford (1974) reviewed the effect of LOSC on the construction industry's conduct of industrial relations. In a survey of 200 directly employed construction workers 76 per cent of the sample felt that LOSC adversely affected the bargaining power of the trade unions at site level.

Since this survey the debate has moved on and because of the political and economical encouragement that Conservative governments gave to labour-only subcontractors during the 1980s, coupled with the decay of 'union power', the result is unlikely to be replicated. In the words of Drucker and White (1996), the 'entrepreneurial climate of the 1980s gave some legitimacy to the position of the self-employed'. Winch (1998) endorses this argument by noting that the culture of the industry perpetuated 'individualism' which 'made the autonomy of labour-only contracts attractive'.

This army of the self-employed often operates in gangs which may have a reasonable length of service with a particular contractor. What is evident is that such individuals need managing and the provision for this management frequently comes from the gang-bosses. Part of this management will be that traditionally associated with industrial relations – hours worked, wages, conditions, etc. Consequently, the industrial relations management is highly decentralized and conducted in small self-contained cells. This would suggest that the management of industrial relations has been displaced and moved to those who perhaps have least training to manage the process. This transfer of responsibility, as a by-product of the subcontracting system, has made main contractors vulnerable to having their production stopped by one trade which impinges upon the work of another. In such an unintegrated industrial relations structure the industrial power may move not to unionized labour but to key craftworkers who are vested with the authority to damage progress of a job.

In such elements, the formal rules of the industrial relations machinery are unlikely to be of any great value and project managers are more likely to rely upon personal persuasion and the building of shared agendas with labour-only bosses to resolve potential industrial relations problems.

The informality of site practices, however, have to take place within the formal regulatory system to which most of the major players within the industry subscribe.

4.5.2 The legal framework

The industrial relations machinery is set within a legal context. One of the benefits that employers and trade unions gain from participating in the national bargaining round is that it gives access to the disputes resolution machinery. While strikes have only occasionally been experienced in the industry, it is noticeable that strikes act as a barometer for the general conditions of trade. As work becomes short the labour force became more docile.

Increasingly the law has been used to fashion how industrial disputes are settled. The trend of industrial stoppages has been declining since the early 1980s, but over the same period applications to employment tribunals increased markedly. The message is that during the 1980s if you had an issue with the boss you went on strike, nowadays you take them to court. The law has encouraged this change and two pieces of legislation have been influential.

The Employment Rights Act (Dispute Resolution) Act 1998

This act was a consolidating piece of legislation which brought together the disparate pieces of law which governed the conduct of industrial disputes. Industrial tribunals were rebadged as employment tribunals and the structure and jurisdictions of the tribunals were set out. Other methods of dispute resolution were mapped out and this included the workings of the Arbitration, Conciliation and Advisory Services (ACAS) scheme. Boundaries of compensation were also defined. An average award for unfair dismissal in 2000–2001 was £2700.

The Employment Relations Act 1999

This piece of legislation was developed from the Government's White Paper, Fairness at Work. The Act comprises five major planks of the governance of industrial relations.

- New statutory procedures for the recognition and derecognition of trade unions for collective bargaining to apply when unions and employers are unable to reach agreement voluntarily, and a requirement for employers to inform and consult unions recognised under the statutory procedure on their training policies and plans.

- Changes to the law on trade union membership, to prevent discrimination by omission and the blacklisting of people on grounds of trade union membership or activities.
- Changes to the law on industrial action, in particular to certain aspects of the ballot requirements and to enable dismissed strikers to complain of unfair dismissal in certain circumstances.
- New rights and changes in family-related employment rights, aimed at making it easier for workers to balance the demands of work and the family.
- A new right for workers to be accompanied in certain disciplinary and grievance hearings.

4.5.3 Managing industrial disputes

It has been shown that strikes and other disputes do not dominate the construction industry but where they do occur it is frequently on large sites. There are likely to be significant physically and politically as well as in their visibility and are likely to be long-lasting and so enable unions to recruit and retain members.

The complexity of major projects lies in the fact they fuse together many different trades which may have membership to different unions, with each union competing for membership and influence. Invariably jobs will have a multi-interactive basis and may be covered by different national agreements, follow different industrial relations policies, have different 'custom and practice' traditions.

This situation leads to a variety of rates of pay for operatives of the same trade employed by different contractors on the same site. The potential problems in this situation have been spotted by large clients and they have frequently specified that the industrial relations are managed before work commences on site. The usual way this is put in place is for the client to 'nominate' the site as being special and be subject to a procedure agreement which would cover the site. This nomination, having been agreed by the CIJC, causes the union to stimulate the drawing up of a Project Joint Council which is made up of the unions, the principal contractors and is often chaired by the client. This Project Joint Council is charged with developing a procedure agreement which supplements or even supplants the national agreement. The advantages of such project based agreements are:

- negotiation at the start of a job can be anticipated in order to reduce questions of demarcation and productivity
- overtime can be controlled
- the amount of shift time can be determined
- the facilities to be provided for shop stewards can be decided
- a policy can be formulated on any selection for redundancy

Site agreements have become more popular for large sites and have helped to calm industrial relations. The difficulties experienced in the 1970s in major construction engineering and petrochemical projects have largely been avoided in the major projects of the 1990s. Certainly the zeitgeist will have played its part – weakened trade unions, an admiration for individual enterprise rather than social solidarity – but the role of large site agreements has also contributed to industrial peace.

Perhaps the best example of the benefit of site agreements was shown by the agreement used during the construction of the Channel Tunnel. This was an unusual agreement in that industrial relations on this project followed a pattern of joint regulation, and strong trade union representation was encouraged and labour-only subcontracting was discouraged. Fisher (1993) highlighted that industrial relations in the Tunnel project was a success. The reason for this success was that the management representatives saw the need for the Working Rule Agreements to be upheld. John Donogue, the industrial relations manager with Trans Manche Link (TML), saw a need to engender a philosophy of 'justice and care' applied to the 6,500 workers employed on the project.

The management of industrial relations was driven by the acceptance by TML and the unions of one site agreement based upon the National Working Rule Agreement supported by Codes of Practice which were applicable to all TML sites. The site-based agreement was issued to every employee and subcontractor. This procedure ensured that subcontractors could not undermine the principles of the site agreement. By managing in this way, labour-only subcontractors were barred from the site. More importantly, TML by giving support to the unions enabled the unions to build an effective organization such that almost 100 per cent of the employees joined either UCATT or the TGWU. Norman Willis (1992), the former TUC General Secretary, summed up the success of this approach when he observed:

> There can be no doubt that the Channel Tunnel project is the most important infrastructure investment under way in Europe, perhaps in the world today, comparable to the Great Wall of China and the Suez Canal. What is remarkable is that on this huge project there has hardly been a day lost through strikes. Any problems have been dealt with on the site. There have been excellent relations between the site officials and full-time union officials.

The greatest crisis to industrial peace came not from traditional topics such as bonus and pay but from the management of safety. The only major stoppage concerned the safety issue: it was customary for men to walk out for 24 hours after a fatal accident. The management of safety on the Tunnel will be referred to in Chapter 5.

4.6 Empowerment and worker participation

Over the past two decades, the issue of participation has largely disappeared as an industrial relations issue in most industries in the face of renewed management assertiveness fostered by government-stimulated managerialism. Yet the overriding influence on all industries and firms has been the drive for competitiveness. In their continuous search for superior performance, firms have recognized the valuable contribution that employees make and, more importantly, their potential for achieving sustainable competitive advantage. The management of industrial relations has been geared to managing such advantage as can be gained. Goss (1994) argues that human resources have become the major source of sustainable competitive advantage for most firms, based on the premise that in an increasingly competitive market, both global and local, all firms can obtain the same materials, equipment and other factors of production at essentially equivalent cost, therefore the main competitive difference between them is their human resource and how well it performs. The growing importance of employees to the survival and success of firms is also reflected in the development of human resource management (HRM) specialists as opposed to industrial relations experts. A large number of firms, across a wide range of industries and countries, have successfully implemented employee participation to improve their competitiveness, producing results such as improved quality, innovation, greater flexibility, faster adaptability, but, more tangibly, increased productivity and lower production costs (Cole 1979, Massarick 1983, Porter 1990, Coffey & Langford 1998). These firms have recognized that the key to achieving these results is the commitment of all the employees in the firm, which is attained through employees being involved in decision-making which in turn requires their genuine participation.

Despite the success of employee participation in other industries, there are no discernible examples in the construction industry, an industry that would appear to be well suited to its application. The lack of application poses the question of whether there are fundamental reasons in the organization, operation, work or workers of the construction industry that prevents the successful utilization of employee participation.

The question of empowerment and participation is often taken as a comprehensive package but Ramsay (1996) sees four strands, namely:

- task and work group involvement
- communication and briefing systems
- consultative arrangements
- financial participation

To an extent, participation remains contextual in its definition, but must include certain defining characteristics of participation which have been identified by Tannenbaum *et al.* (1961) as decision-making, a conclusion also arrived at by Blumberg (1968) who deduced from his extensive study that decision-making was the critical component of participation.

Ramsay (1976) extended this view by noting that the most effective means of providing this participation was involvement in some negotiation procedure. This linked participants to the area of greatest interest to them - their jobs.

The definition of participation with decision-making as its central characteristic requires further development; for participation to be successful it must be genuine, which in turn requires the decision-making to be genuine. The significance of genuine participation is that it is essential to the success of participation, a point confirmed by White (1979) whose study of schemes of participation concluded that continued success was a direct result of the genuineness of the participation. Allport (1949) identified that 'only authentic participation taps central values, calls for the person's ego to be engaged with the issues'. Genuine participation is defined as decision-making that is not subject to the review of approval of the authority of any other individual or group. Verba (1986) originated the concept of 'pseudo-participation' to describe schemes of participation that purported to be genuine but which in reality were not, a concept with which Pateman (1970) concurred stating, 'that participation must be in something and in the case of industrial relations it must be in decision-making'. Many of the reported schemes of participation and employee involvement do not meet the test of genuine decision-making and cannot be considered as participation.

A further characteristic of participation that must be defined is its form, with respect to whether it is direct participation or indirect participation. Stated in simple terms, direct participation occurs where each worker affected by a decision personally takes part in the decision-making. In contrast, indirect participation relies upon representation in which workers' participation in decision-making is undertaken through a representative. The distinction also serves to define the scope, or span, of the decision-making, direct participation is limited to decisions that only affect the work of the individual and/or members of the group making the decision, while indirect participation affects the work of a much wider group. Indirect participation, because of its need for representation and its wider span of decision-making is invariable formally designed and organized, and generally originates from the management of the firm.

This formality was once represented by 'hard forms' of participation such as power-centred works councils but now worker directors have moved to softer forms of empowerment and other techniques for developing employee commitment (Ramsay 1996).

A search for reported schemes of indirect participation in construction revealed there to be no such schemes that could be considered genuine, or that were operated at the worker level. A study of participation and construction was undertaken by Rosenfield *et al.* (1992), who studied the application of quality circles which should be ideal for use in construction. He contended the only quality circle schemes that could be identified involved first-line supervisors and their managers. No quality circle scheme which involved workers on site could be found.

Glancy (1994) sees a number of techniques such as quality circles, semi-autonomous work groups and more recently total quality management (TQM) as vehicles for implementing task participation initiatives. However he doubts the durability of such techniques in advancing participation.

Indirect participation in the construction industry is perhaps the next wave in the HRM function.

The difficulties of implementing such programmes should not be underestimated. The structure of the industry, with high levels of employee mobility and a high proportion of self-employed, diminishes the opportunities for participation. Fisher (1997) agrees that in most conditions joint consultation can work. This consultation can bring together management and workplace representatives (whether unionized or not), first consultations usually focus upon workplace issues concerning the work tasks. This is the conclusion of Coffey and Langford (1998) who studied the willingness of construction workers in the electrical and mechanical trades. They found that workers 'positively and unequivocally want to participate in the organization of their work'. In managing an external environment, the buzz word of the 1990s has been partnering yet this concept can be extended to partnering between employer and employee for the benefit of both.

4.7 The industrial relations machinery

It has been stressed earlier that there is a diversity between the national agreements and local conditions pertaining on site. This diversity is a reflection of the formal and informal systems of industrial relations, which are held to coexist within industry as a whole. In building, the formal system of industrial relations is characterized by the existence of the Working Rule Agreement for the Construction Industry with its assumption that its influence is industry-wide and is capable of imposing common standards upon a diverse industry. Central to this acceptance of the formal system is the understanding that most, if not all, matters can be covered by collective agreements, with pay being determined by industry-wide settlements. Disputes arising from the formal system are derived from differences in interpretation of the national agreement.

In contrast, there is the informal system, which depends upon the wide autonomy of construction managers and trade unions to determine appropriate industrial relations standards. This autonomy recognizes that collective bargaining at the site level is just as important as bargains at the national level. At the local level, the issues which can be considered are extremely wide and will incorporate such issues as discipline, redundancy, work practices, etc. and will have a strong basis of 'arrangements' and 'understandings' acting to police the system.

It can be seen that the informal system can undermine the regulative effect of the Working Rule Agreement (WRA), but many have argued that it is more important than the formal system. However, it must be stressed that total dependence upon the informal system is courting danger, since it is likely to lead to very diverse conditions between company sites and an over-reliance upon expedient methods of dealing with immediate industrial relations policy. This is especially so when it is dealing with a multiorganization setting where some union members may be directly employed, others unionized self-employed and non-unionized workers combining together on one project.

Here collective agreements are carried out at national and site level. At site level, bargaining took place between contractors and union representatives long before the first formalized industry-made agreement was drawn up. The history of collective bargaining dates back to the formulation of the National Joint Council for the Building Industry in 1926. This body brought together employers' representatives and trade unions for the purposes of establishing working rules for the building industry. A parallel body was the Civil Engineering Construction Conciliation Board which focused upon negotiating the civil engineering agreement.

In 1998 a new body, the Construction Industry Joint Council (CIJC), was established. In June 1998 it oversaw the integration of the building and civil engineering working rule agreements into one single agreement covering the whole of the construction industry. Like previous joint councils, it comprises a number of employers' federation and trade union representatives. Figure 4.5 illustrates the structure of the CIJC. Within this structure the key players are the chair and vice-chair and the two joint secretaries, one from the employers the other from the trade unions.

In comparison with the highly decentralized machinery that had gone before, the structure is strongly centralized with no scope for regional or local involvement (except in Scotland which has its own regional council).

The functions of the CIJC are to draw up the Working Rule Agreement. Like those before it, the WRA is a substantive agreement which attempts to set down minimum standards for the industry as a whole.

Areas which the WRA covers include the following:

- wages
- working hours
- overtime rates
- daily fares and travelling allowances

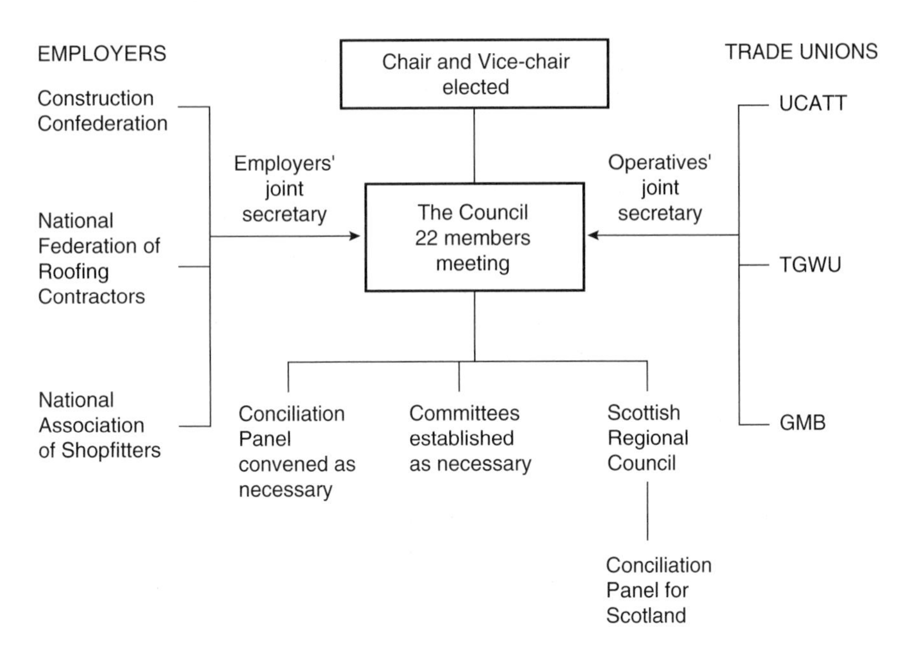

Fig. 4.5 The Construction Industry Joint Council.

- shift working and special conditions
- subsistence allowances
- holiday entitlements
- sick pay
- trade union facilities and negotiation procedures
- termination of employment procedures
- occupational pensions (under the Building and Civil Engineering Benefits Scheme)
- disputes procedure

In addition to the national agreement, locally made procedural agreements are encouraged, but the NWRA notes that any supplementary agreement for a site must not conflict with the WRA. This is reinforced by reference to the disputes procedure where the WRA takes precedence (Working Rule 29). However, the site-based deals may be used to settle disputes and grievances and this may go some way towards explaining the relative unemployment of the formal disputes procedure which has been embedded into the WRA (Working Rule 22).

Greater latitude is allowed for the bonus arrangements. Bonuses account for a substantial portion of a building worker's gross pay and are a much larger proportion of the wage in building than in all other manual trades. This element of the wage is left to local negotiation and is merely an acknowledgement of 'the decline in the intent to which industry-wide agreements determine actual pay' (Donovan 1968).

The WRA acknowledges that its remit cannot extend to covering bonus arrangements and so limits its advice to one sentence, 'It shall be open to employers and employees on any job to agree a bonus scheme based upon measured output and productivity for any operation or operations on that particular job' (WR2 Bonus).

An important aspect of procedural agreements is the principle of 'status quo ante' (the situation which applied before). This means that changes in the working conditions, rates for the job, bonus schemes, manning levels, etc. should not be made without prior consultation with the trade unions. It is now TUC policy to insist upon a 'status quo ante' clause in all new procedural agreements.

With national firms, there is a tendency for procedure and substantive agreements to be made between national officers of the union and the employers, and where such arrangements are made there is a tendency to write in details of working arrangements and site conditions. By doing this, areas of possible differences are removed and as people employed know what has been agreed there is less likelihood of a dispute when the contract has commenced.

However, the question of disputes should not be overstated. While the building industry has the conditions for unrest – dangerous working conditions, low basic pay supplemented by fragmented bonus arrangements and the fundamental unease in relationships between the employed and the employer – strikes have not been a dominant feature of the industry.

In the main, the absence of disputes can be partially explained by the comprehensive disputes procedure laid out in the WRA (WR22 Grievance Procedure). The procedure is paraded as a step by step route through the disputes resolution machinery. Put diagrammatically, the route is shown in Figure 4.6.

The working rule agreement also details the procedures to be used at the conciliation panel including the rules for presenting evidence.

The WRA highlights that stoppage of work or 'go slow' or lockout shall not take place until the grievance procedure is exhausted.

Summary

The widespread use of labour-only subcontracting in the industry has meant that the formal system of managing industrial relations as represented by the Construction Industry Joint Council is underused. The informal system, which constitutes understandings and arrangements made between management and arrangements made between management and labour, is more likely to be influential but the formal system provides the framework for these understandings. Consequently, issues of wage rates, hours worked, site facilities, etc., are likely to be negotiated

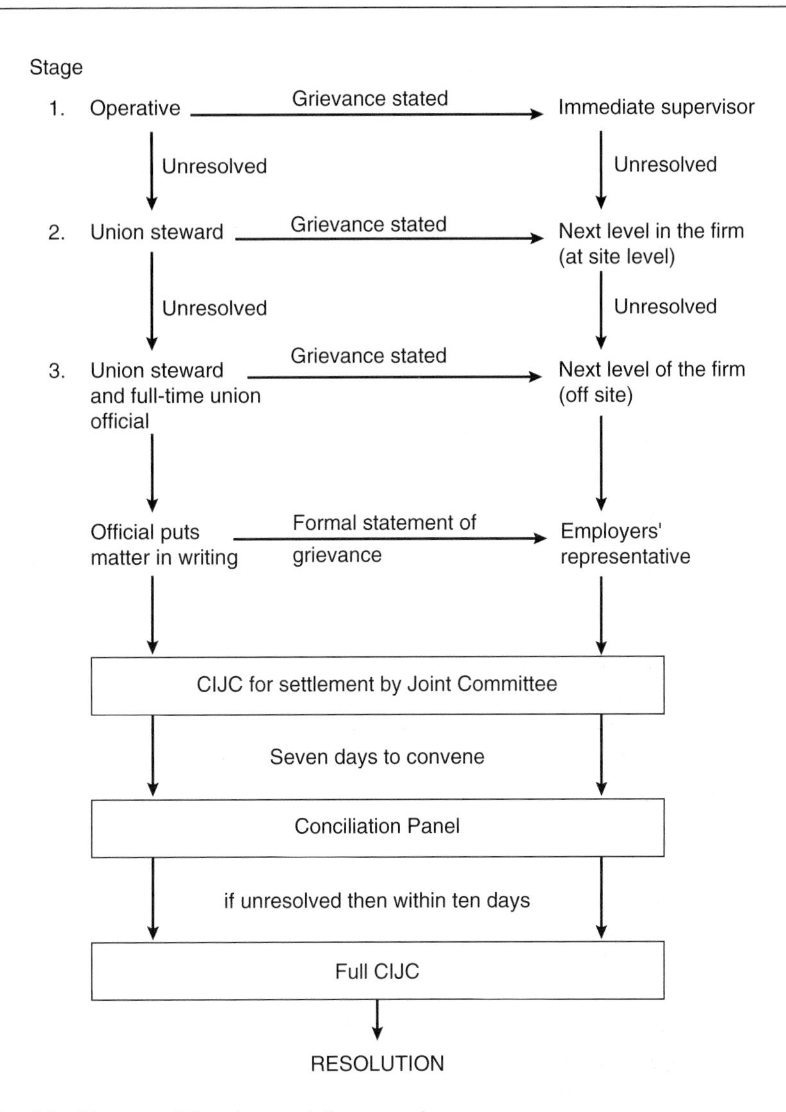

Fig. 4.6 Diagram of disputes resolution procedure.

by subcontractors and managers but the national conditions will be used as a floor below which no one should drop.

The institutions which participate in the industrial relations machinery have strong representational and lobbying roles in respect of government policy but both employers and unions would like to see greater numbers of employers and workers in employers' associations and unions.

In many ways the growth in the size of companies has meant that the traditional advisory role of the employers' confederation has passed to specialists employed by the company; the traditional role of the shop steward in negotiating on behalf of the workforce has passed to the gang-boss of specialist tradespeople.

Questions

1. In comparison with many other industries, construction has a good record in industrial relations. Discuss the factors which give rise to this situation.
2. In the future it may be necessary to train multiskilled operators. Speculate on the responses from the employers and the trade unions to this prospect.
3. Evaluate the role of the Construction Confederation in the construction industry. How is this role changing?
4. Large civil engineering projects are often the flashpoint for industrial relations in construction. Explain why this is often the case and how the situation could be improved.
5. Clients have a responsibility to improve industrial relations in construction. Discuss how the client can have a positive impact upon site industrial relations.
6. Are trade unions in construction governed democratically? Argue your case with reference to the structure, and organization of the trade unions.
7. You have been appointed as the industrial relations manager of a large construction company. Your first task is to introduce yourself to the senior site staff and to explain the role you expect to fulfil. Prepare a paper, to be sent to senior site staff, which identifies your role and its relationship to the work of sites.
8. The authority of the Construction Confederation is often under threat from the constituent employers' associations. Explain why this may be the case and what can be done to increase the solidarity of the confederation.

References and bibliography

Allport, G. (1949) *Personality: In a Psychological Interpretation*, Constable.

Burgess, K. (1975) *The Origins of British Industrial Relations*, Croom-Helm.

Clack, G. (1967) *Industrial Relations in a British Car Factory*, Department of Applied Economics, University of Cambridge.

Clarke, L. (1992) *Building Capitalism*, Routledge.

Coffey, M. and Langford, D. (1998) 'The prosperity for employee participation by electrical and mechanical trades in the construction industry', *Journal of Construction Management and Economics*, Volume 16 (pp. 543–552).

Cole, G.D.H. (1953) *Attempts at General Union*, Macmillan.

Cole, R. (1979) *Work, Mobility and Participation. A Comparative study of American and Japanese Industry*, University of California Press.

Drucker, J. and White, U. (1996) *Managing People in Construction*, Institute of Personnel and Development.

Donovan, T.N. (1968) *Royal Commission on Trade Unions and Employers' Associations 1965–1968*, HMSO.
Fisher, J. (1997) 'Industrial Relations and the Construction of the Channel Tunnel', *Journal of Industrial Relations*, Volume 24, Number 3 (pp. 211–223).
Glancy, J. (1994) 'Task Participation: Employees' participation enabled or constrained in personnel management', *A comprehensive guide to theory and practice in Britain*, Sisson, K. (Editor), Blackwell.
Goss, D. (1994) *Principles of Human Resource Management*, Routledge.
Hilton, W.S. (1968) *Industrial Relations in Construction*, Pergamon.
Langford, D.A. (1974) *The Effect of Labour-Only Subcontractors upon the Construction Industry*, unpublished MSc thesis, University of Aston, Birmingham.
Marx, K. (1867) *Capital*, Volume 1 (1970 Edition), International Publishers.
Massarick, F. (1983) 'Participative Management', *Work in America Institute: Studies in Productivity, Number 28, Highlights of the Literature*, Pergamon Press.
NFBTE (1978) *An Outline History of the NFBTE 1878–1978*. National Federation of Building Trades Employers.
Pateman, C. (1970) *Participation and Democratic Theory*, Cambridge University Press.
Porter, M. (1990) *The Competitive Advantage of Nations*, MacMillan.
Postgate, R. (1923) *The Builders History*, Labour Publishing Company.
Ramsay, H. (1976) 'Participation: the shopfloor view', *British Journal of Industrial Relations*, Volume XIV, Number 2, July.
Ramsay, H. (1996) *Engendering Participation*, Department of Human Resource Management, occasional paper Number 8, University of Strathclyde.
Rosenfield, Y., Warszawski, A. and Laufer, A. (1992) 'Using Quality Circles to Raise Productivity and Quality of Working Life', *Journal of Construction Engineering and Management*, Volume 118, Number 1, March.
Tannebaum, R., Weschler, R. and Masserick, F. (1961) *Participation by Subordinates in Leadership and Organization* (pp. 80–100), McGraw-Hill, New York.
Verba, N. (1986) '1961' in *Towards a New Industrial Democracy: Workers Participation in Industry*, Routledge.
White, K.J. (1979) 'The Scanlon Plan. Causes and Correlates of Success', *Academy of Management Journal*, 22 June (pp. 292–312).
Willis, N. (1992) *The Link* (monthly magazine of Trans Manche Link (TML), March.
Winch, G. (1998) 'The Growth of Self-employment in British Construction', *Journal of Construction Management and Economics*, Volume 16, Number 5, September (pp. 531–540).

5 Health and Safety in Construction

The necessity for construction firms to adopt sound safety policies has seldom been questioned and the Construction (Design and Management) Regulations 1994 enforced the industry to accommodate changes in the regime of managing health and safety. The organizational changes which have been encouraged in the industry over the last decade have made policing health and safety more diffuse. Among the changes influencing the safety climate will be:

- New procurement routes have fragmented the traditional main contractor control over large numbers engaged to work on a site.
- Client expectations for faster building have accelerated the pace of construction – a factor which may encourage less safe practices.
- Larger projects which are higher are more likely to involve the movement of large prefabricated components.
- There are now increasingly complex organizations for managing projects and the safety systems.

The current debate sees trade union and employer unanimity regarding the need to manage health and safety effectively. This was not always the case. Trade unions have always called for a wider use of statutory measures whilst employees have frequently, but not invariably, preferred to deal with the situation without resorting to the external pressure that the law can apply.

Whatever the nature of this discussion, which focuses upon ways of achieving safer sites, there is a universal recognition that there is a moral and economic necessity to maintain safe working practices on the construction site. Unfortunately, the construction industry has become stereotyped as an accident-prone industry, in fact only the mining and fishing industries have higher fatalities per 1,000 workers employed. There is some relief from this bleak picture in that the number of fatal accidents experienced in the industry was falling in the 1990s, although fatalities relating to the numbers employed rose sharply during 1997–98 but fell in 1998–99. Depressingly, the accident rates experienced closely correlate to the level of activity within the industry – when workload is high accidents increase with construction output.

Against this background the legal, economic and moral aspects of safety in construction will be discussed.

5.1 The magnitude of the problem

Some 889 people were killed in the construction industry over the period 1987–99. In addition, every year, approximately 2,000 workers are injured seriously enough to keep them away from work for three or more days. Construction accidents mostly happen to men going about their daily work and, consequently, those involved with site safety must not only concern themselves with large, technically complex sites but also with the more routine problems of fragile roofs, site transport, scaffolds, etc. The Health and Safety Executive's figures for the sources of fatal accidents confirm that the commonplace operation is the most hazardous. The pie chart in Figure 5.1 indicates the sources of non-fatal major injuries in 1999/2000 and Figure 5.2 shows the trends for fatalities and major injuries over the period 1995/96–1999/2000.

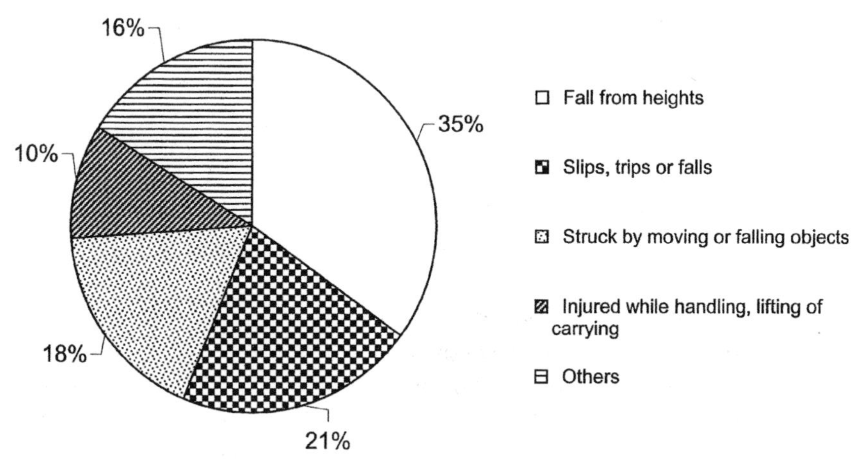

Fig. 5.1 Non-fatal major injuries by kind of accident 1998/1999.

As can be seen, by far the greatest risk arises from falls and, perhaps more surprisingly, the vast majority of accidents occur to experienced tradesmen and building workers engaged in simple traditional activities. Part of the problem arises from the nature of the work, but attitudes towards safety should compensate for this. A site worker has greater autonomy in carrying out work than, say, a factory worker, and therefore there must be a corresponding degree of responsibility for his or her own safety and the safety of others. However, the trends in accident performance give some encouragement in the absolute numbers involved. Figure 5.3 demonstrates that there has been a decline in fatalities and notifiable injuries, but it must be pointed out that the annual returns fluctuate considerably. What is less encouraging is that over the ten years up to 1999, the fatalities have averaged at around eight per 100,000 people at risk, which is typically four to five times greater than that for the

	1995/96	1996/97	1997/98	1998/99	1999/2000
Fatal injuries to employees and self-employed (10s)	9.7	9.3	8.9	7.4	8.5
Major injuries to employers per 100,000 employees	220	400	375	400	380

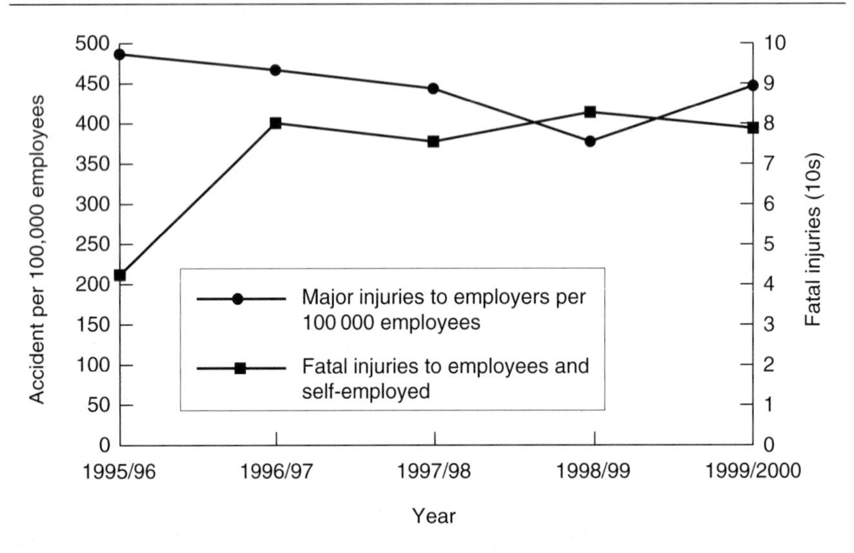

Fig. 5.2 Fatal injury rates and major injuries (1995/6–1999/2000) per 100,000 workers.

manufacturing industry. With respect to disabling injuries, the chances of such an event are almost twice that for the rest of industry. Many explanations have been advanced for this disparity. The demand for higher productivity which is being met by new building methods and mechanization has been cast as one explanation. The challenges set by the Latham Report (1995) for construction times to be reduced by 25 per cent implied a need for urgency in construction programmes. This challenge may be met by greater integration of design and construction, and managerial improvements in the different processes involved in making a building grow. It is perhaps unfortunate that Latham did not urge safer construction as part of the culture change which delivered better buildings to clients, although John Egan's (1998) task force sought safety improvements to the industry. Notably part of the missing agenda is giving construction workers healthier and safer working conditions.

These structural issues, coupled with the casual nature of employment within the industry, have been put forward as the main explanations for poor safety performance. The Swedish equivalent of the Building Research Establishment (BRE) has identified stress as a major contributor to accidents, the stress being caused by the accelerated tempo of work carried out on sites with construction times being too short and the trades working tightly behind one another, creating stressful, and, consequently more hazardous, working conditions. The progressive attitude of the Swedes towards accidents is reflected in their good record in construction safety, having an accident rate less than half that occurring in the UK. However, Britain's accident prevention work is probably more effective than its other European neighbours and it is developing even further as the Construction (Design and Management) Regulations become better understood.

5.2 Health and safety legislation

There are two major pieces of legislation which govern health and safety law. They are:

- The Health and Safety at Work Act 1974
- The Management of Health and Safety at Work Regulations 1999

Allied to these are several statutory instruments governing safety. Relevant ones will include:

- The Construction (Health, Safety and Welfare) Regulations 1996
- The Construction (Design and Management) Regulations 1994
- The Provision and Use of Work Equipment Regulations 1998
- The Personal Protective Equipment Regulations 1992
- The Construction (Head Protection) Regulations 1989
- The Lifting Operations and Lifting Equipment Regulations 1998

5.2.1 The Health and Safety at Work Act 1974

Craig and Miller (1997) have argued that few areas of the law can have been transformed in such a relatively short period of time as the area of health and safety at work. The radicalism evident in the Health and Safety at Work Act 1974 represented an entirely new philosophy of managing health and safety. The Act itself prescribes general duties as listed below but is supported by a raft of codes of practices and regulations governing individual industries and actions.

One of the main areas of the Health and Safety at Work Act was to sweep away the vast array of legislation and regulation which existed in the late 1960s (over 500 pieces of legislation and regulation were in force). It was initiated with the intention of creating a single legal and administrative structure under which all workers and the public at large were to be protected. It is an enabling statute imposing a general care on all people associated with work activities and so is a change from previous health and safety legislation which was more concerned with physical harm.

Generally, the Act consists of new, or at least more specific, obligations on managers, supervisors, and worker representatives and has four particular aims:

1. To secure the health and safety and welfare of persons at work.
2. To protect persons other than persons at work against risks to health and safety arising out of, or in connection with, the activities of persons at work.
3. To control the keeping and use of explosive or highly flammable or otherwise dangerous substances and generally prevent the unlawful acquisition, possession and the use of such substances.
4. To control the emission into the atmosphere of noxious or offensive substances.

One of the principal objectives is to involve everybody at the workplace – management and workpeople – to create an awareness of the importance of achieving high standards of health and safety, and the primary responsibility for doing what is necessary to avoid accidents and occupational ill-health lies with those who create the risks. With the Act came specific duties that employers should perform. In particular, Section 2 of the Health and Safety at Work Act specifies that employers are to ensure the health, safety and welfare of all employees. From the generalized duty, employers have detailed responsibilities:

- to develop systems of work which are practicable, safe, and have no risk to health
- to provide plant to facilitate this duty, and this general requirement is to cover all plant used at the workplace
- to provide training in the matter of health and safety; employers must provide the instruction, training, and supervision necessary to ensure a safe working environment
- to provide a working environment which is conducive to health and safety
- to prepare a written statement of safety policy and to establish an organizational framework for carrying out the policy; the policy must be brought directly to the attention of all employees

However, the employees also have specific duties, namely:

- to take care of their health and safety and that of other persons who would be affected by acts or an omission at the workplace;
- to cooperate with the employer to enable everyone to comply with the statutory provisions

Obviously, these general duties imposed upon employers and employees

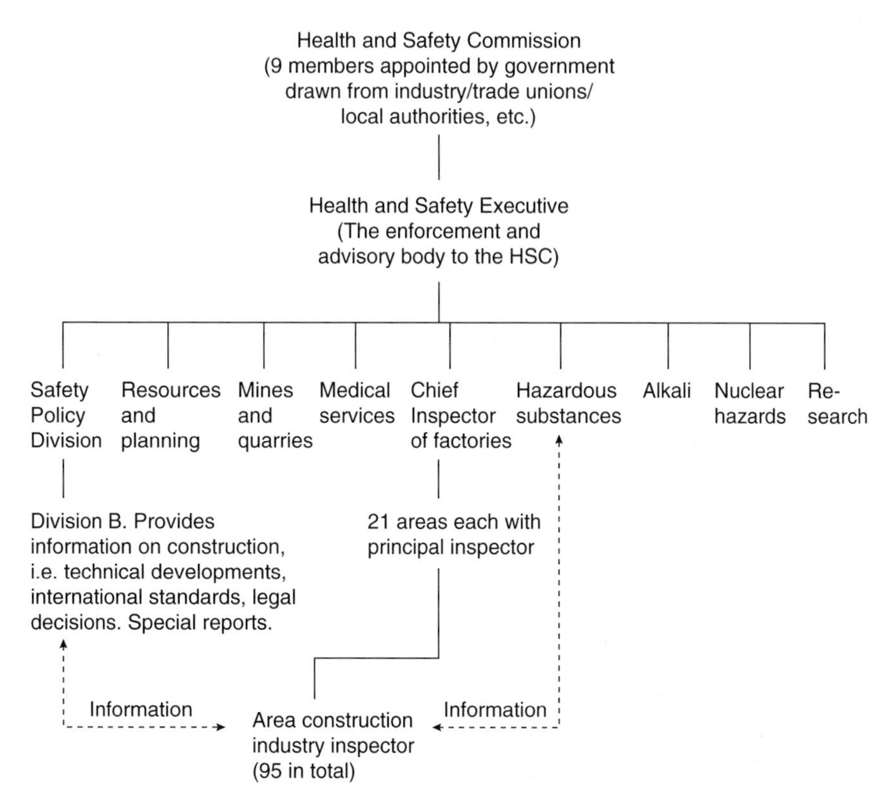

Fig. 5.3 The structure of the Health and Safety Commission.

have to be backed up with an effective inspectorate. The structure of the inspectorate is shown in Figure 5.3.

Clearly, the inspectors are the persons who will be directly involved in checking the safety of particular sites. As such, they are granted the power to enter sites, at any reasonable time, for the purposes of inspection. The inspection procedure may include the taking of photographs, samples, etc., and the inspector may request site managers to provide documents or any other information he may require in carrying out his duties. The inspector has a duty to inform the workforce on a site of any matters which are likely to be deleterious to the workers' health and safety, but the inspector is also obliged to give the same information to the employer. If the inspector finds that a site is in breach of the requirements of the Act, then three possible remedies exist:

- the issue of improvement notices
- the issue of prohibition notices
- the issue of 'seize and destroy' instructions
- prosecute a person who does not comply with the above

The *improvement* notice means that if a site is in contravention of a statutory provision, an inspector may serve the improvement notice, stating his opinion concerning the breach of regulation, the particulars of the breach, and the demand that the contractor rectify the situation within a given period. The *prohibition* notice arises when the inspector is of the opinion that construction activities being carried out involve the risk of serious personal injury. The inspector must state why, in his opinion, a prohibition notice is necessary. The prohibition notice will instruct a person not to carry on with an operation until the necessary steps have been taken to improve safety. These notices may be immediate or deferred. With *seize and destroy* notices, the inspector may seize and destroy any article or substance which causes an imminent danger to persons. These penalties have statutory authority but the underpinning philosophy of the Health and Safety at Work Act 1974 is to move the balance of safety control towards positive prevention of accidents rather than punitive action after the event. This is perhaps a purely pragmatic approach since the thousands of sites in Britain are covered by only small numbers of site inspectors from the HSE, and consequently the chances of a breach of law being discovered are low. Finally the HSE has powers of *prosecution* if these orders are ignored.

The machinery for the detailed prevention of accidents is embodied by codes of practice. Under Section 16 of the Health and Safety at Work Act the Health and Safety Commission has the power to draw up or approve codes of practice for specific working procedures. Observance of codes of practice is admissable as evidence in any criminal proceedings for an offence under the Health and Safety at Work Act. Firms may draw up their own codes of practice, but it is important that these are approved by the Health and Safety Commission. A full explanation of the regulations as presented in Holt (2001).

5.2.2 The Management of Health and Safety at Work Regulations 1992

These regulations provide specific directions on how to carry out the duties. A principal responsibility is the duty to undertake a risk assessment of the hazards to which employees or the self-employed are exposed. The purpose of this risk assessment has been described by the Health and Safety Commission's Code of Practice as being 'to help the employer or self-employed person to determine what measures should be taken to comply with the employers or self-employed person's duties under the relevant statutory duties' therefore, in essence, the risk assessment guides the judgement of the employer or self-employed person as to the measures they ought to take to fulfil their statutory obligations.

This risk assessment process is a vital communication document in the management of health and safety. From the assessment, the workforce should be told three facts:

1. the risks that have been identified
2. what actions are to be taken to prevent the risks
3. risks associated with shared workplaces

5.2.3 The Construction (Health, Safety and Welfare) Regulations 1996

These regulations subsume the specific regulations concerning construction work. These regulations were first incorporated into the Factory Act which was repealed upon the introduction of the Health and Safety at Work Act. Now the main body of construction-specific regulations are to be found in the Construction (Health, Safety and Welfare) Regulations 1996. The 1996 regulations consolidated and updated a range of regulations which were put in place during the 1960s. The thrust of these regulations is that employees, the self-employed and construction managers have statutory duties to ensure safe working practices on site. In general they require employers to provide:

- safe access and egress from places of work
- safety railings to prevent falls
- competent tradespeople to erect and dismantle scaffolding
- designated areas where hard hats must be worn
- rules concerning demolition
- support for open excavations
- rules for handling explosives

A new area of regulation introduced in the 1996 regulations was the management and routing of traffic on site. Secondly, the Management of Health and Safety at Work Regulations of 1999 required each site to have a named person responsible for evacuation of the place of work. The construction regulations require the site manager (or a named deputy) to be responsible for controlling emergency situations. More generally, the regulations require contractors to prepare risk assessments for their employees and for other persons who may be affected by their activities (such as members of the public). These risk assessments will need to indicate:

- The seriousness of the risk.
- The type of assessment made. These may be routine or generic assessments for situations which occur on every site or bespoke assessments which are project-specific.
- The integration of the contractor's risk assessment with those of other contractors. In this evaluation, the assessment should indicate where they will have no effect, limited effect and major effect on other contractors.

5.2.4 *The Construction (Design and Management) Regulations 1994 (CDM Regulations)*

These regulations emanated from European legislation. In 1987 the European Commission launched the third programme on Health and Safety at Work and later the Social Charter and Action programme in 1989. These programmes saw self-regulation under the guise of risk assessment as being at the heart of the European approach to managing safety. This approach was made into a European Union directive as the Temporary or Mobile Worksites Directive of 1992. This was adopted by the British Parliament as the Construction (Design and Management) Regulations of 1994 (CDM Regulations).

The CDM Regulations shifted the emphasis for the control of health and safety. Under the Health and Safety at Work Act the responsibility has lain with individual firms to provide a safe working environment and individual managers to police this environment.

As construction is a fragmented activity, this meant that a myriad of employees (frequently as trade, specialist or subcontractors) had responsibilities for safety management on different parts of the job. Responsibility was organizationally and geographically dispersed. The CDM Regulations moved the responsibility from the firm to the project.

In the CDM Regulations, the planning of safety is as important as cost or time planning of projects and so may be handled in the same way. Moreover, all parties to the construction process have responsibilities for safety and their duties and the necessary components of safety planning are outlined below.

The client

The client has a statutory duty to appoint a 'planning supervisor' for the purposes of preparing a safety plan and monitoring the implementation of the safety plan on site. (The work of the planning supervisor is discussed below.) The client needs to provide the planning supervisor with details of the project in terms of function, scale, location, etc. The client also has responsibilities in selecting design consultants who are able to offer designs which are not only safe to use but are capable of being built safely. Finally the client must appoint a contractor who is competent and can build the project in a safe way. This may involve safety records being inspected as a qualification for selection as a tenderer.

The planning supervisor

The planning supervisor is responsible for:

- creating a safety plan
- developing a safety file
- ensuring that designs are safe

The CDM Regulations state that the planning supervisor should be appointed as soon as is practicable after the inception of the project.

The safety plan

The safety plan needs to be prepared before arrangements are made for the contractor to carry out or manage construction work and the information required in the safety plan should include:

- a general description of the work
- the timescale of the project
- details of risks to the health and safety of the workforce
- details of how resources are allocated in managing safety

This framework plan must be developed by the principal contractor and each element of the construction will need to be subject to a risk assessment. In many ways this will need to be presented as a method statement with an emphasis on how the chosen method for constructing each part of the works is good safe practice. The planning supervisor should advise the client in the adequacy or otherwise of the safety plan.

The Health and Safety File

The file starts at the inception of the project and commences with details of the design. The file is built up as the project progresses and can include variations to the design and will be added to by the contractor as the works proceed. Regulation 12 of the CDM Regulations puts in place mechanisms for this file to be handed over to the client at the end of the project.

Ensuring that designs are safe

The planning supervisor is charged under the CDM Regulations to ensure that the designer designs in a way which avoids risks to health and safety. However, the need to design out hazards is bound by the caveat 'so far as reasonably practical'. The overall design process should not be dominated by the need to avoid all risks during construction but the intention is to encourage decisions at the design stage which reduce risks during construction or assist contractors in devising safe methods of work.

In the design development stage, the designer will need to communicate design ideas to the planning supervisor and cooperate with the planning supervisor to ensure that safe designs emerge for the project.

The designer

Regulation 13 of the CDM Regulations implies that an integral part of the design function is to balance design considerations and the effect upon the health and safety of construction workers. This will involve the designer carrying out a risk assessment. The duty of the designer is limited to risks which the designer could reasonably foresee; that is to people constructing, maintaining and repairing a building. The risk assessment work will involve the designer examining methods by which a structure might be built and analysing the hazards and risks associated with these methods in the context of design choices. Those risks identified should be tackled at source by reducing or controlling their effect. It should be noted that some risks will have to be tolerated and the purpose of the risk assessment is intended to draw attention to areas of risk in the design which may be accepted or rejected and so stimulate a search for alternative structures.

Where the contractor is to provide a detailed design, the designer should indicate the principles of the design and describe any special requirements for the purpose of construction. Where a hazard is identified, the details must be passed to the planning supervisor for inclusion in the health and safety file. However, the CDM Regulations do not require architects and engineers to dictate construction methods or exercise a health and safety management function over contractors as they carry out construction work.

Principal contractor

The CDM Regulations insist that the principal contractor shall undertake the coordination of Health and Safety. This will involve the policing of subcontractors to ensure that rules and regulations are adhered to. The principal document developed to enable this to take place is the *safety plan*.

It is the principal contractor's responsibility to ensure that training takes place and that subcontractors are informed of all risks. This communication function is a two-way street since the CDM Regulations encourage workers, be they directly employed or subcontractors, to discuss all matters which directly affect their health and safety.

Naturally the principal contractor has responsibility for monitoring health and safety, with at least daily, if not twice daily, inspections for breaches of safe working practices.

5.2.5 The Provision and Use of Work Equipment Regulations 1998

These regulations place a responsibility upon contractors to ensure that plant and equipment to be used on site is safe and used properly. In instances where plant is shared (e.g. tower cranes), the installation and management should be overseen by the principal contractor. In certain circumstances the responsibility for equipment may be devolved or let as a separate work package – but whatever the arrangements the details of the control, coordination and management of shared plant should be specified in the safety plan. Regulations regarding the wearing of hard hats and the provision of protective clothing were subsumed into these regulations when they were updated in 1998.

5.2.6 The Personal Protective Equipment Regulations 1992

These regulations require the employer to provide, without charge, clothing to protect workers from inclement weather. Equally, employees are required to *use* this personal protective equipment.

5.2.7 The Lifting Operations and Lifting Equipment Regulations 1998

This relates to the plans for reducing accidents whilst lifting equipment or materials. Issues such as training in the correct ways to lift items need to be considered. The regulations require a risk assessment to be made before lifting operations commence.

5.2.8 The Confined Spaces Regulations 1997

These regulations relate to work in spaces which may give rise to 'specified risks'. In construction, examples of confined spaces include trenches, sewers, pits, etc. The hazards are said to arise from the nature of the place of work and so increase the risks to safety and health.

5.2.9 The Reporting of Injuries, Diseases and Dangerous Occurrences Regulations 1995

These regulations require all accidents to be reported to the HSE. Records of these accidents must be kept by the employer for a period of three years.

To sum up the legislation framework, the principles underpinning the protection of workers can be identified as follows:

- *If possible, avoid the risk completely* by using alternative methods or materials.
- *Combat risks at source* rather than use measures which leave the risk in place but attempt to prevent contact with the risk.
- *Wherever possible, adapt work to the individual* particularly in the choice of work equipment and methods of work. This will make work less monotonous and improve concentration, and reduce the temptation to improvise equipment and methods.

- *Take advantage of technological progress* which often offers opportunities for safer and more efficient working methods.
- *Incorporate the prevention measures into a coherent plan* to reduce progressively those risks which cannot altogether be avoided and which takes into account working conditions, organizational factors, the working environment and social factors. On individual projects, the safety plan (Regulation 15) will act as the focus for bringing together and coordinating the individual policies of everyone involved. Where an employer is required under Section 2 (3) of the Health and Safety at Work Act 1974 to have a health and safety policy, this should be prepared and applied with reference to these principles.
- *Give priority to those measures which protect the whole workforce or activity* and so yield the greatest benefit, ie. give collective protective measures (such as suitable working platforms with edge protection) priority over individual measures, (such as safety harnesses).
- *Employees and self-employed need to understand what is required of them,* e.g. by training, instruction, and communication of plans and risk assessments.
- *The existence of an active safety culture affecting the organizations responsible for developing and executing the project needs to be assured.*

5.3 The cost of safety

The moral question for safety has been promulgated and, as we have seen, there is a legal requirement for the provision of safe working environments for building workers and passers-by. But added to these imperatives will be an economic incentive to ensure safety. In an era when controlling costs and time in projects is a significant factor, it is impossible to ignore the economic argument.

Assigning monetary values begs a rhetorical question: 'Can a meaningful cost be applied to accidents?' For material losses in which no injury occurs, the accounting of loss can be easily assessed; where human loss is concerned, the costing becomes difficult and burdened with ethical pitfalls. A life or a human facility cannot be credibly evaluated in terms of money. Yet money is the resource which is used to compensate injured parties or relatives (in the event of a fatality).

The courts, insurers and government ministries charged with managing health and social security have suggested the amounts paid as compensation are very variable. There seems little pattern of awards paid by the courts for compensation in accidents. Consequently it is a fine judgement as to whether it is better to risk litigation with its high risk and high costs.

Most compensation payments are paid by contractors' insurers but this should not be a disincentive to improve safety. Insurance companies will base premiums on safety performance and poor safety records will be reflected in the premiums paid. However, there are doubts as to whether the premium-loading for poor performances is adequate since insurance companies will use a portfolio approach and spread risk across all policy holders rather than discriminate between good and poor safety performances.

One approach is to treat the direct costs of accidents and the costs of preventative measures taken to avoid them as the total accident cost and then seek to minimize this cost. In construction, the costs of accidents will fall as safety measures increase; this is presented graphically in Figure 5.4. If this relationship is accepted, then we have two components: accident cost and prevention cost. As we reduce risk, the accident cost will also be reduced, but in order to reduce risk we must spend money on accident prevention. Therefore, we have an intersection of graph lines in Figure 5.4. Now, by adding together the accident cost and the prevention cost, we shall develop a dish shape and the lowest part on this curve will indicate the optimum expenditure on safety prevention.

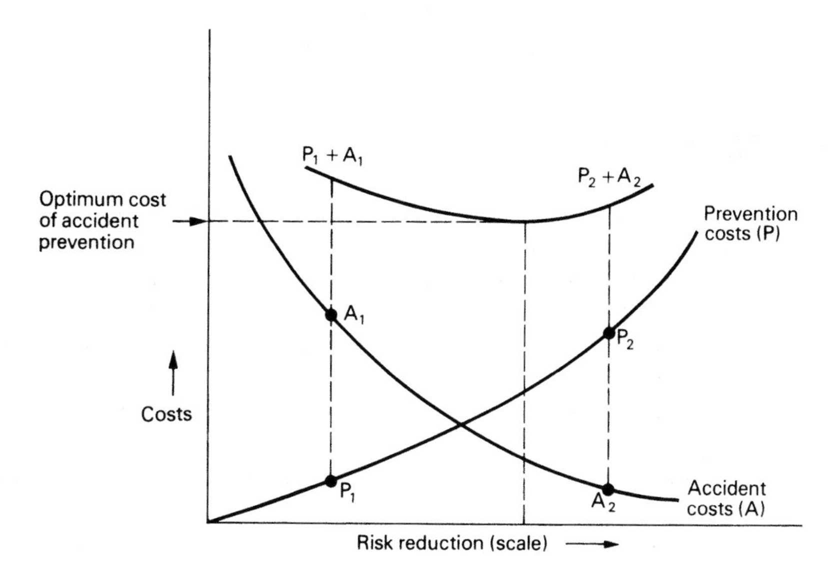

Fig. 5.4 Determining optimum costs of accident prevention.

The difficulty with this approach is that it does not take into account the subjective nature of many accident costs. Sinclair (1972) has attempted to quantify accident costs by breaking accidents down into three parts: fatalities, serious injuries (over four weeks off work), and other injuries. The cost of accidents can then by expressed as:

$$C_A = R_D x (A_{OD}) + R_S(A_{SS} + A_{OS}) + R_O(A_{SO} + A_{OO})$$

Where C_A = annual accident cost per worker
R_D = annual risk of death per worker
R_S = annual risk of serious injury per worker
R_O = annual risk of other injury per worker
and average A_S = subjective element of cost
A_O = objective element of cost

With subscripts D, S, or O for death, serious injury, or other injury, as above.

Sinclair has applied this method to large groups of workers, and with industry-wide figures, the cost of each accident has some relevance when compared against the cost of prevention.

The cost of prevention can be based upon more objective data. Companies can abstract from records the costs of safety in terms of the provision of safety administration, protective clothing and equipment, insurance, extra manning for safety reasons, estimated loss of production due to unavoidable hazards on sites, etc. The value of the costs of prevention per worker can be compared against the costs of accidents per worker. Sinclair has argued that when the preventative costs of accidents equal the costs of accidents, the optimum cost of prevention has been reached. This approach has its critics, since there will be a time lag between the institution of preventative measures and their effect, but it can be used as a yardstick by which accident prevention programmes can be evaluated.

A simpler, and probably more practical, approach was put forward by Capp (1977) who analysed the actual cost of accidents in industry. The cost assessments are based solely on the working days lost by injury which are reported to the Health and Safety Inspectorate. The study was carried out in 1970 when approximately 23 million days were lost through accidents or industrial disease, with the average accident/incident ratio of 34/1000. However, these figures are based upon reported accidents which resulted in more than three days off work yet, as many surveys have pointed out, considerable numbers of firms in construction do not report all accidents. One such survey was carried out by the Government's 'Labour Force Survey' in 1994, when the number of claims for industrial injury benefit relating to people injured in construction work was compared with the number of injuries actually notified by employers to HM Factory Inspectorate during the same period. The study revealed that some 38 per cent of legally notifiable cases were not being reported.

Hence, the real cost of accidents may be far greater than that revealed by calculations based on official statistics. Nonetheless, it is useful to evaluate the available data. Several assumptions are made:

- The loss to a company is the total return for an employee's labour at an average daily rate of £70 with companies involved in accidents having to pay 200 per cent of this figure; 20 days average absence.

- Two hours has been assumed as the time lost for treatment and incidental disruption and assistance to injured persons.
- The average cost of repair or replacement resulting from each incident of property damage is £300.
- Insurance premiums are increased by 20p per £100 of wages.
- Assume an accident rate of 33 accidents per 1,000 workers at risk in a year (1994 figures).
- 240 days are worked each year.

From the above assumptions, the calculated cost of £433.22 per employee to a company with 1,000 employees is presented in Table 5.1.

Table 5.1 The cost of safety

Predicted accident rates	*Category*	*Actual number*	*Loss*	*Value (£)*
1	Reported accident Loss to company: 200% of wage rate	33	33 × 20 days = 660 days lost per 1,000 workers at 200% = 1,320 days at £70 per day	92,400
10	Minor accidents	330	330 × 2 hours = 660 hours lost per 1,000 workers = 73 days at 200% = 146 days at £70 per day	10,220
30	Property damage	990	990 × £300	297,000
	Insurance costs		20p per £100 wages for 1,000 workers $\frac{0.2 \times 16{,}800}{100} \times 1{,}000$	33,600
Total cost per 1,000 employees				433,220
Cost per employee (per annum)				433.22

To the figure of £433.22 wasted per employee is to be added the consequential costs such as delay and lost production and, therefore, the figure is, by nature, conservative and estimates of the rates of consequential loss have been gauged at three or four times the direct losses rates.

5.3.1 *Indirect costs*

The table clearly omits the indirect cost of injuries. Lingard and Rowlinson (1994) assert that indirect costs are four times that of the direct costs. Levitt and Samelson (1987) and Hinze (1997) have created a list of the sources of indirect costs which is presented below.

Injured worker costs

- lost productive time on the day of the injury
- reduced productivity upon returning to work

Transportation costs

- transporting injured worker to obtain treatment
- arrangements and attendance by other personnel

Crews costs

- lost productive time while assisting injured worker
- reduced productivity caused by injured worker
- reduced productivity caused by replacement worker

Costs of hiring a replacement worker

- administrative costs of locating and hiring
- training and orientation costs
- reduced productivity due to low familiarity with work

Other crew costs

- direct impact on the activities of another crew
- lost productivity while watching or talking about injury
- reduced morale of company employees

Supervisory costs

- assisting the injured worker
- investigating the accident
- preparing the accident or injury report
- time spent with media for dramatic incidents
- time spent with regulatory inspectors and insurance representatives

Others

- damage to company image
- loss of opportunities to qualify for tendering due to poor safety record
- scare people who want to enter into the industry, thus increasing labour shortage problem

Another approach is to offer incentives for good performance rather than penalize bad performance. Some clients, in an attempt to encourage safe working practices, have included fully specified safety-related items in a bill of quantities or other bid document. These items are then certified and the contractor is paid for them when they are performed. Failure to perform would result in the payment being withheld.

However, it is important to stress that the economic aspects of safety should not be seen as the prime motivator for construction safety. The central issue is the intangible element of a secure and healthy workforce for construction. Cost considerations and calculations of cost optimization can only be seen as a guide to determining priorities and improving health and safety in construction.

5.4 The role of the safety officer

Contractors should give careful thought to the role of the full-time safety officer. Two basic concepts exist about the role:

- safety officers should be advisers to site management
- the safety officer undertakes the safety responsibilities on sites

Whichever *modus operandi* is adopted, the safety officer will have responsibilities for:

- the development of a safety culture
- the delivery of a safety policy
- the provision of safety training

5.4.1 Safety culture

A vital ingredient in generating safe working is the safety culture which pervades a construction organization. Anderson (1997) sees safety culture as a mixture of individual and group values, attitudes, perceptions, competencies and patterns of behaviour that determine the commitment to, and the style and proficiency of, an organization's safety management. In short, the safety culture is the distillation of beliefs that members of an organization share about safety. This culture cannot be manufactured and installed; it has to grow organically, it will take time to take root. The reward is that those who work in the firm do not merely comply with safety rules and regulations but have internalized the need for safe working. The features of a safety culture can be identified:

- leadership and commitment to safe working from the top which is genuine and visible
- there is a long term strategy
- there must be a policy of high expectations, conveying a sense of optimism about what is possible which is supported by appropriate procedures

- a sense of 'ownership' of safety standards with widespread involvement in the policing and training for safety
- targets established for safety performance and measured regularly to compare performance with the targets set (naturally these targets should be realistically set)
- good safety behaviour should be a condition of continuing employment and considered in annual appraisals
- a management information system which includes the evaluation of safety as well as commercial information

In this way, construction firms can move towards a safety culture – it is an effective, resilient and painless way of improving the safety climate in an organization. The safety officer will be an important agitator for a strong safety culture.

5.4.2 *Safety policies*

As has been seen, the Health and Safety at Work Act 1974 contains a strong element of self-regulation. It seeks to stimulate firms into having an efficient organization for health and safety. This approach of self-regulation has obviously created problems for firms operating in the construction industry. Management has to contend with a variety of problems which will vary from site to site during the progress of the job. Climate, regional differences in the labour force, time of year, type of contract, method of payment, type of employment practices, and labour mobility all mean that construction management will have a difficult job to control safety on the site.

The demands put upon the industry by Latham (1995), with calls for 25 per cent reduction in time and 30 per cent reduction in the cost of building, impose fresh challenges to managers who have to meet new demands while maintaining a safe working environment.

Successful firms within the industry have solved their difficulties without affecting productivity and have effectively integrated the safety and health of the workforce into the mainstream of their organizational operations. Progressive management will argue that the safety and health of workers must be as equally well-organized as the commercial aspects of their company.

However, there is a wide diversity in attitudes to safety within the industry. It is noticeable that the number of fatalities is disproportionately higher in smaller organisations despite the biggest proportion of the industry's workforce being employed by larger contractors. Perhaps it is not surprising that the larger organizations have a better safety record for they have the resources to attend to safety and can monitor safety by employing specialist staff. The more casual operators within the industry are less likely to have such specialists. Despite this difference, employer concern is vital since it is widely recognized that a company's safety policy lives or dies by the support it receives from top management. This commitment to safety from the top is important and it must be emphasized that a safety policy developed by management must be more than a statement of good intent. Senior management must show a commitment to safe and healthy working conditions throughout the company. The Health and Safety Commission have identified a reasonable test of sincerity: they suggest that top management should be seen to support site management when decisions are taken which relegate profit below a concern for safety.

In general, the safety policy must be backed up by organization and arrangements which will secure the maximum effect. A typical organization structure for the control of safety is shown in Figure 5.5.

It is not intended to describe the detailed arrangements for the organization for safety, because each company will have its own style and manner of operations and the intention behind the self-regulation principle is that safety policies should reflect a compatibility of safety procedures and general organization. However, some general rules can be developed:

- the delegation of responsibility for safety down to the workforce
- the identification of key personnel to direct the safety effort in specific areas of work (e.g. plant maintenance, keeping of records, etc.)
- the development of job descriptions which emphasize that site managers are accountable for safety on site
- the monitoring of safety by the safety officer of a company and the submission of reports to senior management in such a manner that they have a picture of what is happening on site in terms of health and safety
- a strong emphasis on the development of safe systems of work
- the importance of good communications between sites and head office over safety matters

The last point, concerning communications, needs close attention. In particular, firms may need to ensure that safe systems manuals are transmitted to sites where they will be needed. Additionally, information will need to flow between sites and head office which will make senior management aware of the site conditions and will alert sites to the senior management's attitude toward safety. Finally, a good communication system can assist in the circulation of information concerning new hazards within the company and the industry in general. The transmission of information concerning safety has been one of the traditional roles of the employers' associations, and the CEC plays an important part in providing information on matters of health and safety to their members. The trade unions allied to the construction industry can also provide information.

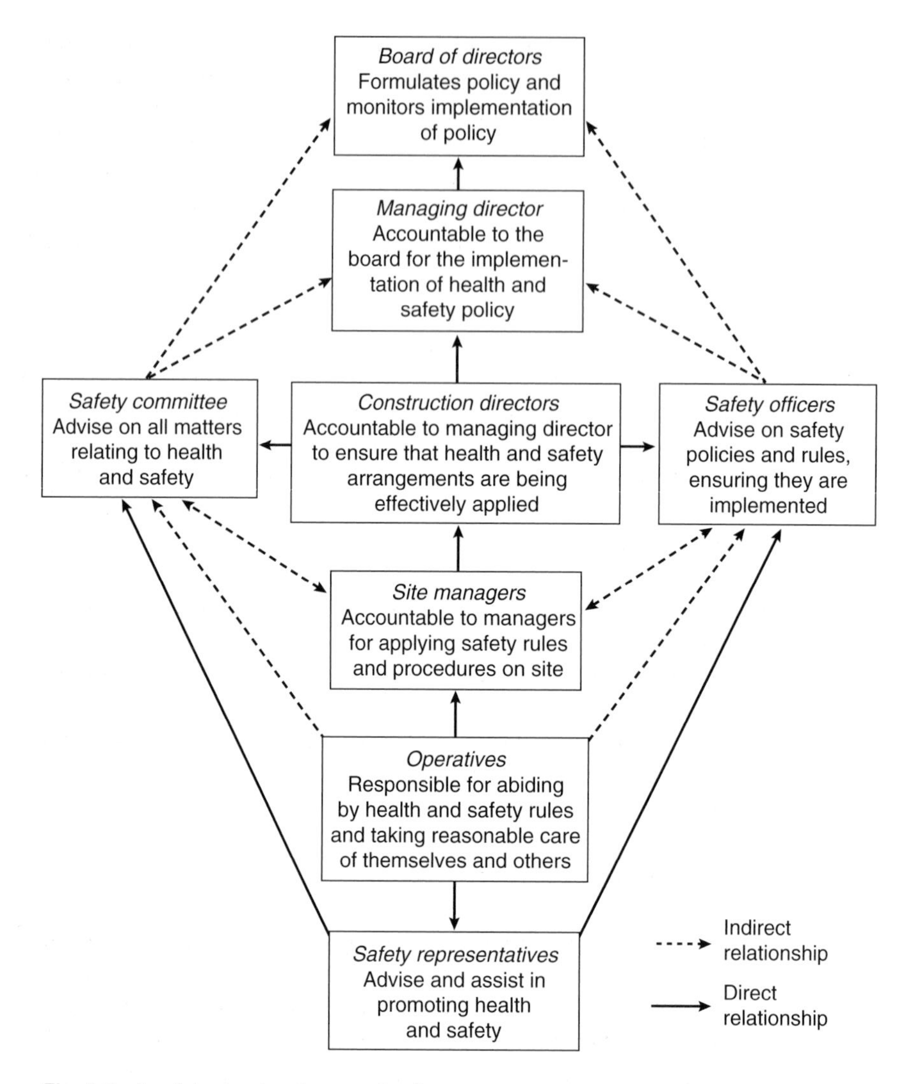

Fig. 5.5 A safety structure for a contracting company.

5.4.3 *Safety training*

Many of the larger contractors have a well-developed programme of safety training, but as yet there is no systematic, comprehensive training provision within the industry. Much of the training carried out is done in-house and is occasionally supplemented by the Construction Industry Training Board (CITB).

Safety training should begin with the new employee and continue throughout the time he or she is with the company. The type of training, frequency, material presented, and presenter of the material will obviously vary with the type of employment and trade involved. However, safety training should be given to all employees regardless of previous experience. The focus of this in-house training should, of course, be the company's safety policy, but specific information may be imparted verbally with a discussion of the company's attitude to safety. Such items as the statutory requirement for the wearing of protective clothing for particular tasks, the location of first-aid equipment, and the person(s) qualified to administer first aid are pieces of basic information which can be transmitted verbally.

Even experienced workers can benefit from additional on-the-job safety training. One of the most effective means is the use of periodic safety meetings with the use of visual aids to dramatize the effects of poor adherence to safe procedures. However, such meetings should only be long enough to present the desired information, and are often best undertaken in an informal setting where workers can contribute observations concerning safety on site. Some of the matters that can be discussed at such meetings could include:

- information on accidents that have happened elsewhere on similar sites (the company safety officer should keep site managers informed about such accidents and any conclusions that have been drawn from them)
- precautionary measures necessary on any new section of the work
- reviewing first-aid procedures (see review of Health, Safety and Welfare Regulations)
- pointing out (preferably without mentioning names) any unsafe practices that have been noticed

The spread of Vocational Training and competency testing through National Vocational Qualifications for construction trades has presented more opportunities for safety training to be presented as part of such training programmes. Certainly the Trade Unions have a view that safety training should be strongly featured in the courses that train tradespeople.

Part of this training will be technical and procedural but much benefit can be obtained from attitudinal and behavioural changes. The most recent efforts at safety improvement have been focused upon behaviour-based safety management. The objectives of such improvement programmes are to increase the incidence of safe behaviour by site workers. Such research has had widespread acceptance of its potency, with research on the theme of behaviour modification being reported from the UK, Hong Kong and Finland. The underpinning theory of behaviour modification programmes is that the motivation of workers to avoid accidents is a key factor in the quantity and type of accidents that occur. In short, if workers are motivated to work safely then they will do so and moreover if this motivation comes from within themselves then performance will improve. In order to test this theory researchers have set up a number of site-based experiments to empirically validate that behaviour can be changed by positive intervention into the safety management programme and that this positive intervention leads to benefits in terms of safety performance.

Most of the experiments conducted have followed a set pattern. This may be characterized as follows:

1. Measure the safety climate by observations of the site. The breaches of safety protocol contribute to a safety score. Low scores indicate potential hazards lurking on site.
2. Low scores stimulate a safety training session and publicity in site offices, workplaces, etc. emphasizing this training.
3. This safety climate of site is then remeasured to test for any change in the safety climate.
4. Selectively withdraw the improvement programme and test for changes in safety climate.

Obviously the intention is for the interventions to improve safety. The results of the experiments conducted are not uniform. In Hong Kong, Lingard and Rowlinson (1998) reported on four areas of safety performance:

- site housekeeping
- access to heights
- the use of bamboo scaffold
- the use of personal protection equipment

The three-step process outlined above was implemented with the 'use of personal protection equipment' being used as a control. In a 34-week experiment on seven sites the research scaled the safety climate in each area of performance. For example, issues such as the safety of material stacks would contribute to the 'housekeeping' dimension: the proportion of ladders which were lashed to the scaffolding would be indicative of the strength of the safety climate in this aspect of safety performance. The outcomes of this safety audit were represented on charts and displayed. The impact of the safety training was measured and in Hong Kong the results of the intervention are summarized in Table 5.2.

The researchers note that these results need to be interpreted cautiously since the predominant method of payment for Hong Kong construction workers is piecework or, when directly employed, a bonus is payable for faster than average work. Consequently the researchers present the caveat that the attractiveness of reward for quick work is greater than the value of obtaining a higher safety profile.

Earlier experiments in the UK by Duff *et al.* (1994) generated safety

Table 5.2 Behaviour modification and safety

Safety element	*Effect of behaviour modification*	*Effect when behaviour modification was withdrawn*
Housekeeping	improvement on all seven sites	deterioration on six of seven sites
Access to heights	improvement on three sites deterioration on two sites	no significant deterioration
Bamboo scaffolding (only four sites studied)	no significant improvement	no significant deterioration
Use of personal protection equipment	effect not measured	effect not measured

improvements by using behaviourally based management techniques. On six sites, a regime of goal-setting and performance feedback was established. The theory behind this technique is that goals are a powerful regulator of human activity and that performance targets to meet the set goals (provided that the goals are specific, legitimate, achievable and accepted by those who are charged with achieving the goals) stimulate improvements. Feedback acts as important information and shapes performance.

The relationship between goals, performance and feedback can be exposed as a classical contract loop as shown in Figure 5.6.

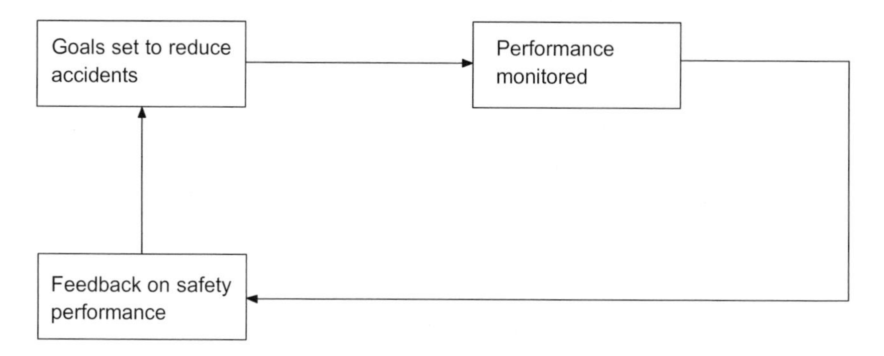

Fig. 5.6 The relationship between goals, performance and feedback.

Duff *et al.* (1994) set up an experiment to test this theory. They started by identifying 99 common causes of accidents and tested the importance level of risk and severity of injury associated with each hazard. The 99 hazards were condensed to 24 and then were classified into four composite measures of safety. Again scaffolding, access to heights, housekeeping and personal protective equipment were the measures selected. From this the safety 'temperature' was scaled for each site by an activity-sampling method. From these, safety temperature goals for safety improvements were set and transmitted to the workforce, the performance was monitored and feedback charts prepared and displayed on site. Simple graphs which showed the target level and actual performance on a week-by-week basis were provided. Supporting this feedback were training packages: here operators were exposed to 15-minute slide presentations which focused upon the dos and don'ts for each category under examination (scaffolding, access to heights etc).

The results showed that these interventions dramatically improved the safety performance. In order to validate this finding, the researches withdrew feedback and training support but continued to measure safety performance. As expected the withdrawal of the feedback diminished safety performance but it did not decline to the level found at the start of the experiment. The cycle of intervening and withdrawal demonstrated that safety performance could be improved by modifying the behaviour of construction workers. The pattern of performance could by typified as shown in Figure 5.7

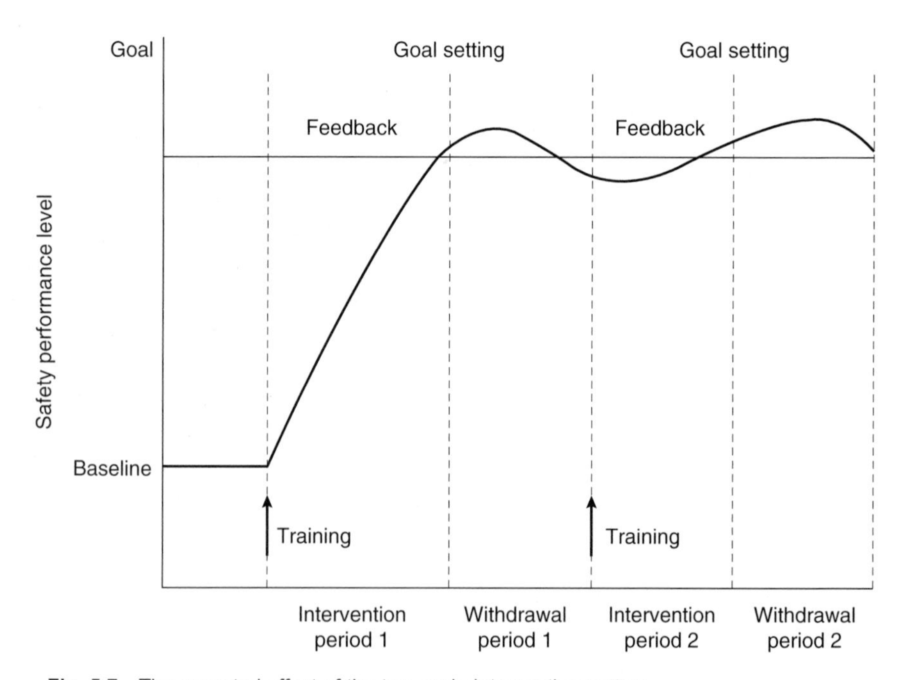

Fig. 5.7 The expected effect of the two-cycle intervention pattern.

The type of interventions were presented in three formats, goal setting, feedback and training, and these three interventions were manipulated to test for the most effective. The researchers concluded that 'goal setting and feedback was better than feedback alone and that the addition of training did not provide benefits'. Although these results are not conclusive, they do suggest that behaviour modification programmes have greater impact than safety campaigns or safety training.

The researchers acknowledge the commitment of management and operators to safety performance as a key variable. On sites where management was cooperative and supportive the improvements made were the greatest; where hostility and suspicion were experienced then performance improvements were weakest. The operators were invariably supportive of the work and enthusiastically awaited performance feedback.

5.5 Changing attitudes to safety

5.5.1 *Monitoring safety performance*

It is widely accepted that management and workers jointly contribute to the inherent risks associated with construction work. Both accept the high injury rates as a 'fact of life' with the annual accident figures at company and industry level acting as a catalyst for action or inaction. If accident figures for a particular year are better than the previous year, complacency sets in; if they are worse, then safety becomes more central to the concern of the company and the industry. There may also be a normalizing process associated with accident rates: to stay within the boundaries of the previous year's returns is often seen as an acceptable philosophy. Inevitably, counting the number of accidents will continue, but a more positive approach may be possible by counting the number of lives 'saved' instead. With this approach in mind, the Construction Central Operations Unit of the Factory Inspectorate encouraged the measurement of safety performance beyond the comparison against the previous year's statistics. Such an approach would systematically classify a hazard for its significance and principle cause, with the result of identifying whether an accident arose through the failure of site supervision or from inadequate training or information. For instance, the maintenance arm of Railtrack (1996) creates matrices which use two dimensions to evaluate safety. They are:

- likelihood
- severity – safety and loss

They have scaled each dimension as follows:

Likelihood	*Severity*
1. Very unlikely	1. Injury – no time loss
2. Unlikely	2. Minor injury – less than three days off
3. Likely	3. Injury – more than three days off
4. Very unlikely	4. Major injury, damage or loss
5. Certain	5. Fatality, catastrophic damage or loss

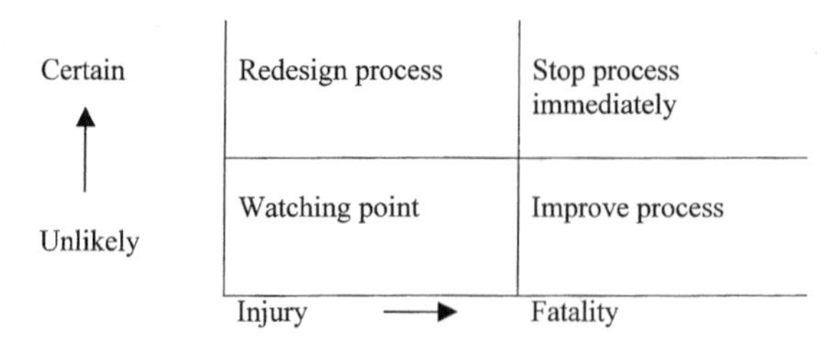

Fig. 5.8 Using two dimensions to evaluate safety.

These factors may be combined to advise on appropriate actions. Figure 5.8 is not the Railtrack model but may be used as a general guideline.

A more quantitative analysis can be determined by the use of 'critical incidence techniques', whereby information is collected from experienced personnel on hazards, near misses and unsafe conditions and practices. The technique involves interviewing workers regarding involvement in accidents or near accidents. Tarrents (1977) has noted that people are more willing to talk about 'close calls' than about serious accidents in which they were personally involved, the implication being that if no loss ensued, no blame for the accident would be forthcoming. In effect, the critical incidence technique accomplishes the same as an accident investigation, by identifying the type of hazards that could result in injury or damage. It has been estimated that for every mishap there are 400 near misses and, consequently, by sampling all persons in a firm, a large sample size can be used as indications of areas in which improvements are necessary.

This approach is one of many which attempt to measure site safety performance. Firms can use their available data to review safety performance over the preceding twelve months. Central to this approach is the systematic classification of every hazard, with indications of its significance and principal cause. In this way it is possible to note whether the hazard arises from a failure in site supervision, senior management, or through inadequacies in training, instruction, or information.

However, this approach should not be seen in a wholly negative manner as positive aspects of site safety should be recorded as well as the misdemeanours. The identification of responsibility for an accident has its drawbacks in that a safety officer requires the cooperation of the line management if his or her job is to be done effectively. By allocating responsibility, the safety officer may destroy a good relationship and hence the motivation of line management to improve safety on site. Assessment schemes can be unpopular with line management, because they are seen as threats to their self-esteem, and consequently line management can become obstructive in order to cope with these threats. In particular, line management may start to question the measurement criteria, with the consequence that the importance of safety inspection is demoted. Also, the source of the assessment – the safety officer – becomes another source of external pressure, rather than a cooperative colleague.

These difficulties can be overcome by using the assessment as a measure of positive values and not negatives ones, with each hazard being assessed systematically for its liability to cause accidents and the extent of the risk it poses. If a numerical scale is used, then this should be adjusted where satisfactory protective features are noted on site. This aspect is important, since site managers are being assessed on what they have achieved in terms of safety and health. It emphasizes that safety can be managed in the same way as other aspects of the resources found on a construction site. Care should obviously be taken to ensure that an assessment discussion is not carried out in an authoritarian way, with the company safety officer sitting in judgement on the site management team. Equally important is the correct structuring of assessments so that site managers feel that they are receiving fair treatment. Of particular importance in this respect is the alerting of site management as to when assessments will be taking place, and information on the criteria of assessment, pointing out the health and safety issues that are of prime concern. Finally, the assessment should begin by giving site management the opportunity to discuss the difficulties they have experienced in attaining high standards of health and safety. Naturally this process of cooperation is not meant to obscure the facts – if a site is poor, then this must be said – but the basic principle is to encourage positive attitudes rather than hostility and resentment.

5.5.2 *Safety and incentives*

Working hours and incentives also have a bearing upon accidents and management may wish to review policy on these issues in order to improve health and safety. It has long been accepted that efficiency falls when excessive overtime is worked, but accompanying the fall in efficiency is a rise in accident rates. In construction, the use of overtime is widespread, and the reasons for this are clear: workers like the additional income and management can gain better utilization of existing workforce. However, the moral to be drawn is that managers who are concerned with safety and accident prevention need to look closely at overtime working, and it may be necessary to limit actual working hours to the European working hours directive.

Construction workers are also highly dependent upon incentive payments to supplement the basic pay. Piecework is widely used in construction through the labour-only subcontracting system of employment. It has often been suggested that labour-only subcontracting does little to improve safety consciousness or to encourage safe working methods at site level. This may be the effect of haste, the lack of stability within the labour-only workforce, and the financial gains to be reaped from an early finish, all of which tend to lessen the importance that should be attached to safe working conditions. These views are seldom borne out by the facts.

The primary incentive scheme for directly employed workers is a production bonus scheme. This type of incentive would appear to do least harm. Various reasons have been put forward for this. Many people have often argued that incentives do little to directly motivate but benefits arise from the intangible effects of better organization or production and this in itself will create a safer working environment. Another view is that incentive bonuses are so often a lottery and have little to do with workers' efforts. As the authors of *2000 Accidents*, Powell *et al.* 1971, observe:

> 'In the general accident situation bonus pay is unlikely to correlate with accidents at work because there are few cases where it properly reflects the human work content of the task.'

However, abandoning bonus payments in the construction industry would be universally condemned when it constitutes an important element of the take-home pay. Nonetheless, management must be aware of the additional hazards which may be created by the use of bonus schemes and to evaluate how far incentives for safer performance can be incorporated into their operation. But incentives are not limited to finance: there can be competitions between sites to encourage better safety performances. Such schemes have the benefit of encouraging safety awareness on site and fostering a spirit of cooperation between management and workers on safety matters. Competitions do, however, have their detractors, who argue that they encourage non-reporting and non-recording of accidents. As one union official drily commented when his firm had won a safety award: 'It's all walking wounded. That's what we call it.'

The general conclusion is that while incentives to achieve greater output and greater safety should not be discouraged, they should be carefully examined to ensure that they do not contain features which militate against genuine safety and accident prevention.

Summary

It has been argued that contractors have a moral, economic, and legal responsibility to ensure that working conditions on site are healthy and safe. But provisions for safety must commence before the construction phase of a project; architects and engineers must have the technical knowledge to design buildings which can be safely constructed, as well as a commitment to safe working conditions for site workers. Indeed, they are legally required to do so by the CDM Regulations.

To generate safety-consciousness within construction organizations, a firm lead must be taken by top management. It is recognized that finance and lost production are convenient measures of accidents, but accidents should generate an emotional response, and if this emotion is genuine it will carry conviction. A firm's safety policy which is founded upon compassion will more often succeed, since it will be impervious to shifts and changes in fashion and production schedules and, consequently, will be less easily diluted. Senior management can do much to implement the policy by adhering to the requirements of the Health and Safety at Work Act 1974, the CDM Regulations and associated regulations. In particular, senior management can encourage the introduction of safety committees which can monitor progress in respect of safe working methods, and, more importantly, can act upon recommendations and reports from the safety committee(s) and also discipline any person in breach of relevant safety laws and codes of practice. Obviously, adequate tools, tackle, plant and protective clothing must be provided and these items can be framed within a budget for the development of health and safety.

Finally, management should not see safety merely as a hindrance to productivity, but as a component of an efficient mechanism of production.

Questions

1. Outline the principal features of a safety policy for a construction company.
2. Discuss the potential sources of conflict between a company safety officer and a site manager. How best are such differences resolved?
3. Assess the impact of the 'six-pack' regulations upon the construction industry. Are such legislative interventions necessary for the improvement of health and safety in construction? Argue your case.
4. It has been suggested that accident prevention costs can be optimized. What are the advantages and disadvantages of this approach to construction site safety?
5. Explain how a safety audit can be carried out on construction sites and identify potential problems in carrying out such a task.
6. Are safety committees useful agencies for monitoring safety on construction sites? Amplify your answer with reference to the work of safety committees.
7. Discuss the utility of techniques of behaviour modification in improving construction safety.
8. Can the 'critical incident technique' be usefully employed on construction sites? Draw up a method statement of how this could be carried out.

References and bibliography

Anderson, H. (1997) 'The problem with construction', *Safety and Health Practitioner* (pp. 29–30).

Capp, R.H. (1977) *Engineers' Management Guide to the Elements of Industrial Safety*, Institute of Production Engineers.

Craig, V. and Miller, K. (1997) *The Law of Health and Safety*, Sweet & Maxwell.

Duff, R., Robertson, I., Phillips, R. and Cooper, M.D. (1994) 'Improving safety by the modification of behaviour', *Journal of Construction Management and Economics*, Volume 12, Number 1.

Egan, J. (1998) *Rethinking Construction*, DETR HMSO.

Hinze, J. (1997) *Construction Safety*, Prentice Hall.

Holt, A. (2001) *Principles of Construction Safety*, Blackwell Science.

Latham, M. (1995) *Reconstructing the team*, HMSO.

H M Factory Inspectorate (1994) *Labour Force Survey*, HMSO.

Levitt, R. and Samelson, N. (1987) *Construction Safety Management*, McGraw-Hill.

Lingard, H. and Rowlinson, S. (1998) 'Behaviour-based safety management in Hong Kong's construction industry: the results of a field study', *Journal of Construction Management and Economics*, Volume 16, Number 4 (pp. 481–488).

Lingard, H. and Rowlinson, S. (1994) 'Construction Site Safety in Hong Kong', *Journal of Construction Management and Economics*, Volume 12, Number 6 (pp 501–510).

Powell, P., Hale, P., Martin, J. and Simon, M. (1971) *2000 Accidents*, National Institute of Industrial Psychology.

Railtrack paper (1996) 'Managing Safety in Railtrack', Manage Risk or Manage Failure Conference, Institution of Civil Engineers.

Sinclair, T.C. (1972) *A Cost Effectiveness Approach to Industrial Safety*, HMSO.

Tarrents, W.E. (1977) *Utilizing the Critical Incident Technique as a Method of Identifying Potential Accident Causes*, US Department of Labor.

The Confined Spaces Regulations 1997, HMSO.

The Construction (Design and Management) Regulations 1994, HMSO.

The Construction (Health, Safety and Welfare) Regulations 1996, HMSO.

The Health and Safety at Work Act 1974, HMSO.

The Lifting Operations and Lifting Equipment Regulations 1998, HMSO.

The Management of Health and Safety at Work Regulations 1999, HMSO.

The Personal Protective Equipment Regulations 1992, HMSO.

The Provision and Use of Work Equipment Regulations 1998, HMSO.

The Reporting of Injuries, Diseases and Dangerous Occurrences Regulations 1995, HMSO.

6 Managing People

The management of people is a key element of a construction manager's job. The construction industry has seen unprecedented calls for changes in organization, performance and ways of undertaking projects and this has meant that the pressures upon individuals and groups have increased. Most of these pressures have been externally imposed through legislation, such as the Housing Grants, Construction and Regeneration Act 1996 (the 'Construction Act') or European Community directives on safety management, as well as by clients seeking to obtain better performance from the industry (Egan 1998, Latham 1994). Other changes will be caused by the industry's response to the bandwagons which beset all industries: TQM (total quality management), BPR (business process engineering), environmental management etc. Some of these fads have fallen in the area of people management and one can think of 'empowerment', 'right sizing', 'de-layering' and so on as being initiatives which have implications for how people are managed.

This chapter seeks to explore common and timeless processes which underpin the people-management function of the construction manager's job. These functions will include:

- planning human resources
- managing human resources

Given the changes in the structure of employment, with trade or specialist contractors providing much of the human resource for the physical process of construction, the focus will be on managing the managers – those people who control the supply of the administrative, professional, technical and clerical staff in a construction organization.

Part A
Planning Human Resources

It must be recognized that any human resource plan for the industry, or individual firms within it, must be flexible. In an industry where demand is derived from the fortunes of other industries, shifts in the construction strategy of major clients (including government) will obviously have repercussions upon the human resource needs of firms and the industry in general.

Such conditions do not allow for tight planning but even guidelines are preferable to the oscillation of recruitment and redundancies which follow from human resource strategies which are uncoordinated with the business cycles of the industry.

6.1 Objectives of human resource planning

In short, the objectives of human resource planning are to ensure that the organization:

- recruits and retains the people it needs to undertake the work: these people are to be of the right quality and in sufficient numbers to conduct the work of the organization
- anticipates shortages or surpluses in the work place
- uses its staff effectively

For many years the construction industry relied upon strategies of casual employment to manage the operative labour force. Employment contracts were for 'services' rather than for service. During the 1990s the Inland Revenue and the Department of Employment sought to tighten the rules of self-employment and employers were encouraged to provide direct, rather than subcontracted employment. Despite this innovation, the numbers of workers employed by the industry varied immensely and this was mirrored by the trends in operative and also administrative, professional, technical and clerical staff.

Figures 6.1 a–d graphically illustrate these relationships. Not only does employment fluctuate with workload but the high level of insolvencies experienced by construction companies exacerbates the problem. Figure 6.2 shows the level of insolvencies over the period 1971–1996 in all industries (in England and Wales) and construction insolvencies. While construction contribution to GDP is in the order 8 per cent, the extent of insolvencies ranges between 17 per cent to 10 per cent in the years considered.

Cynics may argue that these figures demonstrate the futility of attempting human resource planning for the industry, but equally it may be suggested that the instability of the labour force is a reflection of the industry's failure to undertake long-term forecasting of human resource needs. Clearly, capital-intensive industries are better placed to forecast accurately their labour requirements, and in construction the margin for error is likely to be large but a pragmatic managerial approach would suggest that a 'guesstimate' is better than nothing, and by doing such exercises we can reduce as far as possible the area of uncertainty in overall

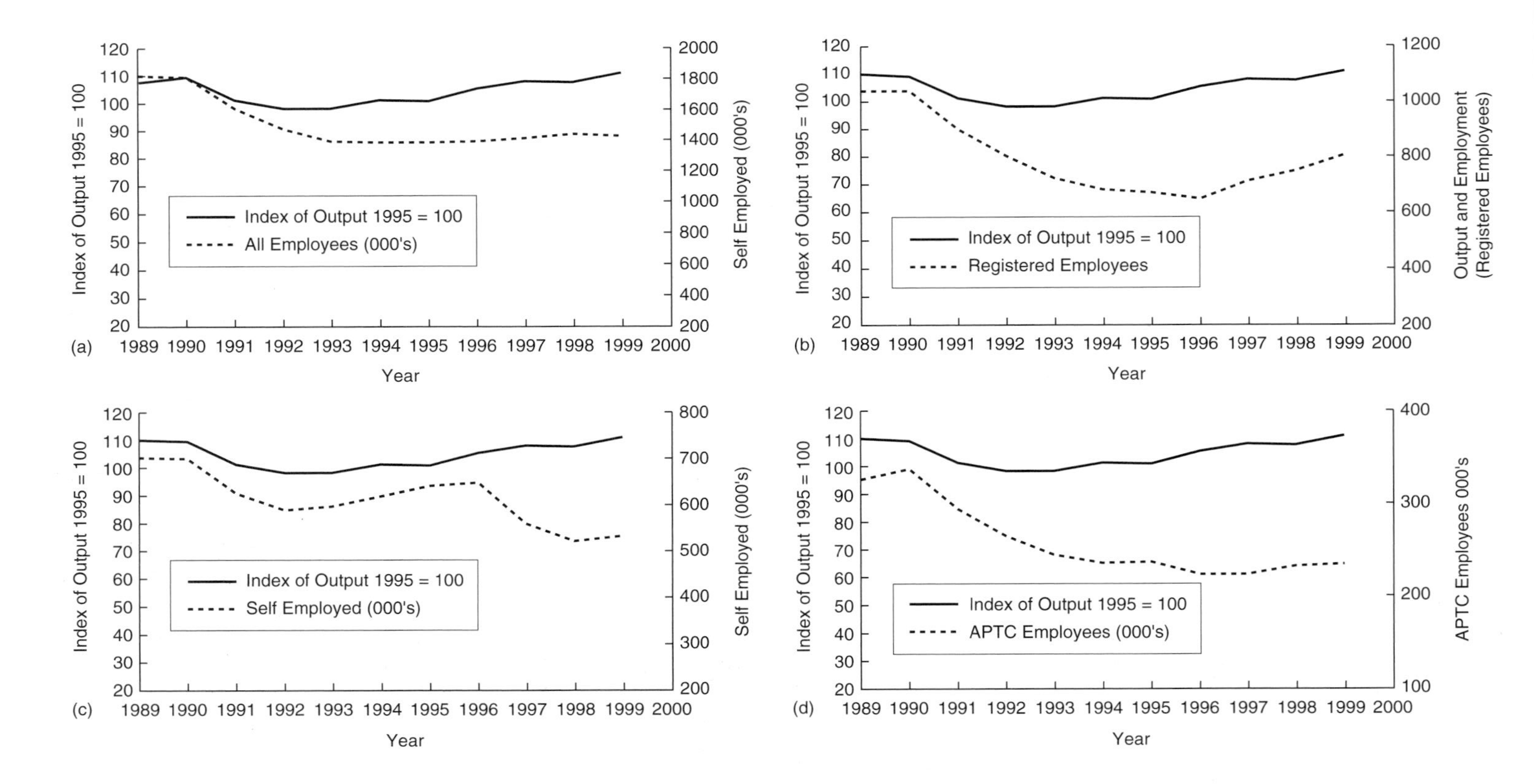

Fig. 6.1 Output and employment. (a) All employees; (b) registered employees; (c) self-employed; (d) APTC.

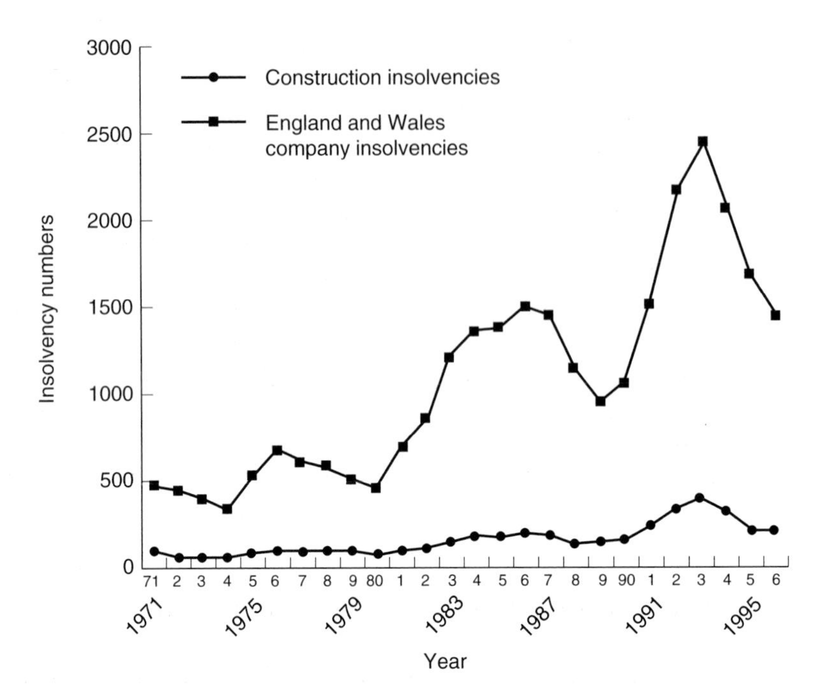

Fig. 6.2 Insolvency levels.

business planning. Therefore, to the industry and the firm, human resource planning has several advantages. A human resource plan can:

- reduce personnel costs because of management's ability to anticipate shortages and surpluses of human resources and make appropriate corrections
- serve as a basis for making use of employees' abilities and, consequently, the industry and firms can optimize their use of human resources
- be used to establish the best cost balance between plant and human resources
- determine recruitment levels for various grades of staff
- anticipate redundancies and avoid unnecessary dismissals
- determine optimum education and training levels and management development programmes
- act as a tool to evaluate the effects of alternative policies

It is important that the human resource plan be integrated into the overall business and economic planning process and is a component of a business strategy.

Despite these advantages, many companies will be reluctant to commit resources to human resource planning because of the uncertainty of the workload and any development of a human resource plan must commence with an analysis of the work carried out in previous years to see if an underlying trend can be observed. For instance, the proportion of times that a pre-qualification for a project is converted into a contract award or the occasions that a preferred bidder status ends up as an award of a PFI project may be valuable information. The profile of projects attained is also useful information when looking for trends. Are they in the public or private sector, building or civil engineering, traditional or industrialized building? Are they design and build or management contracts? From this analysis one can move towards an assessment of demand for various types of staff, with an identification of the skills required to service anticipated workloads and whether forecasts should be made for the whole of the organization or merely sections of it. Also specification of the degree of accuracy required and the period over which forecasts are to be made will be essential.

6.2 Procedures for human resource planning

If human resource planning within a firm is to be successful, a set of procedures must be followed. These can be identified as:

- analysis of current staff resources
- analysis of changes in human resources
- analysis of staff turnover
- effects of changes in the conditions of work
- analysis of external factors influencing the supply of staff
- integrating human resource plans into the company organization

6.2.1 Analysis of current staff resources

Here it is useful to retain records to show the profile of the existing staff. Items worth recording would be the sex, age, education, promotability, salary, length of service, etc., of all employees. From these data, profiles of the staff structure of the firm can be made. Histograms are a useful tool to demonstrate problem areas, for example, management may be concerned that existing staff are growing old together and therefore a histogram can graphically show the age distribution of employees (see Figure 6.3).

Similar histograms can be drawn for the organization's employees, for example, what is the proportion of construction managers with degrees or professional qualifications against those from a trade background? These analyses can throw into sharp focus impending difficulties for the future of the organization, for example, a skewed age distribution may mean that a succession policy is not being developed, or a preponderance of graduate site managers suggest that promotion prospects for other types

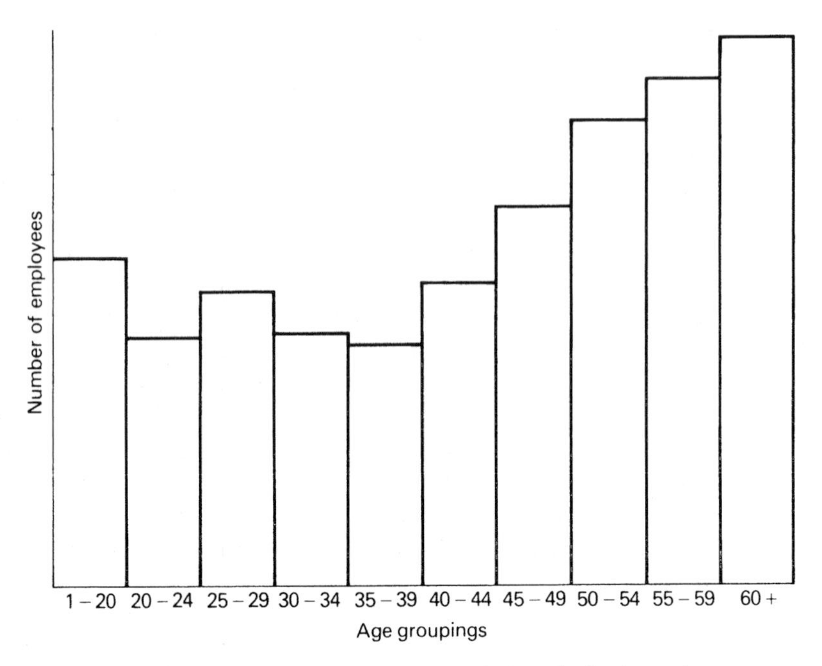

Fig. 6.3 Graphical methods of showing the age distribution of a firm's employees.

of entrant are inadvertently being limited; alternatively, if there are few non-construction graduate/professionally qualified staff, is the firm preventing access to jobs which require a knowledge of sophisticated management techniques?

Demand forecasting is an inexact science and the more distant the future the more difficult it becomes. In the short term a planning model may be used which relies on evidence of the past. For the medium term, scenarios may be postulated; for the long term one is usually left with hope. Figure 6.4 maps these three phases.

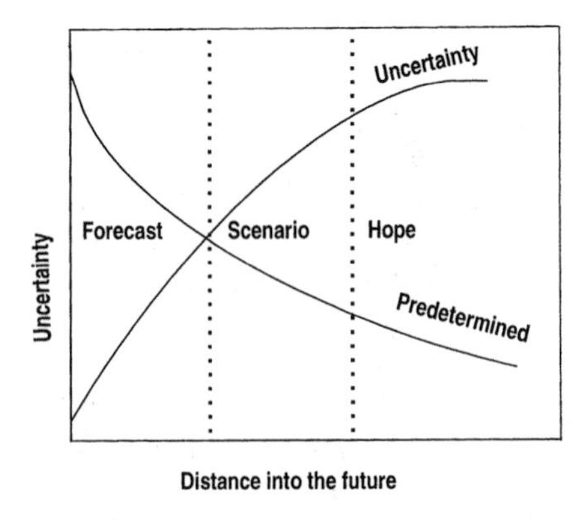

Fig. 6.4 The balance of predictability and uncertainty in the business environment. After Van de Hiedjen (1997).

6.2.2 Analysis of changes in the labour force

It is sometimes useful to assess changes in the composition of the staff complement for time to time. Again, this can be best demonstrated by using histograms (see Figure 6.5). Such data can demonstrate the possibilities of growing imbalances: perhaps the numbers of administrative and management staff are growing at a faster rate than the growth in turnover, or perhaps the firm is not fully utilizing its training programme. Of course, the changes in the structure may have been planned and may be a response to technological factors or to the type of work the company is carrying out, but more often than not certain categories of staff will drift without a staff budget. It may be necessary to establish departmental or functional structures which can be used to control the number of employees. Information such as this can assist in clarifying a staffing policy by focusing attention upon career paths for individuals, e.g. can some of the management support staff be transferred to line management if necessary?, what are the qualifications for advancement from operative to management level?

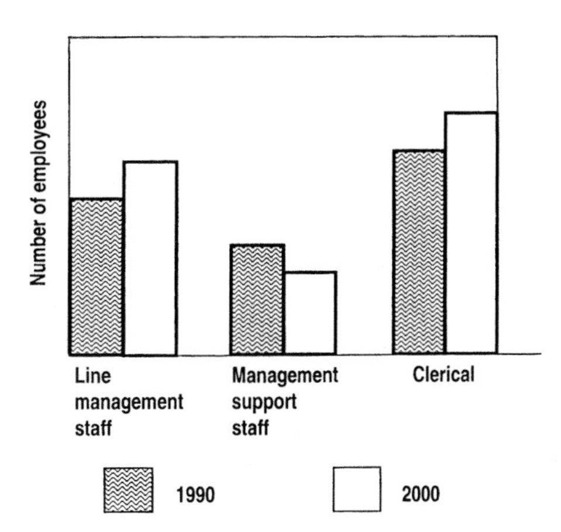

Fig. 6.5 Changes in the composition of the labour force.

6.2.3 Analysis of staff turnover

It is common for managers in construction to be concerned about staff turnover within a firm and there are several ways of measuring this turnover. A simple manner of evaluating matters is to apply the following equation:

$$\text{Staff turnover} = \frac{\textit{Number of leavers in one year}}{\textit{Average number employed in the same year}} \times 100$$

This approach gives a fairly crude evaluation since, in such an industry as construction, ingress and egress of blue collar workers to and from the industry is commonplace and the mobility of professional staff may be high. A more refined analysis may be determined by the 'cohort theory', whereby the pattern of leaving can be determined. This can be presented graphically, as in Figure 6.6.

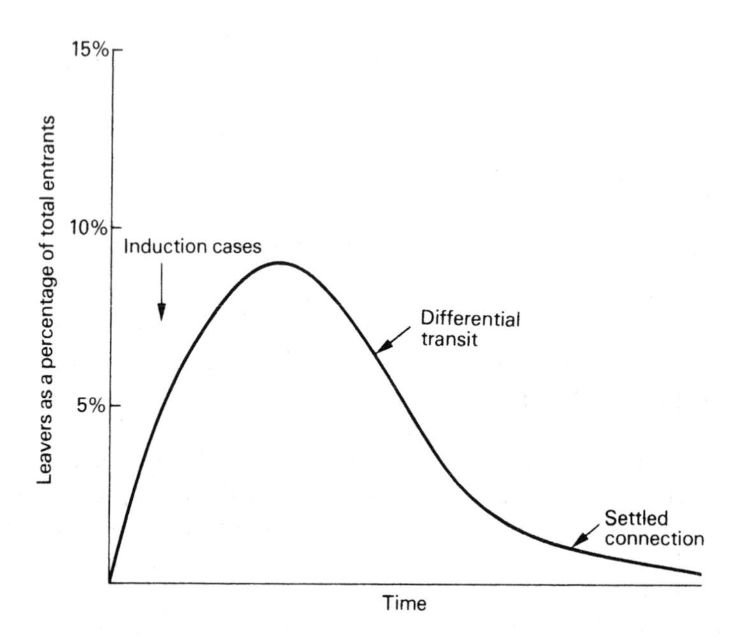

Fig. 6.6 The cohort theory of labour turnover.

The early peak in turnover can be explained by those persons who leave early because they seldom stay long in any job. This will be followed by those who leave after finding out about the organization, and have found that they do not like it. After these two groups have departed, there will be a settled connection between the individual and the firm. If the persons who have established a settled connection begin to leave in numbers, then it may suggest that something is awry within the organization. Obviously turnover may be more evident in certain parts of the organization than others. Site staff may be more fickle than those at head office or vice versa. High turnover in particular areas may suggest a problem with earnings or style of management.

It is advisable for a supervisor to interview those persons leaving the firm to determine their reason for leaving, for often the stated reason may hide real or imagined grievances. Discussion may help to rectify any dissatisfaction. Many building firms may not feel that it is necessary to maintain tight records of starters and leavers, but it is essential if a firm is to have a good human resource policy. Should staff turnover be excessive, then records will show the extent of the problem and corrective action can be taken if required.

However, it would be dangerous to assume that a high staff turnover indicates internal problems. Hyman (1976), in a study of blue-collar worker turnover in two engineering plants, suggested that economic conditions were important variables in the level of labour turnover. Factors within the firm determine a worker's willingness to seek new employment, but external factors influence their ability to find it. Additionally, salary is not a prime determinant in staff turnover; people do not act in accordance with classical economic rationality (i.e. seek out high salaries at all times). Security of employment and career development can be far more influential.

6.2.4 Effects of changes in the conditions of work

Here the human resource plan of a company will be influenced by changes in corporate objectives and environmental changes. The corporate objectives of a construction firm will have been formulated by the perceived opportunities and the firm's capability to respond to these opportunities over a period of time. The choice of these objectives will be influenced by the availability of work, finance, machinery and human resources. With respect to environmental changes, it may be clear after analysis that there are alternative opportunities to which the resources of the company can be applied. For example, during the early 1990s, many firms operating in the traditional construction market moved into design and build work in order to sustain turnover and get closer to the client's decision-making process. Others sought opportunities in Facilities Management in order to be close to clients' strategic thinking about the management of a property portfolio. It is, therefore, necessary for a firm to monitor market and business changes. Such data can be used as a guide to the type of human resource changes they are likely to face.

However, more mundane factors will also influence matters, for instance, if the retirement age is changed by the government, maximum working hours stipulated by the European Community are enforced, or minimum wage levels are changed. All these factors will have an influence upon the human resource requirement of a construction firm.

6.2.5 Analysis of external factors influencing the supply of human resources

So far, we have concentrated upon the internal supply of staff that needs to be available to meet expected demand conditions within the firm. By matching future requirements against existing resources, the human resource specialist will be able to judge the type of recruitment programmes required. But, as has been seen, to make such judgements in isolation from the economic and social environment would be rash. There is little point in planning business expansion if the skill to service new workloads is not going to be available. Similarly, construction programmes need to take into account the likely human resource supply to particular contracts. A vital part of a manager's job is therefore directed towards analysing the factors which will have a bearing upon the supply of people to the firm in general and individual projects in particular. Some of the factors which will have a bearing upon supply are:

- population density in the area of a project
- local unemployment levels in the principal trades and professions
- current competition from other companies in the area and the likely future competition
- local transport facilities
- availability of short-term housing within the area
- the impact of government training agencies and 'new deal' arrangements
- the impact of legislation, e.g. retirement ages, working hours and minimum wages
- specialist and trade contractor arrangements

It will clearly not be possible for managers to measure precisely the effect of local and national supply factors upon a firm since employment practices within firms will vary and the best a company can do is to carry out an intelligent appraisal of the way the market for particular trades and professions is working.

Trade organizations and employers' associations have a brief to monitor such matters and are able to give advice.

6.2.6 Integrating human resource plans into company organization

The whole of the human resource planning procedure can be expressed diagrammatically as in Figure 6.7, from which we can see that people in construction are a key resource and should claim equal attention to that given to production and profits. This view has given rise to increased attempts by construction companies to plan their human resource requirements. Those companies which have undertaken an element of such planning have often recognized the benefits in terms of higher efficiency and productivity as a result of better utilization of human resources and the elimination of waste in recruitment and training.

However, in such a labour-intensive industry, the best results for this exercise are gained from integrating a company's human resource plan with overall company objectives; therefore, the human resource specialist in construction will need to forecast the total available work in a particular market and a company's anticipated turnover in selected markets in order to translate these business predictions into human resource requirements. Secondly, the human resource specialist will need to point out to top management the constraints that the availability of staff will

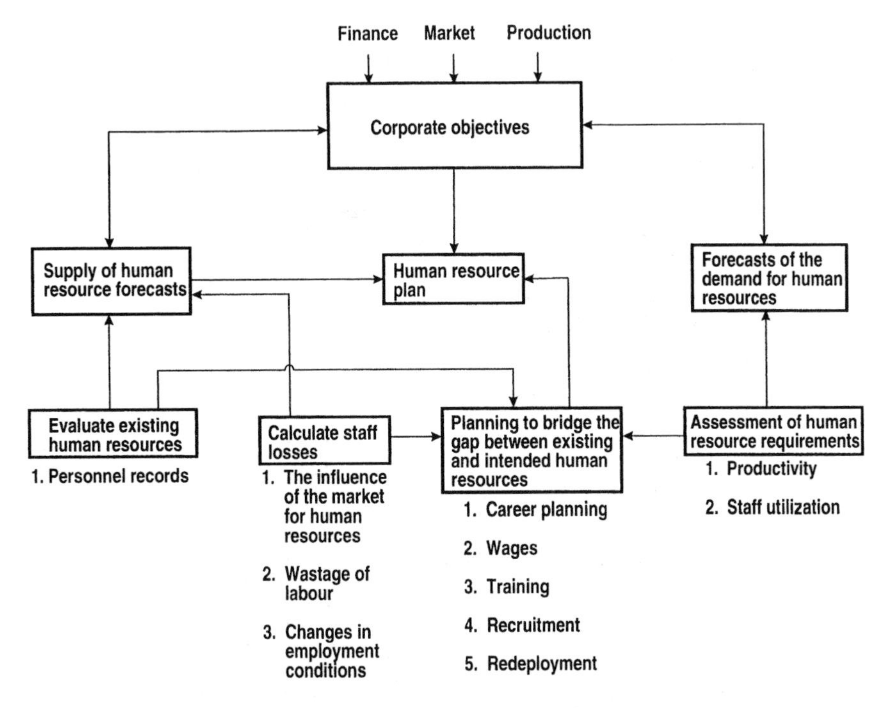

Fig. 6.7 Human resource planning within corporate objectives.

place upon company objectives. Such forecasts can be based upon five-year intervals, with monitoring at stages between such dates, although the unpredictability of the construction market means that a large margin for error should be built into the plan. Nonetheless, guidelines, albeit loose ones, are better than none at all.

Armstrong (1996) sees the integration of human resource planning into the business process as consisting of a number of distinct steps. These are:

- Demand forecasting – estimating future staff needs by reference to the corporate plan.
- Supply forecasting – estimating the supply of staff in particular trades and professions in the context of current and future supply. The estimate will include inevitable wastage.
- Forecasting requirements – analysing the balance between demand and supply so as to be able to predict deficits or surpluses.
- Productivity and cost analysis – this will include any process re-engineering to eliminate wasteful practices in the use of human resources.
- Action planning – the preparation of plans to manage recruitment or set in chain a programme of reduction of human resources.
- Budgeting and control – setting human resource budgets and monitoring them against the plans.

Figure 6.7 illustrates the integration of human resource planning with corporate objectives.

The CITB (1997) has made operational this approach by forecasting construction employment over the period 1997–2001. The CITB built these forecasts from a macroeconomic model which took into account the expected growth in construction output, the future expected changes in interest rates and wages rate. The global number of construction employees forecast was then divided into 23 occupations allied to the construction industry pro rata to the existing division of labour between the crafts. Using historical data it was able to forecast the total employment for the future (Table 6.1). These data were then broken down into four groupings: Building Trades, Specialist Building Trades, Civil Engineering and Building Services. The forecasts are presented in Figures 6.8–6.11.

Table 6.1 Total construction employment, Great Britain, 1991–2001. Source: CITB Employment Model July 1997; Construction Forecast and Research, July 1997 and Cambridge Econometrics, June 1997; actual: Department of the Environment Transport and the Regions.

	Year	*Output, annual growth rate (%)*	*Total unemployment, direct and indirect (1000s)**
Actual	1991	−7.0	1698
	1992	−4.0	1521
	1993	−0.2	1411
	1994	3.2	1384
	1995	−1.0	1375
	1996	1.1	1370
Forecast	1997	3.8	1386
	1998	3.8	1410
	1999	2.9	1440
	2000	2.7	1455
	2001	2.7	1475

* The model currently uses the narrower definition of construction employment of DETR which is based on VAT returns. This is consistent with the measure of construction output. The estimate of construction employment from the Labour Force Survey is some 20 per cent higher, however the trend is similar.

This analysis showed that there is a shortfall of craftspeople in 1998 and 1999 but this corrected itself in 2000 and 2001. Figure 6.12 shows the shortfalls. The biggest difficulty will be in finding carpenters, electrician managers, plumbers and bricklayers. This is an invaluable model for the industry.

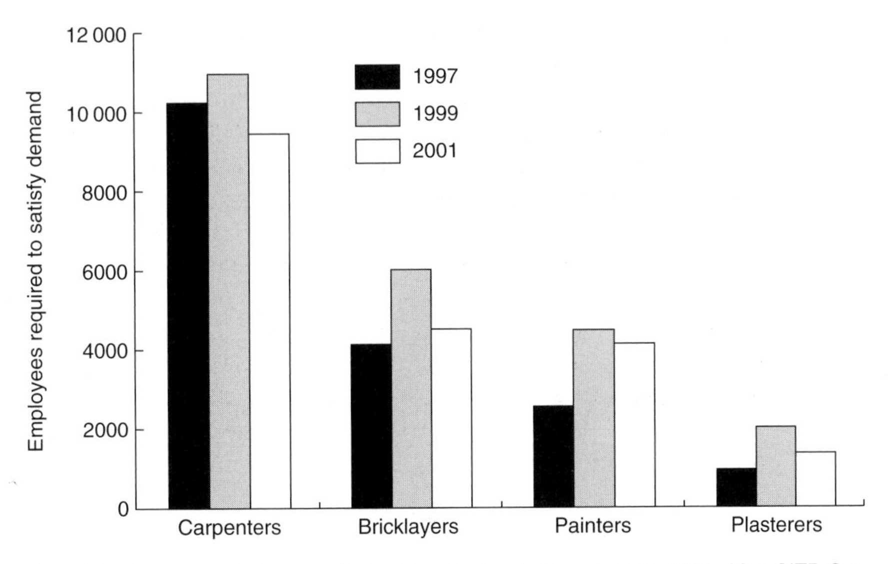

Fig. 6.8 Employment forecast: the building trades 1997, 1999 and 2001. After CITB Construction Employment and Training Forecast (1997).

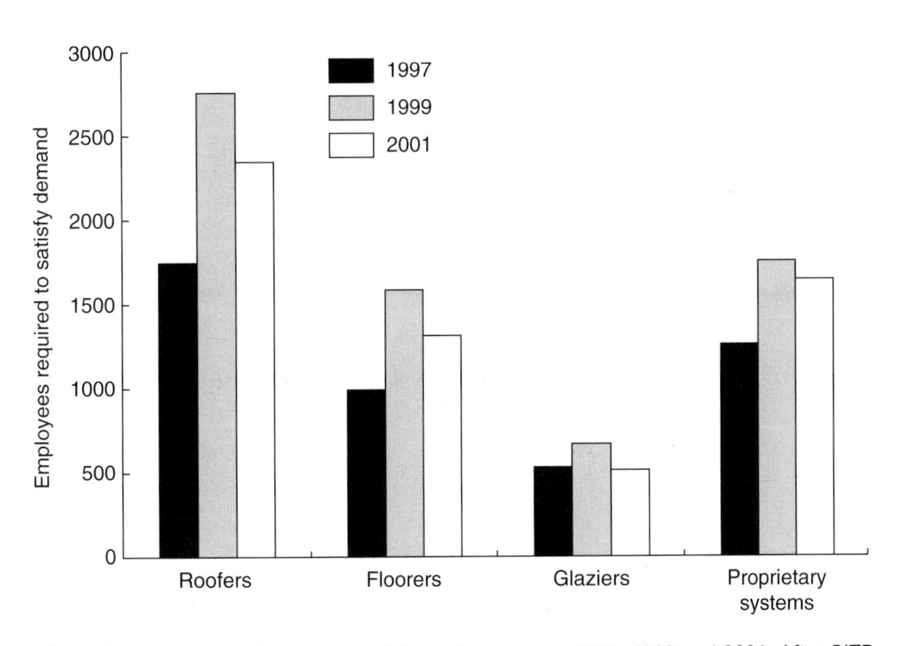

Fig. 6.9 Employment forecast: specialist building trades 1997, 1999 and 2001. After CITB Construction Employment and Training Forecast (1997). NB. Proprietary systems include ceiling fixers, loft insulators, thermal insulators, etc.

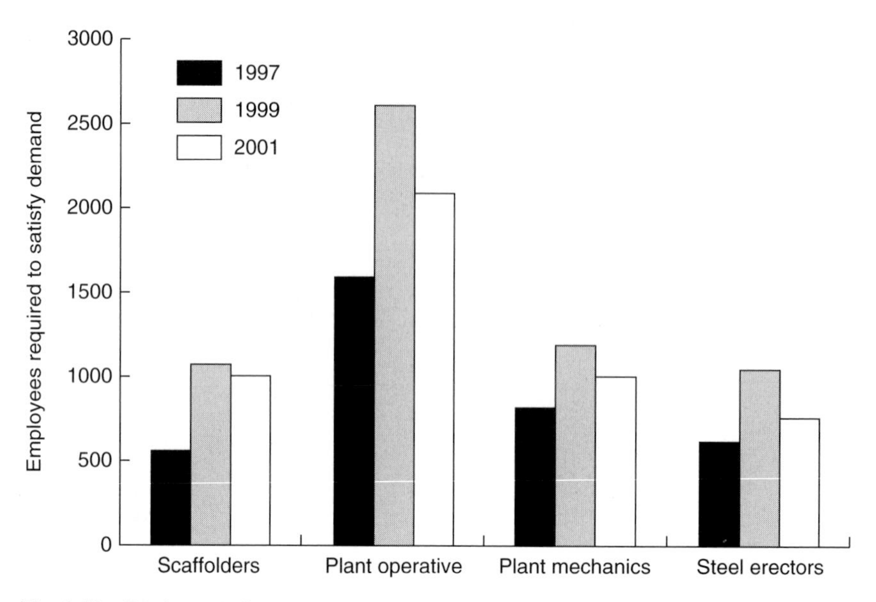

Fig. 6.10 Employment forecast: civil engineering trades 1997, 1999 and 2001. After CITB Construction Employment and Training Forecast (1997).

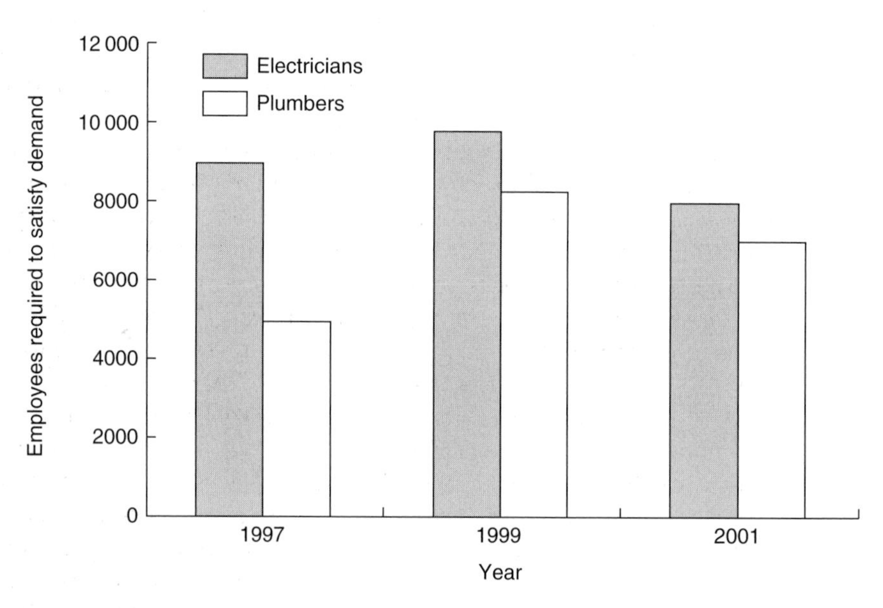

Fig. 6.11 Employment forecast: building services 1997, 1999 and 2001. After CITB Construction Employment and Training Forecast (1997).

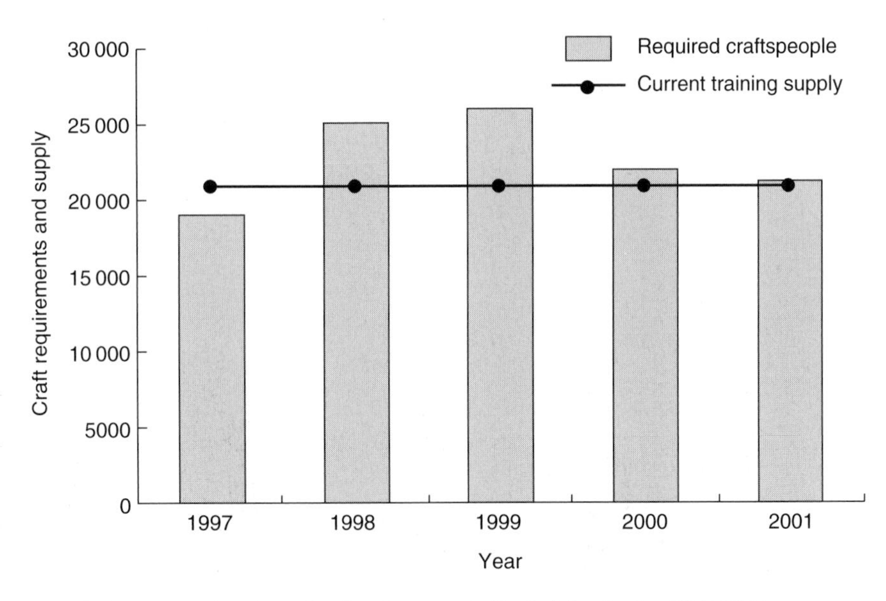

Fig. 6.12 Demand and supply of craftspeople in the building trades 1997–2001.

Part B
Managing Human Resources

If youngsters are attracted to the industry then they are faced with a myriad of choices about the role they wish to undertake. At the level of the professions, most of them have a single door through which entrants must pass to gain entrance. For example, if you wish to be a doctor then you take a degree in medicine then specialize in one of the 'colleges': general practice, surgery, etc. All will be members of the British Medical Association. If you want to be a lawyer then you go and do a law degree and then qualify for the Law Society. Compare this to the confusion that faces someone attracted to construction; they have a choice of being one of four sorts of engineer, three types of architect, have to select between four surveying groups and any number of variants of building/building engineering employment. While the CITB has attempted to make progress in the information available to aspirant professionals by creating the Construction Careers Service, the presentation of the industry as a whole is unlikely to make rapid advances until wholesale changes are made to the structure of the professions in the industry. A unified professional body serving the industry with an articulated first degree followed by specialist postgraduate studies would help present the industry as an attractive proposition to the brightest school leavers. Recruitment can be taken at several levels – to the industry, to a trade or profession or to a firm. The attraction of the industry to potential professionals is dependent on a dialogue between the individual armed with information and perceptions they have of the industry and the industry's presentation of itself to a wider world. Baker and Gale (1990) researched the image held of the construction industry by teachers, career advisers and school students and discovered all have, at best, a superficial knowledge of the industry. In order to attract the brightest candidates this information gap must close.

6.3 Recruitment to the industry

One strongly held perception was that the balance of supply and demand for construction labour is difficult. Unemployment in construction may be high generally yet in some parts of the UK contractors have difficulty in recruiting certain skills. The reasons for this are that unemployment is often regional in nature and people cannot move owing to the higher cost of housing in the areas where jobs are available. Recruitment to the industry should not be a problem for such an exciting industry, but nonetheless it is a matter of concern, particularly the quality of recruits that are attracted. Is this because the picture of the industry presented to recruits characterizes construction as unattractive, dirty and dangerous?

This image of the building industry is off-putting and obviously has a direct effect upon recruitment, and in turn the quality of recruitment will affect the image that the wider world has of the industry. So there is a necessity to market the industry so that it is attractive to school leavers. Superficially the industry sells itself with glossy photographs of large jobs, but school leavers and careers officers may be cynical about the likelihood of working on such projects. To counter these attitudes, the construction professions, the Construction Industry Council, the Construction Industry Board and the CITB have done much to promote the image of the industry, but changing deep-seated prejudices needs persistence and patience. It will be achieved in due course.

6.4 Recruitment to a trade or profession

Throughout the 1990s there have been around 30,000 trainees in construction. This figure has reduced from around 40,000 who entered the industry in the period 1985–1990. New Deal arrangements, with the trade training linked to 'Welfare to Work' schemes, have provided for National Vocational Qualification (NVQ) Level 1 and 2 trade competencies. More often skills are picked up as the person goes along. The prevalence of subcontracting encourages this trend. How recruits are trained once they have been recruited is covered in Section 6.6.

6.5 Recruitment to the company

At the level of the company, its attractiveness can be influenced by the interplay of the company's reputation and the individual's needs. The individual to be recruited may look at a company and consider it against the following list:

- reputation within the industry and as an employer
- training and development opportunities
- longevity of employment required
- location and transport links
- salary and benefits
- opportunities for progression
- opportunities for national or international travel
- the culture of the company
- connection with families or friends

In organizing recruitment, it is often best if this is done centrally by the head or regional office of a company. Line managers seeking to hire staff may be required to complete a proforma which provides the personnel office with details of the job to be filled and the specific operating conditions pertaining to the job. Such a request may be based upon a simple proforma which could contain the following information:

- job title
- list of duties
- pay, conditions and hours of work, usually determined by conditions of the Working Rule Agreement (WRA)
- 'essential' and 'desirable' attributes of the person in the context of the job
- the physical environment of the job
- the person to whom the employee will be responsible

Such information can help the personnel office to identify the job and can be used to prepare any necessary advertisements. The central personnel office can supplement this by providing details of the company, social facilities, training opportunities and future prospects. It may be possible to fill the vacancy from within the company, but if not, the details prepared by line management should direct the personnel office to specific sources for likely applicants, be they the careers office, job centre, employment agency or advertisement in local, national, or trade press.

If recruitment of new employees is necessary then it is important to adhere to selection procedures. Naturally, the practice and rigour of selection will vary according to the level to which the appointment is made. Notwithstanding this, standardized basic information will need to be provided. Obviously the level of detail required will vary according to job classification, with more sophisticated forms being necessary for managerial or professional appointments. Some firms have also experimented with selection tests.

The selection tests can provide a benchmark for threshold standards or ensuring an appropriate mix of types of people but, when this has been established, other factors such as drive, energy, commitment, social skills, etc., come into sharper focus as criteria for success.

Following the receipt of an application form the personnel manager may wish to interview applicants. This interview is important and sufficient time should be dedicated to it. A systematic approach to interviewing pays dividends, with questions directed with a view to revealing attitudes and skills which closely match the job requirements. Interviewing applicants is a skilled task and many companies have sought to improve this facet of their personnel management by providing specialist courses for those involved. For a fuller treatment of interviewing see *Human Resource Management in Construction* by Langford, Hancock, Fellows and Gale (1995).

6.5.1 Induction to the company

As has been seen, there is a tendency for a high separation rate at the early stages of employment. Difficulties may be experienced in settling down to a new job within a new organization. Within the construction industry, accommodating a change of job may be less unsettling due to the relatively high incidence of job changes, but if labour stability is to be sought, then an early development of a 'sense of belonging' should be encouraged. In this matter, information about the organization can help. Many construction companies find it beneficial to supplement verbally transmitted information with company handbooks and printed company rules and disciplinary procedures. Such information can be prepared as a package for the new employee and the statutory obligation to provide a copy of the safety policy can be fulfilled at the same time. At no later than thirteen weeks after commencement, the new employee must be given a written contract of employment.

However, the formal induction procedures need to be reinforced by attitudes which welcome the new recruit. It is important that the new employee is well received by the workforce and, if necessary, follow-up interviews should be conducted to monitor the employee's adaptation to the work and organization.

The issue of aligning the new recruit with the prevailing culture of the organization is important. This is best done by emphasizing what is important to the organization in terms of the behaviour of its employees. Most construction firms will have different values, beliefs, and rituals which will have been shaped by the company's history, the type of business it does and with whom, the leading personalities and the industry norms of behaviour. According to Handy (1985) cultures cannot be defined precisely but they can be differentiated and a good fit between the firm's culture and the employee's preference for a particular type of 'atmosphere' at work leads to a satisfying and productive relationship.

The appointment of a mentor or 'shepherd' to the new recruit can assist in the transmission of the cultural norms of the business.

6.5.2 *Statistics on turnover, stability, and absenteeism of staff*

Essentially, this function is one of record keeping. The basic problems of staff turnover have been discussed elsewhere in this book, but a vital index for the personnel function will be staff stability. This may be calculated by the following formula:

$$\text{Staff stability} = \frac{\textit{Number of employees with at least 1 year's service}}{\textit{Number currently employed}} \times 100$$

The figures attained from such a calculation will show the rate at which the labour force is being diluted by workers with little experience of the company. In the manufacturing industry, a staff stability index of less than 70 per cent indicates that the company is operating with a significant proportion of inexperienced workers, which will inevitably lower performance and quality standards. In construction this ratio will probably need to be reduced due to the inherent instability of labour and the high level of subcontracting.

The turnover of the core staff should be monitored particularly closely. Using Atkinson's (1984) model of the flexible firm (Figure 6.13) it can be seen that the core group is surrounded by less tenured employees. The wisdom of applying this model to society as a whole is questionable. The idea of flexible workforces has been criticized for its corrosion of trust between employer and employee and between individuals comprising a society. This trust is seen as a social glue, holding a society and an organization together.

Those leaving the firm voluntarily should ideally be interviewed by their immediate supervisor to determine the reason for leaving. Recording the stated reason can assist in identifying employment dissatisfaction which the company may be able to rectify. In the case of a dismissal, the reasons for the dismissal must be recorded and the company must satisfy itself that the legal provisions laid down in the Employment Relations Act 1999 have been followed. In order that justice is seen to be done, internal appeals may be necessary. Given the necessity for tight control, the administration of dismissals is probably best done from the Head Office. Of equal concern is the problem of absenteeism. Again, simple records

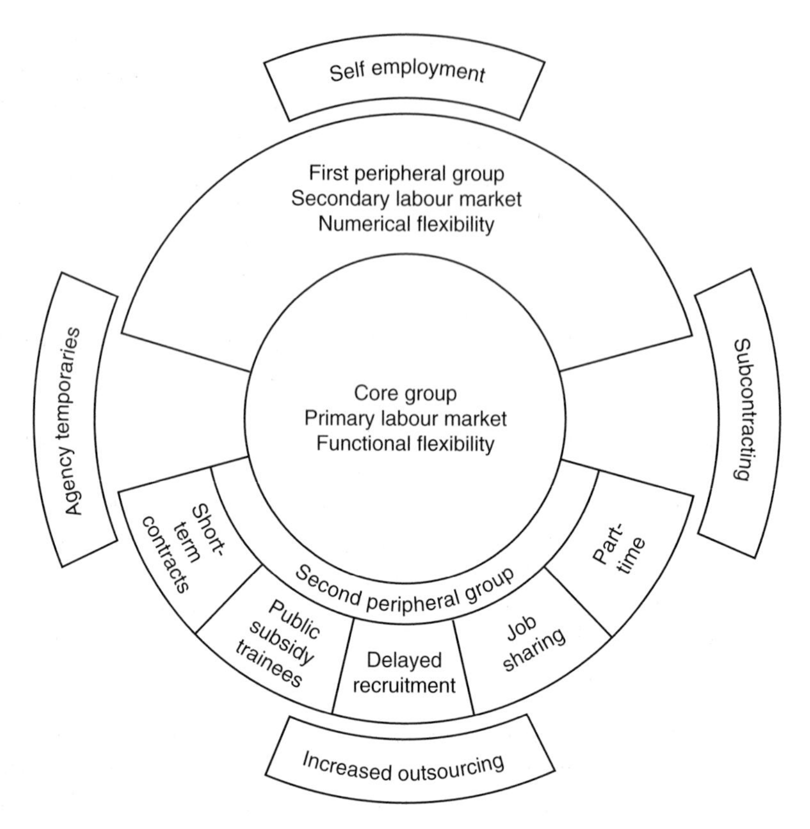

Fig. 6.13 Atkinson model of the flexible firm.

which can show the extent of the problem should be maintained. Different parts of the firm and different sites will have different rates of absenteeism and there are a multitude of factors which will influence matters, including earnings, location of the site, the nature of the labour force, the nature of the work itself, the regional unemployment level, etc. Monitoring of the situation is vital.

6.5.3 Development and administration of redundancy schemes

Construction firms have often argued that the uncertain nature of the business, changing technology, and greater mechanization mean that redundancies will be inevitable. Sharp recessions and government legislation since the mid-1960s have made it necessary to undertake a level of redundancy planning. The legislation concerning redundancy is the Employment Relations Act 1999.

It is widely recognized that the social and psychological effects of redundancy can be devastating to an individual. This factor demands that a company pre-plans any redundancy; if it does not, then good relationships with trade unions, employees, and the public at large may suffer. Moreover, pre-planning is required in order to retain the balance of the necessary skills for completion of existing projects. The fear of redundancy may also have a debilitating effect upon morale and productivity. If redundancy is in the offing, good workers who do not have many years' service may leave of their own volition with a consequent impact upon the balance of the labour force. The forward planning of human resource requirements can assist in defusing this potentially dangerous situation and can reveal if excesses of human resources are likely to occur. If such a position arises, then a construction company can adopt a variety of strategies, among which could be:

- policy of no recruitment other that for 'essential' vacancies
- transferring people from one region to another
- retraining of the existing staff to match requirements
- the phasing of any redundancies over a period of time
- the use of voluntary redundancy
- hiving off a particular service for outsourcing
- considering whether redundancy counselling services should be employed

Such strategies can be part of a redundancy policy which will be a component of long-term planning. Such plans may never be used, but it is better to be prepared for such contingencies as this will avoid any snap decisions. The costs of making people redundant should not be ignored, particularly where the statutory minimum payments are often exceeded by internal negotiations; and although the Employment Relations Act 1999 allows companies to recoup some of these payments, companies will inevitably be committed to extensive expenditure. Furthermore, if redundancies are to be made, it is vital that the relevant government department be notified in advance.

Firms within the industry are seldom troubled by having to make large redundancy payments because of the casual nature of the industry and the legal requirement of two years' service for qualification for redundancy payments, the strategy being to retain long-service employees and release those with shorter service. The increasing use of short-term contracts exacerbates the volatility of employment.

6.6 Training

Training can be discussed under two broad headings:

- apprentice training
- management training

6.6.1 Apprentice training

The UK construction industry is frequently accused of low productivity. Part of the explanation for this low productivity is the relatively labour-intensive processes which dominate the UK construction industry. The Organization for Economic Cooperation and Development (OECD) has argued that this labour-intensiveness flows from an undercapitalization of the construction industry when compared to the Scandinavian countries, the USA and Japan, where workers are highly trained to work expensive machinery to aid the construction process.

However, the construction process has moved from one based largely on craft skills to one where considerable off-site assembly takes place. This has meant that the skill base of construction workers has been eroded. At the end of the 1990s, the construction press were evaluating the extent of the 'skills shortages' in various parts of the UK and reporting that such shortages were driving up labour costs. In a labour market which is largely unregulated because almost half the workforce are self-employed, chaotic fluctuation in the costs of labour is likely to have a significant effect on the costs of construction.

Moreover this organization structure is unlikely to promote training because it is dominated by small, informal groups of self-employed craftsmen. Winch (1998) sees the decline in the number of trainees in the industry as a function of the decline in direct employment and a growth in self-employment. Before analysing training needs, it is useful to have a map of the trade structure of the industry. Clarke and Wall (1998) proposed a trade structure which grouped trades into eight occupational groupings, as shown in Table 6.2.

Structuring the trades in this way has meant that training programmes have centred on traditional trades with a narrow range of tasks at which the individual tradesperson has had to become competent. The large number of trade categories has grown organically and so each different specialism has required an increasingly narrow band of competencies. Trades, instead of overlapping have become compartmentalized. This fragmentation is reflected in the growth of National Vocational Qualifications available in construction. Clarke and Wall (1998) report that while over 50 NVQs are available in construction trades, only 18 per cent of the NVQs awarded in 1996 were in trades outside the 'big five': bricklaying, carpentry, painting and decorating, plastering, and general construction operative. Figure 6.14 indicates the relatively stable proportions of selected trades over the period 1974–1994.

Over the twenty years under consideration, the technology of construction has changed dramatically but the balance of the trades seems impervious to these changes. One explanation would be that the 'other' groups experienced considerable changes, such that those trades engaged in preassembling, say, toilet pods or making off-site components or cur-

Table 6.2 Trade cluster

1. *Woodwork/wood substitute* Carpenter and joiner Woodmachinist	6. *General construction/plant* Concretor Paviour Groundworker Civil engineering operative Plant operator Plant mechanic Demolition worker
2. *'Wet trades'* Bricklayer Plasterer Mason Tiler	7. *Accessing operations* Scaffolder Façade worker Lightning conductor engineer Steeplejack
3. *Roofing* Slater/tiler Roofer Mastic asphalter	8. *Building service* Electrician Plumber Heating and ventilating engineer
4. *Painting and decorating* Painter and decorator	
5. *Prefabricated component fitting* Partitioner/dryliner Ceiling/floor system installer Cladder	

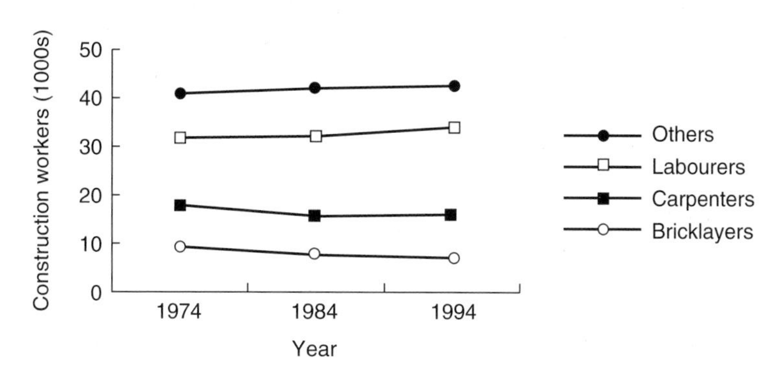

Fig. 6.14 The change in the number of workers in selected trades 1974–94.

tain wall units are unrecorded as construction workers and so the balance of on-site workers remains unchanged.

What has changed is the number of trainees in construction. Table 6.3 charts the decline in the number of trainees.

This diminution of trainees is surprising given that the dominant pattern of delivery has been the NVQ system which is not fully accepted by the industry and the shedding of the intellectual parts of the curriculum and focusing upon the mechanical tasks associated with a trade have encouraged a sharp division between 'those that do' and the cadre of managers who 'tell them what to do'. (The European tradition for trade training still retains a scientific content all the way through an apprenticeship.) The NVQ Level 3 is seen as satisfying a definition of 'having a

Table 6.3 The decline in trainees

Year	*Trainees in construction*	*Trainees in the operative workforce (%)*
1974	81,000	9.2
1984	50,000	5
1994	13,000	4.7

trade' and in 1996 the launch of the 'modern apprenticeships' movement espoused aspirations for tradespeople with skills in numerary and literacy. The principal qualification for a bricklayer is NVQ Level 2 but where the bricklayer needs to undertake specialist work which is considered as complex and engaging in discussions with supervisors then Level 3 may be more appropriate. The definitions of the different levels of NVQs is shown in Table 6.4. Levels 1–3 are craft levels; levels 4 and 5 are supervisory and professional levels.

Table 6.4 Definition of NVQ levels

Level 1:	Competence in the performance of a range of varied work activities may be routine and predictable
Level 2:	Competence in a significant range of work activities, performed in a variety of contexts. Some of the activities are complex or non-routine and there is some responsibility or autonomy. Collaboration with others, perhaps through membership of a work group or team, may often be a requirement
Level 3:	Competence in a broad range of work activities performed in a wide variety of contexts and most of which are complex and non-routine. There is considerable responsibility and autonomy, and control or guidance of others is often required
Level 4:	Competence in a broad range of complex, technical or professional work activities performed in a wide variety of contexts and with a substantial degree of personal responsibility and autonomy. Responsibility for the work of others and the allocation of resources is often present
Level 5:	Competence which involves the application of fundamental principles and complex techniques. Very substantial personal autonomy and significant responsibility for the work of others and for the allocation of substantial resources features strongly, as do personal accountabilities for the analysis and diagnosis, design, planning, execution and evaluation

The idea of the modern apprenticeship marked the end of the concept of learning on the job within the framework of a single firm. Trainees are no longer recruited to a firm, rather they are trained in college where the foundation of a trade can be formally taught. The structure of the industry has also made site-based training difficult. The shift to specialization and away from the general builder has meant that production undertaken on site is as likely to be the assembly of pre-engineered components as much as the application of particular skills.

Moreover the specialization of firms does not always provide for the range of competencies that the trainee needs to acquire to obtain the modern apprenticeship. Consequently the location of training has been divided into:

- college-based training
- training centres
- on-site training

College-based training

Further Education (FE) colleges are the main providers of apprentice training. The FE colleges have expanded their remit to include the practical as well as conceptual parts of the apprenticeship. The colleges teach and test the competencies as part of the NVQ regime.

Training centres

These are commercial enterprises and provide trade training opportunities in workshops. The Conservative Government of 1993–1997 saw training as a commodity and established Training and Enterprise Councils (TECs). These councils were dominated by local employers and their brief was to deliver training within their geographical area. Money was provided by government to the TECs who paid for training to be provided by suppliers of training. Many contractors saw this as an opportunity to profit from the business of training.

On-site training

The availability of site-based training is limited by the nature of the production process in construction. Self-employed workers paid on piecework are unlikely to be enthusiastic about coaching apprentices since it is likely to be an impediment to production. Much of the provision of training is handled through what is colloquially called the New Deal arrangements. Formally known as the 'Welfare to Work' programme, the scheme offers incentives to employers to hire youngsters by subsidising wages and providing a training grant. The employers broadly welcomed the scheme. However, they were chary about whether the arrangements for 26 days in a college-based environment supporting on the job training for a period of six months provided a solid enough foundation for the appellation of 'craftsperson' to be awarded.

In some instances, particularly local government, sites are especially assigned to trainees. In Glasgow, the Direct Services Organization has a large school of apprentices and uses them to build local authority funded projects.

The role of the Construction Industry Training Board (CITB)

The CITB was established in 1964 when the Industrial Training Act 1964 was introduced. Arguably it has played a leading role in stimulating firms to undertake training for the industry. Originally the CITB was established as a partnership between unions, employers and government. Now the balance has tipped heavily towards employers. The managing board is made up of 24 people, only two of whom are trade union representatives. At one time, apprenticeships had to be registered with the National Joint Council for the Building Industry. The new National Working Rules for Construction launched in 1997 carry no such requirement. As ever, Scotland is different to the rest of the UK; there new 'modern apprenticeships' have to be recorded with the Scottish Building Apprenticeship Council.

The CITB raises its money by levying employers. This is unusual and the CITB and the Engineering Construction Industry Training Board are the only two training agencies left with levy-raising powers. The levy is based upon the firm's payroll and there is a different levy for the amounts paid to labour-only subcontracting. (In 1998 the figure was 0.28 per cent for directly employed and 2.29 per cent of the monies paid to labour-only subcontractors; if the payroll of a company was less than £61,000 then the levy was waived.) The CITB raises something like £65 million through levy and this is supported by £25 million from the Government (CITB annual report 1997). This money is split between training those who have just entered the industry and grants to firms who send staff on continuing professional development (CPD) courses. Clarke and Wall (1998) suggest that the levy payment is seen by the industry as a tax rather than a commitment to training. Nonetheless, the Government has confidence in the CITB's ability to deliver training. In 1997 it was awarded the distinction of being given National Training Organization (NTO) status. This status means that the CITB is charged with delivering the Government training policy which includes promoting the Investors in People standard, generating lifelong learning, continuing education and industry and responding to the challenge laid down in the Technology Foresight Initiative.

An initiative welcomed by the industry is the register kept by the industry's workforce. This voluntary scheme is know as the Construction Skills Certification Scheme (CSCS) and is a record of workers who are deemed to have achieved a predetermined level of competence. The scheme works by allowing the industry to accredit tradespeople who do not have NVQs. Its purpose is to improve skills by setting benchmarked targets for future achievement. It is said to benefit the individual by recognizing skills and allowing for greater awareness of health and safety. The benefit for employers is the confidence of having skilled staff which also reassures clients when entrusting contracts to the industry. Indeed, one move has been to involve the need for evidence of CSCS pre-qualification criteria. The Government is interested in promoting scheme and incorporating it into standard forms of contract which would significantly strengthen the culture of craft training in the industry.

6.6.2 *Management and professional training*

This section considers how managerial and professional staff are trained. It identifies three vehicles of such training:

- professional development
- management development
- the use of learning networks

Professional development

Graduates in the industry have frequently been encouraged to develop their skills so that they may become corporate members of a chartered institution. This usually involves engaging in a structured training scheme. Each profession will have a set of presumptions as to which skills are to be developed in the time between graduation and corporate membership. Table 6.5 illustrates the range of development needs required by four of the major institutions associated with the industry.

Part of the requirement will be for an employer to appoint a supervising engineer/surveyor/builder who is responsible for the graduation progress towards corporate membership. Usually the arrangement is for regular reports to be filed with the supervisor. The completion of the components in the training requirements can be acquired by exposure through work experience or through formal, accredited short courses. These may be run from within the company or provided by training suppliers such as professional bodies, universities or training companies.

Management development

Beyond professional membership, most companies will encourage staff to develop their managerial skills. Langford and Newcombe (1992) defined management development as:

> ...the process whereby the (construction) organization's managerial resources are nurtured to meet the present and future needs of the organization. This process involves the interaction of the needs of the organization and the needs of the individual manager in terms of development and advancement.

This view does not limit management development to the formal activities of education and associated training courses but informal, incidental

Table 6.5 Requirement for corporate professional membership

Graduate entry formal training scheme	*Quantity surveying*	*Building*	*Civil engineering*	*Building services*
	RICS diary and log book	CIOB professional development	Company schemes ICE agreement	Company scheme CIBSE approved
	Record of experience Regular contact with training officers and supervising engineers/surveyors			
Training course				
Safety	•	•	•	•
Temporary works		•	•	•
Concrete		•	•	
Quality control and quality assurance	•	•	•	•
Planning, estimating and costing	•	•	•	•
Plant		•	•	•
Negotiating skills, subcontract administration	•	•	•	•
Soils and foundations, piling			•	
Contractual – ICE, JCT, GC Works, Standard Methods of Measurement, claims	•	•	•	•
Communication skills	•	•	•	•
Team development	•	•	•	•
Construction skills	•	•	•	•
Levelling and setting out		•	•	
External courses	•	•	•	•
Professional membership	ARICS	MCIOB	MICE	MCIBSE

and opportunistic learning helps to put into a work or personal context the formal lessons learned. Figure 6.15 illustrates the different strands of development and how it may take place.

Brandon (1993) sees the strategy for developing managers moving from the formal, school and university-driven education. This education will develop knowledge skills, values and understanding of life and the

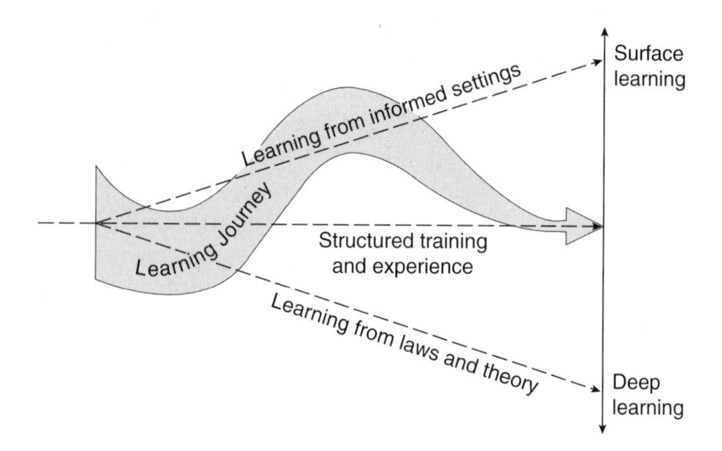

Fig. 6.15 Different outcomes of learning.

society to which the individual makes a contribution. After the acquisition of 'an education' then individuals will undertake training to modify attitudes, add transferable skills or develop the knowledge of a new technique. So, training is a planned activity with the intention of improving performance, not necessarily for developing cognitive skills. It adds to the individual's and the organization's capability. Management development is a synthesis of the two, it integrates the pedagogy of education with the focus of training; its usefulness to the organization is achieved by adding the capability of the individual. Figure 6.16 illustrates this conjunction.

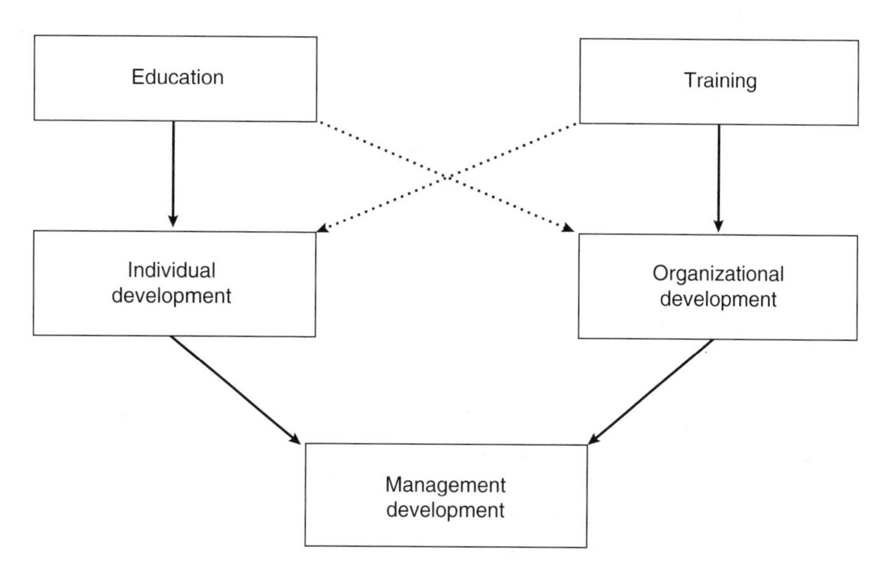

Fig. 6.16 The integration of training and education.

The way management development is handled by firms in the industry will vary with the commitment that a firm has to training and development and the learning culture within the firm. However, three generic plans can be identified (Ashton *et al.* 1995). The characteristics of these plans are mapped out below.

Phase I

- New managers have a low commitment to development.
- The development activity is not seen as part of the company culture.
- Opportunities for development are restricted to selected personnel who are 'sent' on external courses.

Phase II

- Commitment to development grows at senior management level.
- Budgets for training are made explicit.
- In-house training starts to develop.
- Audits of managerial talent begin to identify management development needs.

This phase may be characterized by tensions regarding who is to be developed to do what. Moreover, organizational and personnel needs may come into conflict.

Phase III

- Visible development activity declines and training events are not prominent.
- Management development is an implicit activity and part of what is carried out as routine.
- The culture and features of a learning organization are evident.

Management development is more than a set of training courses, it is part of the culture of the firm. At its highest level, the development of managers is integrated into every functional area of the business. In this setting 'mentoring' is likely to be important. The passing on of experience of coaching younger managers accesses the learning organization in a wholesome way. The challenge for the next ten years will be to create an industry which is attractive enough to recruit young men and women who can benefit from the wisdom of the battle-hardened veterans of the recessions of the 1970s, late 1980s and early 1990s.

A fuller discussion of management development for the construction industry can be found in Langford *et al* (1995) *Human Resource Management in Construction* and Langford and Newcombe 'Management Development in Construction' from Stocks and Male (1992) *In Competitive Advantage in Construction*.

Learning networks

The Foresight Programme for Construction (Office of Science and Technology 1995) identified that one of the engines of change was the promotion of learning networks through improved and more appropriate education and training. This was emphasized in the Egan report *Rethinking Construction* (1997) which called for a 'commitment to people, training and development' as a 'driver for change'. This call imposed new demands upon organizations to engage more staff in formal continuing professional development programmes (CPD). One response to this need for continuing education has been to create learning networks. CIRIA (1997) identified several types of networks which could engage construction professionals in learning; these are:

1. *The formal or organized networks*: This is often based around a specialist trade or professional grouping. For example, all the professional bodies run seminars or meetings of members to discuss particular topics of relevance to the profession. These can include technical, managerial or commercial developments influencing practice. Other networks are based on looser affiliations such as the Construction Productivity Network or the Construction Industry Environmental Forum which are agencies funded by government to promote learning in the areas of construction productivity and environmental concerns as they affect construction. Membership is open to all those who work in the industry. Other learning networks have been used by academics to strengthen ties between construction researchers. Again public money drawn from the Engineering and Physical Sciences Research Council (EPSRC) has facilitated such networks.
2. *Networks which develop as companies work together*: supply chains in the construction process create networks of people which can be used for purposes of learning. This learning is, of course, undocumented and is not likely to be a formal transfer of knowledge. It is knowledge gained by infusion rather than explicit transfer. Nonetheless, it can be powerful as different companies will have different cultures and how a company 'learns' can be an important discriminator when competitive advantage is evaluated. The concept of the learning organization has become an attractive way of defining a company's culture. Pedler *et al.* (1991) have portrayed the learning organization as one which contains several features, as Figure 6.17 illustrates.

   This learning organization has several features. It has a strategy which formally evaluates how the company and its staff can learn, and

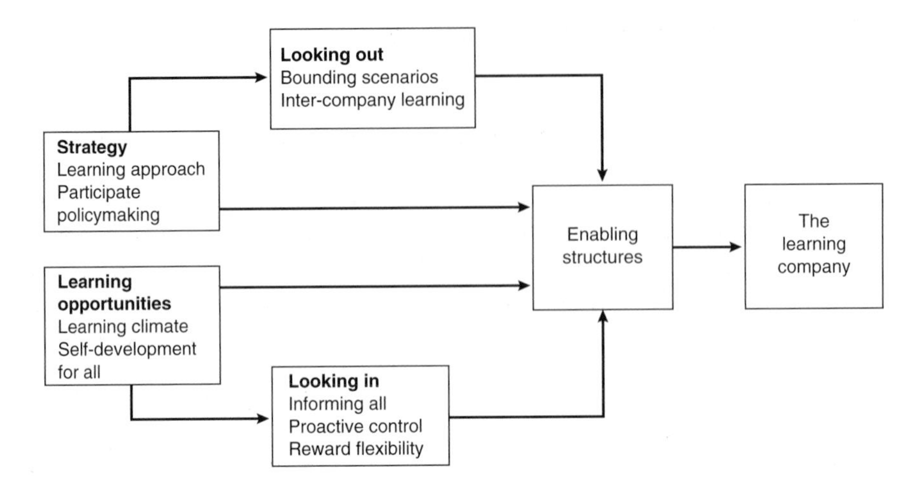

Fig. 6.17 Features of the 'learning company'. After Pedler *et al.* (1991).

part of this learning is by involvement of staff in the strategy formulation. Once the strategy is defined it is implemented in an inclusive way.

The learning organization creates opportunities for everyone from the lowest level in the company through to the managing director or senior partner.

The emphasis of the learning company is bifocal. At one level it should have a strong awareness of how it can learn from other companies and so is likely to value participation in industry networks. It will also be aware of how it can benefit from the experiences of other industries. At the level of the company, it will be flexible about reward structures and so be bound by fixed links between the job undertaken and the reward given. Merit pay is likely to feature strongly. Communications are likely to be ubiquitous and exist in a climate of fluid organizational structuring, unhampered by a culture of excessive formality.

3. *Informal networks*: This type of network involves the network of contacts that people make socially as a direct result of their chosen career. It is built up of contacts made at previous employees' mutual interest groupings based around a shared passion (i.e. the 'Building' half-marathon or the Chartered Institute of Building (CIOB) golf challenge). These networks are not formalized and the learning may be highly personalized but it frequently accesses market intelligence: soft, vital channels of information which can be used for personal or corporate benefit. Often it is both.

6.7 Investors in People standards

One of the more adventurous themes of human resource management of the 1990s has been the consequence of the Investors in People movement. The Investors in People standard is intended to improve business performance through the effective development of people. It is an initiative which overarches all trade, management and professional training. The award of the Investors in People standard entitles companies to carry the Investors in People logo. The award of the standard is given to companies who have developed targets for improving the knowledge and competencies of the people in the business. In a broader sense it may be used as an agent of change. Within a company the standard is based upon four principles which are underpinned by 23 indicators. Companies have to satisfy the Investor in People assessors that the 23 indicators are in place within the company. The principles are shown in Table 6.6.

Beckinsale and Dulaimi (1997) note that the takeup of the Investors in People standard has been relatively slow in construction organizations. They cite this as evidence of an unwillingness on the part of construction organizations to embrace the concepts and practices of learning organizations or that training and development of staff is seen as of secondary importance. The CITB acknowledge the lack of interest by the construction industry since they articulate a modest target of 20 per cent employers with over 25 employees to be in Investors in People registration by 2003 (Fergus 1999).

Summary

This chapter has discussed some of the broad issues of how people are managed in the context of the construction industry. It considered how formal plans for staffing an organization may be drawn up and implemented. Particular emphasis was placed on auditing current staff resources and mapping the current provision against future needs. This planning needs to take place in the setting of a wider appreciation of the strategic direction of the firm, so that the skills available to the company matched its ambitions in the market place.

The planning phase was followed by implementing the plan and the issue of how well-qualified people can be attracted to the industry featured such considerations as career information and the improvement of the image of the industry. Once inside the industry, it is seen as important that staff are properly inducted into a company and provided with a career structure and opportunities for personal and professional development. The labour structure was discussed and the employment of training regimes and training programmes by this structure.

Finally, the chapter has covered some issues of how the firm may

Table 6.6 The principles of Investors in People Standards. Source: A Framework for Business Improvement, Investors in People in the Construction Industry, Construction Confederation 1999

Principle 1	*Principle 2*	*Principle 3*	*Principle 4*
Commitment *An Investor in People makes a commitment from the top to develop all employees to achieve its business objectives.*	*Planning* *An Investor in People regularly reviews the needs and plans the training and development of all employees.*	*Action* *An Investor in People takes action to train and develop individuals on recruitment and throughout their employment.*	*Evaluation* *An Investor in People evaluates the investment in training and development to assess achievement and improve future effectiveness.*
Assessment indicators 1.1 The commitment for top management to train and develop employees is communicated effectively throughout the organisation. 1.2 Employees at all levels are aware of the broad aims or vision of the organization. 1.3 The organization has considered what employees at all levels will contribute to the success of the organization, and has communicated this effectively to them. 1.4 Where representative structures exist, communication takes place between management and representatives on the vision of where the organization is going and the contribution employees (and their representatives) will make to its success.	**Assessment indicators** 2.1 A written but flexible plan sets out the organization's goals and targets. 2.2 A written plan identifies the organization's training and development needs, and specifies what actions will be taken to meet these needs. 2.3 Training and development needs are regularly reviewed against goals and targets at the organization, team and individual level. 2.4 A written plan identifies the resources that will be used to meet training and development needs. 2.5 Responsibility for training and developing employees is clearly identified and understood throughout the organization, starting at the top. 2.6 Objectives are set for training and development actions at the organization, team and individual level. 2.7 Where appropriate, training and development objectives are linked to external standards, such as National Vocational Qualifications (NVQs) or Scottish Vocational Qualifications (SVQs) and units.	**Assessment indicators** 3.1 All new employees are introduced effectively to the organization and all employees new to a job are given the training and development they need to do that job. 3.3 Managers are effective in carrying out their responsibilities for training and developing employees. 3.4 Managers are actively involved in supporting employees to meet their training and development needs. 3.5 All employees are made aware of the training and development opportunities open to them. 3.6 All employees are encouraged to help identify and meet their job-related training and development needs. 3.7 Action takes place to meet the training and development needs of individuals, teams and the organization.	**Assessment indicators** 4.1 The organization evaluates the impact of training and development actions on knowledge, skills and attitude. 4.2 The organization evaluates the impact of training and development actions on performance. 4.3 The organization evaluates the contribution of training and development to the achievement of its goals and targets. 4.4 Top management understands the broad cost and benefits of training and benefits of training and developing employees. 4.5 Action takes place to implement improvements to training and development identified as a result of evaluation. 4.6 Top management's continuing commitment to training and developing employees is demonstrated to all employees.

develop its resources by the use of learning networks and so generate progressive learning organizations for the construction industry.

Questions

1. In such an economically volatile industry as construction, there is no purpose in a firm undertaking human resource planning. Discuss.
2. The failure to attract young people to the construction industry is likely to produce a serious shortage of skilled tradespeople in the future. Outline what you would do to redress this problem.
3. The profile of the construction industry workload is changing, with far more resources being allocated to preassembled construction. What impact will such changes have upon the structure of the labour force and how can such changes be accommodated by the industry?
4. Construction professionals often claim that balancing the interests of work and personal life is difficult. Speculate on the reasons for such claims using published data on working hours and any relevant legislation to support your theories. If this issue is a problem, what can be done to redress the problem?
5. Partnering between clients and contractors, contractors and specialist trade contractors, has been suggested as one means whereby employment in the industry could become more stable. Discuss the drawbacks and benefits of human resource planning in the context of project and strategic partnering.
6. Account for the decline in the use of labour-only subcontractors. How have such groups influenced human resource planning in the industry?
7. Identify the benefits of a structured personnel policy for a construction company. Amplify how such a policy can be used to improve morale and productivity at site level.
8. Prepare a paper for submission to a company's site managers which gives them a basic grasp of the aspects of employment law which have the most bearing upon the construction industry.
9. Discuss the utility and limitations of formal selection tests (e.g. aptitude, attainment and personality tests) as part of the recruitment process for management staff in the construction industry.
10. Assess the importance of maintaining personnel records in a construction company. How can such records be useful in sustaining a personnel policy?
11. Develop a working paper which lays down the criteria for promotion from site manager to a regional manager in charge of several site managers. How are the criteria you feel important to be measured?

References and bibliography

Armstrong, M. (1996) *Handbook of Personnel Management Practice*, Kogan Page Ltd.

Ashton, D. Easterby-Smith, M., and Irvine, C. (1995) 'Management Development Theory and Practice', *Management Bibliography and Reviews*.

Atkinson, J. (1984) *Emerging UK Work Patterns in Flexible Manning – the way ahead*, Institute of Manpower Studies Report Number 88.

Baker, E. and Gale, A.W. (1990) *Women in Construction Management: A report on two pilot programmes*, South Bank University.

Beckinsale, T. and Dulaimi, M. (1997) *The Investors in People standard in UK construction organizations*, Campus construction CIOB.

Brandon, P. (1993) 'Expert Systems – Modelling professional expertise'. Proceedings of CIB W65 conference, Trinidad.

CIRIA (1997) *Ciria News Issue 1*, London.

Clarke, L. and Wall, C. (1998) 'UK construction skills in the context of European developments', *Journal of Construction Management and Economics*, Volume 16 (5) (pp. 553–567).

Construction Industry Training Board (1998) *Construction Employment and Training Forecast*, CITB.

Egan, J. (1998) *Rethinking Construction*, Department of Environment, Transport and Regions, HMSO.

Fergus, D. (1999) 'Training for the Scottish Construction Industry', Construction 2000 Conference, ECN.

Office of Science and Technology (1995) *Technology Foresight Panel on Construction*, HMSO.

Handy, C. (1985) *Gods of Management*, Pan Books.

Hyman, R. (1976) 'Economic motivation and labour stability', from Bartholomew, D. (Editor) *Manpower Planning*, Penguin.

Langford, D., Hancock, M., Fellows, R. and Gale, A. (1995) *Human Resource Management in Construction*, Longman.

Langford, D. and Newcombe, R. (1992) 'Management Development in Construction', from Stocks, R. and Male, S. (Editors). *In Competitive Advantage in Construction*, Butterworth.

Latham, M. (1994) *Constructing the Team*, HMSO.

Pedler, M., Burgoyne, J. and Boydell, T. (1991) *The Learning Company*, McGraw-Hill.

Winch, G. (1998) 'The Growth of Self-Employment in British Construction', *Journal of Construction Management and Economics*, Volume 16 Number 5 September (pp. 531–540).

7 Financing Business Units

One of the major problems facing any business enterprise is that of obtaining finance. This is a problem not merely of quantity but also of type. The situation is compounded by legislation and by the dynamism of the economy but, perhaps more fundamentally, by the requirement to minimize costs.

Before discussing how various types of business may be financed and how suitable financial structure can be selected, certain basic concepts must be appreciated and some assumptions made. Economies are dynamic but, for the purpose of analysis, the situation is frozen so that the effects of changing one variable may be evaluated. Economic person, whether an individual or a firm, is assumed to act rationally to maximize utility, usually considered to be the maximizing of satisfaction for an individual consumer or the maximizing of profit for a firm.

Profit is a term for which there are several definitions. Those concerning accountancy (gross profit and net profit) will be considered later (see Chapter 8). In economics, total profit is usually considered to comprise two elements: normal profit and supernormal profit. Normal profit is that level of surplus of income over expenditure which accrues to the entrepreneur over the long period and is the minimum return required to retain the investment in that use. (The long period is that length of time in which all the firm's costs of production may be changed - the fixed, variable, and semi-variable costs.) Thus, in the long period, normal profit may be regarded as a quasi-cost for, if the firm does not achieve (at least) its normal profit, it should cease trading.

Supernormal (also called abnormal, monopoly or excess) profit is profit earned in excess of normal profit and may be a short-period or long-period situation depending upon the type of competition prevailing.

The margin is a very useful and important concept in economic analysis. It is a concept of increments, the marginal unit being the last unit produced or consumed (or the production or consumption of one additional unit).

The average is a broader concept, usually considering an entire spectrum of activity. An average is usually an arithmetic form of calculation; for example, average total cost is the total cost of the operation divided by the number of units of output produced. Averages naturally tend to follow the trend set by margins. (Averages and margins are encountered most commonly in connection with costs and revenues of firms and industries, often in the context of equilibrium analysis.)

7.1 Types of business unit

The construction industry comprises a wide variety of business units from the single-person enterprise to the large, multinational public company. Each business unit will have its own financial structure and requirements but it remains meaningful to consider categories of business unit as the differences are rarely fundamental within each category.

7.1.1 *Single-person enterprises*

This is the simplest form of business unit or *firm*. As the title implies, the owner of the firm is the firm, at least from the legal viewpoint. The owner has unlimited liability and is therefore personally liable for the debts of the firm, so that if the firm has insufficient assets to meet its liabilities, the assets of the owner may be realized (sold to obtain money) to meet the outstanding liabilities.

Single-person enterprises operate on capital supplied by the owner primarily, especially in the early stages of trading. The capital will have been obtained by saving or borrowing from individuals or a bank; commonly, any such loans being secured against the owner's personal assets (e.g. a mortgage on the owner's house). Naturally, if the firm develops successfully, it may be possible to secure a loan against the assets of the firm.

It is not surprising that single-person enterprises constitute the small firms of an industry, and construction is no exception. That is not to say that single-person enterprises have no employees; they frequently do, but not very many per firm. Further limitations, such as the managerial and organizational ability of the owner, lack of technical expertise and, often, contentment with being the owner of a small firm (sometimes coupled with a 'fear' of expansion) tend to keep the size of these enterprises down. The work they undertake is also at the lower levels of the size scale, being jobbing work, small extensions (usually to houses), occasionally small scale new work (again, usually housing), and specialist trades (such as plumbing, roofing, and electrics) direct to domestic clients or local contractors.

7.1.2 *Partnerships*

Partnerships occur in various forms but under the Partnership Act 1890 a partnership is defined as 'the relation which subsists between persons carrying on a business in common with a view of profit'. A partnership is formed by entering an express or implied contract. Thus, a partnership may be implied by conduct by sharing the profits of a business with another person. However, it is more usual and, indeed, more reasonable, for a partnership to be under a formal partnership agreement. Normally, the agreement will specify such important aspects as the capital contributed by the partners, their roles in the firm, any interest payments in respect of the partners' capital investments, the method of sharing any profits, and so on. Unless the partnership agreement specifies to the contrary, a partner may dissolve the partnership by giving notice to the other partner(s); also the death or bankruptcy of a partner dissolves the partnership *technically*, although, in practice, many partnerships continue to exist and operate. Further, partners are precluded from selling their share in the firm to anyone else without the consent of the other partner(s).

Under the Companies Act 1948, the number of partners in any firm is limited to between two and 20. However, the Companies Act 1985 (ss. 716 and 717) specifically allows partnerships of solicitors, stockbrokers and accountants to have more than 20 partners - no upper limit applies. Further, the Secretary of State is empowered to issue regulations removing the upper limit on the number of partners permitted for other types of partnership - surveyors, estate agents, valuers, consulting engineers and building designers are included in the types of partnerships covered by such regulations. Banks have no upper limit to the number of partners they may have (Banking Acts 1979 and 1987).

Partners are subject to unlimited liability in respect of the debts of the firm but under the Limited Partnerships Act 1907 a limited partnership may be created. A limited partner enjoys limited liability, i.e. the partner is liable for the debts of the firm up to the amount which he has invested only. However, a limited partner has no right to participate in the management of the firm and all partnerships must have at least one general partner (who has unlimited liability). A partnership which has limited liability partners must be registered with the registrar of companies (also the Registrar of Limited Partnerships) for the area of the firm's principal place of business by lodging details of the partnership with the registrar - the firm's name, nature of the business, name of each partner, etc. Unlimited partners are jointly and severally liable for the debts of the partnership.

Normally, the name of the firm will not contain the names of all the partners. Historically, all the partners had to register their names and nationalities with the Registrar of Business Names (Registration of Business Names Act 1916), but this requirement was repealed by the Companies Act 1981. Under the Business Names Act 1985, a partnership which uses a business name - a name under which the firm trades other than a name consisting of the family names of all the partners - must:

- display the names of all the partners on its business stationery – letters, orders, invoices, receipts, etc.
- display the partners' names at all premises where it carries on business and to which customers and suppliers have access
- disclose an address in Great Britain where, for each partner, the service of any document relating to the business will be effective

Partnerships do not pay corporation tax. The partners are assessed for income tax under the Schedule D provisions which, although an individual requirement for each partner, in practice is usually achieved by the firm providing details of the income of each partner from the partnership. However, under the Finance Act 1994, in partnerships created after 6 April 1994, no partnership tax return can be filed, so each partner must deal with the Inland Revenue separately.

The legal parameters relating to partnerships tend to limit the size of this type of firm. The feature of most partners being subject to unlimited liability is a major constraint; hence, it is most unusual for building firms to be partnerships. It is in the professions where partnerships are common. Most professional institutions used to prescribe that their members may not form incorporated associations and so the single-person enterprise or the partnership were the only types of firm available to the institutions' members; the spectre of possible recourse to the personal assets of the practitioner in order to recompense a client for losses occasioned by negligence of members of the practice was considered a major force to uphold professional standards. Naturally, adequate professional indemnity insurance, as commonly required in the codes of conduct of professional institutions, is a more appropriate means of ensuring that sufficient financial recompense will be available. Thus, many architectural, surveying, engineering, quantity surveying, and similar professional organizations are partnerships but, although a few partnerships are very large, employing many hundreds of people, the usual size of firm is small.

7.1.3 *Companies*

Several types of firm may include the word 'Company' in their name; a partnership may be known as 'X and Co.', but a company is more usually regarded as an incorporated association which has a legal identity distinct from its owners and those who participate in its activities. Incorporated associations are registered under the Companies Acts and their owners, in the vast majority of instances, enjoy limited liability. A limited company will, therefore, have the word 'Limited' (or 'Ltd') in the last part of its name, a public limited company's name must end with 'plc' and an unlimited company will have the word 'Unlimited' as the final word of its name (Companies Acts 1980 and 1981).

The promoters (those people who wish to form the company) of a new company must select the type of company to be formed from the following:

- public company limited by shares
- private company limited by shares
- private company limited by guarantee and without any share capital
- private unlimited company with a share capital
- private unlimited company without a share capital

However, as public companies are formed by the conversion of a private company, the usual choice for the promoters is whether the new private company will have liability of its members limited by shares or by guarantee, due to the advantages of limited liability. Normally, the choice is of a private company limited by shares.

A company (an incorporated association) obtains a separate legal identity upon registration with the Registrar of Companies who issues to the company a Certificate of Incorporation.

In order to obtain a Certificate of Incorporation, the promoters of the company must deliver the Memorandum of Association and (usually) the Articles of Association to the Registrar of Companies together with a statement of the company's nominal capital and a list of the people who have agreed to become directors. The Certificate of Incorporation signifies that the company is a body corporate and has perpetual existence, independent of its members. (A company may, of course, be dissolved under prescribed conditions.)

The Memorandum of Association sets out the objects of the company and the type of business to be undertaken. It is a statement of the fundamental principles which govern the company. The Memorandum must state the following:

1. The name of the company. If the company is limited by shares or guarantee, the last part of the company's name must contain the word 'Limited', 'Ltd' or 'plc', as appropriate. Undesirable names will be refused registration, e.g. where the name of the new company is so similar to that of an existing company that the two could be easily confused.
2. The address of the company's registered office. This need not be the company's head office and is quite commonly the office of the company's accountants or solicitors.
3. The objects of the company; the purposes for which the company has been created. This clause will be consulted by those considering trading with the company or by those considering investing in the company. The members of the company may, by obtaining an injunction, restrain the company from undertaking activities outside

its stated objects (*ultra vires*). The fundamental nature of this clause makes it very difficult to alter, and so it is common for the objects clause to be as widely scoped as possible.

4. A statement that the liability of members of the company is limited (for companies limited by shares or by guarantee).
5. A statement of the nominal share capital and the denomination of the shares into which it is to be divided (for companies limited by shares).
6. A statement of the guaranteed capital (for companies limited by guarantee).

The Articles of Association set out the internal organization of the company and its rules for management. It is obligatory for a company limited by guarantee or an unlimited company to deliver the Articles of Association to the Registrar of Companies, although this is optional for a company limited by shares.

The Articles of a company limited by guarantee must state the number of members with which the company proposes to be registered. For an unlimited company the Articles must state the number of members and, if applicable, the share capital with which the company proposes to be registered. In either type of company, the Registrar must be informed promptly of any increase in the number of members of the company.

The signatories of the Memorandum of Association are the original members of the company. It is usual for the membership of a company to be increased by further people taking up shares and having their names entered in the company's register of members.

If a company limited by shares does not deliver Articles of Association to the Registrar, the model articles contained in the Companies Regulations made under the Companies Acts will apply. The model articles will apply even if the company does deliver Articles in so far as the delivered Articles do not exclude or modify them. (The model articles are contained in Table A of the Companies Regulations 1985.) In practice, it is both advisable and desirable for a company to have its own Articles which are complete in themselves, if only to avoid consultation of two major documents to determine the contents of that particular company's Articles.

As the provisions of the Articles of Association are of less fundamental significance than those of the Memorandum, including the size of quorums for meetings and the limits on the number of directors, it is easier for the Articles to be changed. However, one area of importance is the permitted scope of action of the directors, particularly in binding the company contractually.

Every company must hold an annual general meeting (AGM) which all its members are entitled to attend. At this meeting the affairs of the company over the preceding trading year are reviewed in the light of the accounts for that period and the directors' report. This meeting is of great importance where significant issues are often debated and decided, those most commonly arising being the distribution of any profit, the election or re-election of directors and the appointment of auditors. A company may also hold extraordinary meetings.

The directors of a company are responsible for overall control of the operations and the general management. Their duties are quite strictly controlled by the provisions of the Companies Acts. Notably, they have a fiduciary duty to the company to avoid a conflict of their personal interests with the interests of the company, a duty not to take advantage of their position to achieve personal gain and, of course, a duty of care in the exercise of their functions.

As a company is itself a legal personage and is thus distinguished from its members, it has a separate and continuous existence. The life of a company may, however, be brought to an end (the company is dissolved) by the process of winding up. Compulsory winding up will occur upon the happening of certain events, the most common of which is that the company is unable to pay its debts (due to lack of liquidity) and an unpaid creditor has successfully presented a petition to the court to have the affairs of the company wound up. The court will place the affairs of the company in the hands of the official receiver prior to the appointment of a liquidator.

The winding up may be voluntary. Here a resolution for winding up will be passed by the members of the company who will then appoint a liquidator. Usually, this is done for restructuring purposes.

The third possibility is for the winding up to be under supervision. This is where a resolution for winding up has been passed by the company but is subject to a court order for the winding up to be under supervision. The liquidator is, therefore, under some supervision by the court.

Private companies

The Companies Act 1980 defines a private company as '... a company that is not a public company'.

The majority of private companies are limited, thus their members have a liability to contribute to the debts of the company on its being dissolved, up to a maximum of the amount unpaid on their shares (if any) or the amount they have guaranteed. There are, however, strict limits appertaining to the shares of private companies: the number of shareholders must be a minimum of two but with no maximum (the previous maximum of 50 has been abolished), there can be no invitation to the public to take up shares, and the rights of shareholders to transfer shares are restricted.

There are many reasons why forming a private company may be considered to be preferable to forming a partnership. These will usually include the limiting of the liability of the members (a limited partner is precluded from participation in the management of the partnership), the tax incentives (company car, etc.), and the easier raising of capital as more people may become members of the firm.

Within the construction industry, private companies are a very common form of business enterprise. For obvious reasons, they will comprise the smaller companies ranging from the two-person contractor (often subcontractor) upwards, their size being restricted by constraints such as capital requirements. Thus, the current upper size limit for a private company would be that of a local contracting firm.

Public companies

A public company is defined in the Companies Act 1980 as:

> ...a company limited by shares or limited by guarantee and having a share capital, being a company –
> (a) the Memorandum of which states that the company is to be a public company; and
> (b) in relation to which the provisions of the Companies Acts as to the registration or re-registration of a company as a public company have been complied with on or after the appointed day...

Under this Act, the minimum number of shareholders in a public company was reduced from seven to two. Also, the Memorandum must now state the name of the public company to conclude with the words 'Public Limited Company' (or 'plc').

A public company must deliver the written consent of the directors to act as such to the Registrar of Companies and their agreement to take and pay for the shares which qualify them as directors. Even after the Registrar's certificate has been issued, a public company may not begin trading until the directors have actually taken up and paid for their qualifying shares. Further, a plc requires a minimum allotted share capital of £50,000 before it can commence business; each share must be 'paid up' to a minimum of 25 per cent of its nominal value, plus any premium.

In order to raise capital from the public by inviting people to subscribe for shares or debentures, a public company must first issue a prospectus. The prospectus must comprise information relevant to the invited investment including the identities of the directors, the profits which have been made or which are anticipated, the amount of capital required by the subscription, the company's financial record, the company's existing obligations in respect of existing contracts, details of any voting rights, and the dividend rights of each class of shares. It may well be of benefit for the company to include a statement of an expert (e.g. an accountant) in the prospectus to assist in promoting the subscription but this may not be done without the written permission of that expert.

As public companies may offer their shares for public subscription, they are subject to greater control and scrutiny by the authorities than other forms of business enterprise. To 'go public' a firm must not only satisfy the statutory requirements but also the Stock Exchange. There are many advantages for the public company form of enterprise but there are also significant disadvantages.

The major advantages enjoyed by a public company are:

- limited liability of its members
- ease of raising large amounts of capital through public subscription
- shares on sale to the public and not subject to transfer restrictions
- usually more security due to the firm's having a large capital base
- finance may be obtained from a wide variety of sources and, due to the reputation of the firm and the greater security offered, may be relatively cheap
- as these firms tend to be large, they usually enjoy economies of scale in their activities
- large firms usually have sufficient funds, security, and ambitions to undertake research and development work, innovations, and training schemes

The major disadvantages, however, are:

- ownership and management may be divorced
- very large organizations may become 'bureaucratic' and of reduced efficiency
- quite a small shareholding may give one person effective control
- individual members with a small shareholding usually have no effective say in the firm's operations
- large firms sometimes become fragmented with departments reducing cooperation with each other and with head office, often pursuing their own independent goals

It is apparent that the advantages outweigh the disadvantages resulting in a tendency for firms to grow to become large public companies, commonly achieved by takeovers and amalgamations resulting in groups of companies rather than several completely individual companies.

Thus, in the construction industry, the largest firms are public companies, often of a group structure. It is common for the group to include not only building companies and civil engineering (often coupled with overseas) companies but also companies for plant hire, services work, specialist work (such as ground engineering), component manufacture and materials manufacture. Often, in the largest building companies, the structure of the firm is regionalized with each regional division operating as a separate organization. Naturally, these firms tend to concentrate on large projects, the size of firm varying from the large local building firm upwards. Usually, housing has been a separate division in a group, as has rehabilitation and refurbishment works but due to, *inter alia*, the prevailing economic climate and government policies (notably planning regulations), many large building companies have become increasingly involved with rehabilitation and refurbishment projects.

The move into refurbishment, rehabilitation, and repair and maintenance work is a function of the stage of development in the UK, in that most new infrastructures and buildings have been produced and, due to their age and the changing requirements of the community, are in need of repair and adaptation. Thus, the refurbishment, etc. share of UK construction output is estimated to exceed 50 per cent of total work (precise statistics are problematic as refurbishment and rehabilitation work is included in the figures for 'new work'). A further, general trend is for firms to retrench to what they regard as their 'core business'; while such concentration should yield advantages of specialization, it is likely to render such firms vulnerable to declines in demand in those specialist activities (hence, making the businesses of higher risk; diversification of a firm's portfolio of activities is generally, regarded as a risk-reducing strategy).

7.1.4 Cooperatives

In the construction industry, as in most industries in the UK, cooperatives are rather rare. Those which do exist are small scale enterprises and have very varied structures. Commonly, cooperatives experience problems such as obtaining tax exemption certificates as well as scepticism from potential clients due to the unusual nature of the organization in what is generally acknowledged to be a conservative industry. Also, many are unable to give trust and credence to an enterprise in which the profit 'motive' is absent and which operates on an apparently idealistic basis of fulfilling a need and sharing the proceeds of work equally. Cooperatives often subsidize the work they do for less affluent clients from the proceeds of work undertaken for clients in a better financial position, the reasoning being that should the situation be left to the free market, the poor could not afford to have necessary building work done which would result in further deprivation through a depletion of the building stock.

Jo Grimmond, MP, Chairman of Job Ownership (a group promoting Mondragon-type cooperatives) wrote (Grimmond 1980):

> The worker who has no share in the ownership has usually little to gain from the profit motive. Self-respect, the well-being consequent upon belonging to a community, and opportunities for individual enterprise, are most likely to flourish when workers come together in a business they own and control themselves...

Finance and capital are seen as a considerable problem for cooperatives, the sources being the members and loans from banks (usually reluctant) or other institutions and individuals. The formal establishment of a cooperative is valuable in that it gives the members limited liability, a registered cooperative being a legally recognized body in its own right.

7.1.5 Joint ventures

In many respects, all organizations are joint ventures between the people involved; in construction, every project contains elements of joint venturing, although, usually, only informally through the operation of Temporary Multi-Organizations (TMOs). However, some projects are set up as formal joint ventures - either for an individual project or for 'strategic' purposes, such as project type or to enable an overseas-based organization to operate in the 'home' country. Formal joint ventures occur to include particular areas of expertise, to spread risk, to enable particularly large projects to be undertaken, etc. - commonly a mix of such reasons is instrumental in generating a joint venture. Dietrich (1994) asserts that joint ventures are 'characterized by oligopolistic partners', which suggests particular features of the organizations involved and their behaviour, especially due to their market power.

Joint ventures are common in major infrastructure projects, increasing numbers of which are procured under concession arrangements (see 7.2.6) and may be considered as a means by which 'technology transfer' occurs - from a developed country joint venturer to a developing country partner. Frequently, developing countries prescribe that, to undertake projects in that country, an overseas organization must joint venture with a local organization, sometimes with the local organization being the majority owner of the joint venture. A joint venture may involve both horizontal and vertical amalgamations (organizations at the same or at different stages of the supply process).

The owners of a joint venture (usually a joint venture is a separate company which is wholly owned by the joint venturers, although the joint venture may take the form of other business units) invest and share profits of the joint venture. Generally such investment is in proportion to their perceived risk-bearing. In selecting joint venturers and establishing the formal arrangement, the basic issues of transaction cost economics are helpful considerations. The formal arrangements, both intraorganizational and interorganizational are likely to be required to control/limit potential opportunistic behaviour (see, for example, Williamson 1985). Thus, *ex ante* action (to formalize agreements in detail, etc.) may be employed to give control over *ex post* actions (post-contract opportunities to secure individual gains). Clearly it is necessary to balance *ex ante* and *ex post* costs and benefits (notably, financial but regarding relationships too). The formal arrangements are established to combat 'strong cheating' (failing to honour formal obligations); however, 'weak cheating' may still occur (failing to honour informal obligations). What constitutes weak cheating depends on culture (norms of behaviour). Buckley and Casson (1988), in their economic analysis of fifty-fifty joint ventures, note that 'forbearance' (non-cheating) is the honouring of both formal and informal obligations and hence constitutes non-opportunistic behaviour.

As joint ventures arise by the parties achieving combinations of internalization economies, coping with indivisibilities and obstacles to other forms of merger, the level of cooperation depends on the motives of the joint venturers, the nature of the main activity of the joint venture and, perhaps primarily, on the parties' cultures.

Thus, the popular 'invention' of partnering in construction seems to be an exclusively Western phenomenon as opportunism is a notable, frequently encouraged, feature of Western markets (including construction). In other societies, notably those of the East (such as Japan, with which much comparison has been made by Westerners), the drive for partnering is not comprehended as it is (or underpins) the normal, way of conducting business - opportunistic behaviour and cheating (either strong or weak) are not issues! Thus, the necessity for careful selection of joint venturers is paramount, irrespective of the form or purpose of the joint venture.

7.2 Sources of capital

The sources of capital available to any firm are quite numerous but, as has been noted already, public companies have the greatest variety of sources available for their use and the single-person enterprise the least variety. It is also important to seek capital not only from a legally permitted source but from a source appropriate to the type of capital required. The type of capital is dictated by the time period for which it is needed by the firm and the degree of risk involved, the former denoting the possible sources and the latter determining the most economic solution.

For convenience, capital is classified into three types by time period of the requirement; short, medium, and long. The short period is considered to be that length of time during which only the variable costs of the enterprise may change (usually less than one year). The long period is that length of time during which all the costs of the enterprise (variable costs, fixed costs, and semi-variable costs) may change. The medium term lies between the two extremes and is that length of time during which only the fixed costs cannot be changed (usually one year to about seven years). The time period for which the capital is required is thus a very important parameter which affects not only the sources available to the firm but also the cost of the capital, due to time preference. (Note: while in economics periods are determined by the length of time required to change the firm's costs, in finance, periods are denoted by durations in years for which the finance is required. Further, economics considers capital to be real assets: plant and machinery, buildings, etc.; in finance capital is regarded as money.)

7.2.1 Long-term capital

Long-term finance is required to form or expand the long-term capital of the firm. It will be used to purchase fixed assets such as buildings, plant, and equipment, the durable (fixed) assets of the firm. Initially only external sources of capital such as shares, debentures, and mortgages, are available to a firm but as profitable trading progresses, internal sources such as retained earnings, reserves, and depreciation provisions may be used. The internal sources of capital are of obvious importance as no payments must be made by the firm for the use of those funds; however, the opportunity cost of any internal funds must be considered.

Shares

Shares confer a stake in the ownership of the company upon the shareholders (the owner of the shares). Shares have a par, face or nominal value which represents the ownership contribution of each share. Shares may be fully paid up or only partly paid up, in which instance the shareholder may be required to contribute up to the unpaid amount of his shareholding should the company be dissolved. It is important to note that, although such outstanding liabilities do affect share prices, the Stock Exchange share price is no evidence of whether a share is fully paid up – the share certificate and other documents appertaining to the relevant share issue should be consulted for this information.

Shares may be of several types, each with different rights. Ordinary shares, or equities, represent the major ownership and risk-bearing element of entrepreneurship. Holders of ordinary shares are entitled to a share in the profits of the company (a dividend) only after all other liabilities have been met. Ordinary shares usually entitle the holders to voting rights, the votes being in direct proportion to the shareholding (non-voting ordinary shares are often called 'A' shares). Thus, ordinary shareholders, in theory at least, have control of the company but also are the main risk bearers, having only a residual claim on profits.

Preference shares were also common, entitling the holders to a dividend up to a prescribed level prior to any distributions being made to holders of ordinary shares. Thus, preference shares are a safer form of investment than ordinary shares and so the return on investment in the long period tends to be lower.

Cumulative preference shares are rather less common and carry a right for any dividend unpaid to be carried forward for payment out of the profits of future trading periods. Participating preference shares entitle the shareholder not only to a preference dividend but also a further dividend from the company's profits of the trading period should those profits exceed a stipulated amount (such amount being set to permit a reasonable return to ordinary shareholders prior to the participation of these preference shareholders in any further distribution).

Issues of shares are of three types: a new issue for sale, a rights issue, and a scrip issue. A new issue for sale may be made in a variety of ways and may involve various intermediaries between the company and the purchasers. Shares may be issued directly by the company by means of subscriptions or tender from prospective purchasers. More usually, however, the company will issue the shares through a specialist intermediary, most commonly an issuing house (probably a merchant bank; issuing brokers are sometimes used). The issuing house executes the majority of the administration work associated with a new share issue, will offer advice on the form and timing of the issue and, in return for a commission, may underwrite the issue in part or in total thereby guaranteeing the company a minimum amount of finance from the issue.

A rights issue is where a company offers its existing shareholders the option of purchasing new shares in proportion to the existing shareholding, usually at a low price relative to the prevailing market price of the company's shares. Any shares not taken up in this way, by the date prescribed in the offer, will be sold via the Stock Exchange. Thus, a rights issue is a rather cheap way for a quoted company to raise capital as any underwriting, advertising and similar costs associated with a new issue for sale are avoided. The existing shareholders who do not wish to take up their rights may sell those rights.

A scrip (or bonus) issue does not raise any new capital for the company. It is an issue made to existing shareholders in proportion to their shareholding and is free of charge. It is an adjustment to the capital structure of the company, often following revaluation of major, fixed assets.

It must be noted that only issues of shares and calls on any amounts unpaid on shares raise capital for the company. Sales and purchases of shares (or debentures) which occur subsequent to the issue are merely changes in the owners of the company even though the shareholders may make or lose vast sums in such transactions.

Often, shares are sold on issue at a price above their face value. The capital so raised in excess of the face value of the shares forms a 'share premium account' (Companies Act 1985) and essentially is treated as part of the paid-up share capital of the company.

Debentures

Debentures are fixed-interest securities which are issued by a company in return for a long-term loan. Usually, debentures are redeemable after a specified period (commonly of around 20 years). Debentures may be unsecured but more often are secured against the assets of the company as either mortgage debentures, which are secured against specific assets, or floating debentures, which are secured against the assets generally. Normally, debentures are issued in large denominations.

As debentures represent a loan to the company they carry a fixed rate of interest. Debenture holders are creditors of the company, not owners, and as such may seek dissolution of the firm should the interest payments not be made.

Debenture interest is a cost to the company and, therefore, is a deduction to be made from gross profit in determining taxable profit. The rate of interest carried by debentures depends partly on prevailing long-term interest rates as well as upon the type of debenture and, hence, the degree of risk involved. As the risk carried by debentures is usually small, the rate of interest will also tend to be quite low (for instance, when compared to the return on preference shares).

Thus, debentures are best suited to financing companies which have quite stable levels of profit and a large amount of fixed assets (e.g. property companies) and tend to be used instead of preference shares.

Retained earnings

Retained earnings is profit retained within the firm instead of being distributed to the owners. Therefore, retained earnings is a certain and relatively cheap source of long-term capital and indeed is the major source for many firms. Retained earnings does, however, represent an increased stake in the firm by the owners (the ordinary shareholders in the case of a public company) and so, largely accounts for the prices of shares being in excess of their par value in dealings on the Stock Exchange.

Depreciation

Depreciation is a bookkeeping and costing exercise by which the initial cost of an asset is written off over its useful life, i.e. the value of the asset is reduced by a predetermined amount each year. Depreciation is allowed prior to the calculation of profit. It is a means by which the costs of the amounts of the various fixed assets used in production are charged (absorbed into the accounts) and, hence, recovered as part of the sale price.

For internal purposes, a firm may provide for depreciation of an asset on a straight line basis (the depreciation provision per year equals the cost of the asset on acquisition minus any residual or scrap value at the end of the asset's life – all divided by the estimated useful life in years) or on a reducing balance basis (a percentage of the written down value of the asset, the depreciated value of the asset shown in the company's books at the start of a year, is provided for depreciation each year such that the value of the asset is shown at either nil or its scrap value at the end of its predicted useful life). The reducing balance method of depreciating assets is, generally, a more accurate statement of the asset's consumption (or value reduction) pattern. For purposes of corporation tax, however, the method to be used for depreciating any asset is prescribed in the tax regulations and so, it may be necessary to produce two accounts, one for internal purposes and the other for taxation purposes.

Depreciation may also be regarded as a source of capital. If no depreciation were charged on, say, machinery, a greater amount of profit would be available (after tax, of course) for distribution to the owners. Thus, reserves created by the process of depreciating fixed assets represent a stake in the firm by the owners, in a similar manner to retained earnings. However, an actual distribution to the owners of the depreciation provisions would be a depletion of the capital of the firm: 'consumption of capital'.

Depreciation is a relatively cheap and popular source of capital for firms and is essential for capital maintenance. Like retained earnings, depreciation provisions contribute to the market value of a company's shares being in excess of their par value.

7.2.2 *Medium-term capital*

Medium-term capital is the most indeterminate of the three categories, being regarded by some as akin to short-term and by others as akin to long-term capital. From the concept of the medium term stated above, the sources of medium-term capital are short-life debentures, longer-life bank loans and other loans of a life between two and seven years. Retained earnings may also be used to provide medium-term capital.

Bank loans

The prime considerations of the banking system - security of deposits, the required liquidity ratio, and the necessity of obtaining a reasonable return for investors - naturally tend to dictate the lending and investment policies followed. The majority of bank deposits are liable to instant or very short notice withdrawal which acts to limit the extent of all but short-term investment activities by the banks. The security requirements also act to the detriment of the provision of risk capital, the risks faced by banks being of three categories: the risk of total default by the borrower, the risk of inability of the borrower to repay a loan upon the request of the bank and the risk of the borrower being unable to make the repayments at the time(s) and in the pattern agreed. A further limitation applies through the imposed reserve asset ratio which a bank must maintain. (Other restrictions are also imposed upon the banks from time to time but these tend to be of a rather transient nature.)

Thus, although a business might be sound, it is quite difficult to obtain finance from the commercial banks. Should a firm be able to satisfy the bank by demonstrating that the purpose of the loan is commercially sound and that the loan is secure (this will involve the presentation of accounts to the bank of the firm's trading record and a commercial appraisal of the scheme for which the loan is sought), the loan will usually be made for an agreed period of between two and five years. The loan must be repaid at the end of the period and interest will be charged at a commercial rate (a rate above the bank's base rate dependent upon period, risk, etc.). Alternatively, the loan may be repaid by instalments (yearly, half-yearly, or quarterly) and interest charged on the outstanding balance only. It should be noted, however, that banks prefer to make loans for the provision of working capital (such as stocks, which will be processed and sold) which may be considered as self-liquidating (the sale of the finished product providing the funds for the repayment of the loan) and so, should be more accurately considered as short to medium-term financing.

Debentures

Debentures have been considered above in the context of long-term capital provision.

Small firms provisions

In recent years much attention has been devoted to the financial problems of small firms. The Department of Trade and Industry (DTI) offers a loan guarantee scheme for small firms. The firms must have viable business proposals but, due to lack of security, have been unable to secure a loan from 'conventional' sources. The loans are for two to ten years on sums of £5,000 to £100,000 (£250,000 for firms which have been in business for over two years). DTI guarantees 70 per cent (85 per cent) of the loan in return for payment of 1.5 per cent of the outstanding amount per annum. The loan is provided, following formal application, by (mainly) high street banks, which charge commercial rates on the loan. Applicants must be UK companies with a maximum annual turnover of £1.5 million (£3 million for manufacturing companies).

The Industrial and Commercial Finance Corporation was formed in 1945 by various financial institutions, including the Bank of England. The ICFC provides long-term finance for an applicant firm's fixed capital requirement. Usually, the firm will require the finance to enable it to expand by constructing a new factory or purchasing plant and equipment. The firms assisted are small to medium-sized companies and the finance is usually provided as a long-term, fixed interest loan (of seven to 20 years' duration and interest being payable on the outstanding balance only) as preference shares or as ordinary shares or as a combination. The ICFC is also prepared to assist new companies, including start-ups and to provide further services to applicant companies (such as financial advice).

7.2.3 *Short-term capital*

Short-term capital provision and management is vital to the success of a firm. It is this type of capital which is required for day-to-day activities. The sources of short-term capital are both internal and external; the main internal sources being accrued expenses and tax provisions and the main external sources being trade creditors, bank overdrafts and short-term loans. It is the short-term finance which provides the circulating capital for the firm and assists with overcoming potential cash flow problems due to market fluctuations. Notably, the most important source for construction firms is that of the bank overdraft.

Bank overdrafts

A bank overdraft is a process whereby a customer of a commercial bank is permitted to overdraw on that account (to draw in excess of any positive balance) up to an agreed limit for a prescribed period. This is rather similar to a bank loan except that interest is payable on the amount overdrawn only for the period it remains overdrawn and the account is usually repayable on demand or upon the termination of the overdraft period. Overdraft facilities are, however, commonly renewable and so, in practice, may constitute a continual source of short-term capital or liquidity 'insurance' facility.

An overdraft is a relatively cheap form of finance due to its being a short-term facility; a fee for arranging the overdraft facility will be charged by the bank but interest is payable only on the amount of the overdraft actually taken up. Overdrafts are thus very suitable for firms with a fluctuating financial requirement, such as building contractors. It is a widely held belief that almost all building firms operate on an overdraft.

Trade creditors

Initially, it is important to distinguish between trade credit and discount. Discount is a reduction in the price to be paid for a commodity, usually applicable if payment is made by the purchaser within a prescribed period from date of invoice or date of delivery (cash discount) or applicable because the purchaser is a business within a particular trade (trade discount). Most standard forms of construction contract require nominated suppliers and nominated subcontractors to provide prescribed levels of cash discount to the contractor but all other cash discounts and all trade discounts must be passed on to the client (employer).

Credit is the granting of ownership of commodities prior to payment for them.

In construction, it is usual for discount and credit facilities to go hand-in-hand, notably in respect of both domestic and nominated suppliers and nominated subcontractors. Domestic and labour-only subcontractors usually offer the contractor, and the contractor usually affords the client, credit alone. Again, the standard forms of construction contract prescribe the periods of credit to be given to the contractor by nominated suppliers and nominated subcontractors; otherwise the credit terms are set out in the domestic forms of supply and subcontract. During economic recessions, there is financial pressure on contractors and so, suppliers and subcontractors may not be paid until the main contractor has received the relevant interim or final payment from the client. In the UK, the Housing Grants, Construction and Regeneration Act 1996 (following recommendations of the 'Latham Report' 1994) has effectively outlawed 'pay-when-paid' provisions.

The basic concept of trade credit is to bridge the time gap which exists between the purchase of materials and the sale of a finished product (consider the provisions appertaining to nominated suppliers under the JCT Standard Form of Building Contract). Naturally, the credit facilities vary considerably between firms and types of firm.

Control of credit facilities, both those offered to and by the firm, is important, and bears directly upon the capital requirements - indeed, Fellows (1984) demonstrated that operating good credit control is likely to be the most advantageous managerial action for a contractor's cash flow. Ideally, a greater credit facility should be enjoyed by the firm than that which the firm offers to its own customers, considering not only the periods involved but the sums as well (discounts are likely to be an added complication in this calculation).

Short-term loans

Short-term loans are available from individuals, banks and other financial institutions. They are required for the provision of working capital, carry a prescribed rate of interest upon the entire sum and cannot be recalled prior to the due date. Usually short-term loans are obtained from commercial banks.

Tax provisions

A company's corporation tax liability becomes due for payment to the Inland Revenue in the year following that in which the profit, giving rise to the tax liability, was earned. Hence, that finance is available for use by the company for approximately one year. However, at a conference organized by CES Limited in May 1981, Frank Dobson MP asserted that the corporate sector's tax payments to the Inland Revenue in the financial year 1980/81, including bank windfall tax and petroleum tax, constituted approximately 8.7 per cent of the Inland Revenue's income for that year. This indicates that the then prevailing rate of corporation tax (52 per cent) was not such a burden to companies as it might at first have appeared. Nevertheless, payment of corporation tax one year in arrears constitutes an important source of short-term capital.

Value added tax (VAT) is levied by Customs and Excise; the standard rate of VAT is set periodically and has reached 17.5 per cent. Any item sold by a firm is subject to VAT regulations in one of three ways: standard rated, zero rated or exempt. Thus, the purchaser pays no VAT on items which are either zero rated or exempt but the vendor's position is rather different. If an item sold is exempt from VAT, the vendor cannot recover input tax (the VAT paid to suppliers in respect of that item sold); however, if the sale is zero or standard rated, the input tax is recoverable from the Customs and Excise (by way of set-off against the VAT which the firm is liable to pay to Customs and Excise from its standard rated sales).

Any firm with an annual turnover exceeding a prescribed amount must be registered for VAT. Only registered firms may recover input tax. The tax balancing (payment liability and input tax recovery set-off) occurs at either the usual three-monthly or monthly intervals for VAT returns. Naturally, if a firm produces largely zero rated outputs, it should seek monthly returns to facilitate a cash flow from the Customs and Excise. Conversely, if a firm produces standard rate commodities, its cash flow will be advantaged by having three-monthly VAT returns.

New construction work for residential buildings and alterations to produce dwellings or relevant residential buildings are currently zero rated for VAT. All other new construction, refurbishment, alterations and repair and maintenance work is currently standard rated. Thus, unless the majority of a construction firm's work (by value) is standard rated, it should seek monthly VAT returns. It should be noted that the cost of making the additional eight VAT returns per year should be considered prior to electing which period is to apply.

Especially for larger firms, annual accounting for VAT provisions apply: the larger the firm, by annual VAT liability, the more complex the provisions; the largest category (more than £2 million VAT per annum) must make monthly payments by electronic means. Although the monthly payments are calculated on the basis of the previous year's VAT returns, settlement of the current year's VAT liability is quarterly.

For a firm undertaking a majority of standard rated work, VAT may provide a useful source of short-term capital.

Accrued expenses

Accrued expenses are a type of credit. Such expenses comprise wages, salaries, employees' expenses claims, and similar internal credit facilities.

It is evident from the nature of this facility that the amounts and periods of credit will vary, the periods being a matter of days or weeks and the individual sums being quite small. However, especially in times of high interest rates, the overall credit facility will be of significance.

Debt factoring

Finance may be obtained by factoring debts. This is a procedure in which a specialist organization buys the debt owed to the firm at a discount, thereby providing the firm with short-term capital and reduced risk. The size of the discount will depend on the period of the debt until maturity (when it is due to be paid) and the risk perceived by the factoring organization.

7.2.4 Stock Exchange

The Stock Exchange is, primarily, a market where various securities (e.g. government stocks, shares, debentures) are bought and sold. The International Stock Exchange in London (ISE) operates as a continuous dealers' market in which VDUs are employed to display the prices of shares, etc.; deals are struck between traders via telephone links. ISE trading institutions are broker-dealers who can operate on their own account or as agent for another person or institution; a subcategory of trading institutions is those who are registered to act as market-makers or dealers. The activities of the trading institutions are regulated to protect the interests of parties for whom such institutions act as agents in the form of assuring 'best' prices in transactions. The ISE is widely recognized as not only a highly organized but also a near-perfect market.

Apart from acting as a securities market, the Stock Exchange performs several other important functions in connection with the sale and purchase of securities. The Stock Exchange vets any company wishing to go public (i.e. to become a public company) and prescribes certain conditions which must be fulfilled, e.g. a public company must offer at least 25 per cent of its capital for public subscription. The requirements are contained in the Listing Rules. Only companies with an equity share capital which is valued at more than £700,000 are able to obtain a Stock Exchange full listing (the placing of the price of the security on the official list – a privilege granted to public companies which have satisfied the Stock Exchange Council) in order that the company's securities are tradeable on the Stock Exchange. The Stock Exchange Council is rigorous in its examination of all applications for a listing due to its role of safeguarding the investing public.

In December 1980 an alternative to the full listing (which is an expensive exercise for any company and proportionally more so for a small firm requiring to raise a relatively small amount of capital: the cost of a small issue may be approximately five per cent of the proceeds of the issue), the Unlisted Securities Market (USM), was introduced specifically to encourage small companies to raise capital through the Stock Exchange. Any company seeking a USM listing must be registered as a public limited company and have a suitable three-year trading record. Normally, only ten per cent of the company's authorized capital need be sold to the public. The costs to the firm of obtaining a USM listing are also lower (such as the mandatory advertising prescribed by the Stock Exchange) and so smaller companies are encouraged to raise capital through this means. In 1995, the USM was renamed the Alternative Investment Market (AIM).

Only the initial sale of securities raises capital for the issuing company (or government). The majority of Stock Exchange transactions are subsequent to such sales and represent a change only in the ownership of the securities, changes which occur at the prevailing market price. Thus, in evaluating an investment in securities, an investor will consider:

- any amount not paid up (thus representing a possible future liability to pay)
- the market price of the security
- the return on the security (divided or interest)
- the risks involved and status of the company (government securities are generally regarded as almost risk free)
- the economic climate and trends

In a basic economic evaluation, the yield of an investment is important. The yield is the return (average over a period or at a particular time) on the investment compared with its purchase price. Discounted cash flow techniques should be employed in the calculations. Buckle and Thompson (1992) suggest that the current price of a share can be determined:

$$P_t = \frac{E_t P_{t+1} + E_t D_t}{(R - g)}$$

Where: E_t is expectation at time $_t$
P_{t+1} is price of the share at time $_{t+1}$
D_t is dividend paid during the period $_t$ to $_{t+1}$
R is the required rate of discount
g is annual rate of growth of dividends (assumed constant)

7.2.5 *Sale and lease-back*

Sale and lease-back arrangements have been common for some time in the property development industry. The specific arrangements are, of course, peculiar to each individual dealing but may be summarized in general as when a property company which owns a development site sells the freehold to a financial institution (usually a pension fund or insurance company) in return for:

1. a loan to finance a development (usually commercial) upon the land, the interest payable on the loan being deferred until completion of the development, and
2. an agreement to lease the completed building back to the property company

Upon completion of the development, the property company pays rent to the financial institution, thereby providing it with its normal rate of return on such an investment, and itself sub-lets the building to an occupier (usually a commercial enterprise) thus providing itself with an adequate return.

This process permits the financial institution to invest in specific developments while relieving commercial enterprises of the necessity to tie up large amounts of capital in the construction and ownership of their office premises.

It is noteworthy that the process of leasing plant and equipment which is employed by many organizations (e.g. computer facilities, company cars, construction plant - although this is more usually hired than leased) is carried out to avoid the necessity of long-term capital investment by the firm in these items. By employing sale and lease-back arrangements, the capital of the firm is kept available for investment in the firm's specialist activities, those in which it should earn the greatest return.

7.2.6 'Concession' arrangements

An increasingly popular method of obtaining major projects in the public sector is through various forms of concession contracts - Build-Operate-Transfer (BOT); Build-Own-Operate-Transfer (BOOT); Design-Build-Finance-Operate (DBFO); Public Finance Initiative (PFI). Merna (1993) describes a concession or BOOT project as:

> A project based on the granting of a concession by a principal, usually a government, to a promoter, sometimes known as the concessionaire, who is responsible for the construction, financing, operation and maintenance of a facility over the period of the concession before finally transferring the facility, at no cost to the principal, a fully operational facility. During the concession period the promoter owns and operates the facility and collects revenues in order to pay the financing and investment costs, maintain and operate the facility and make a margin of profit.

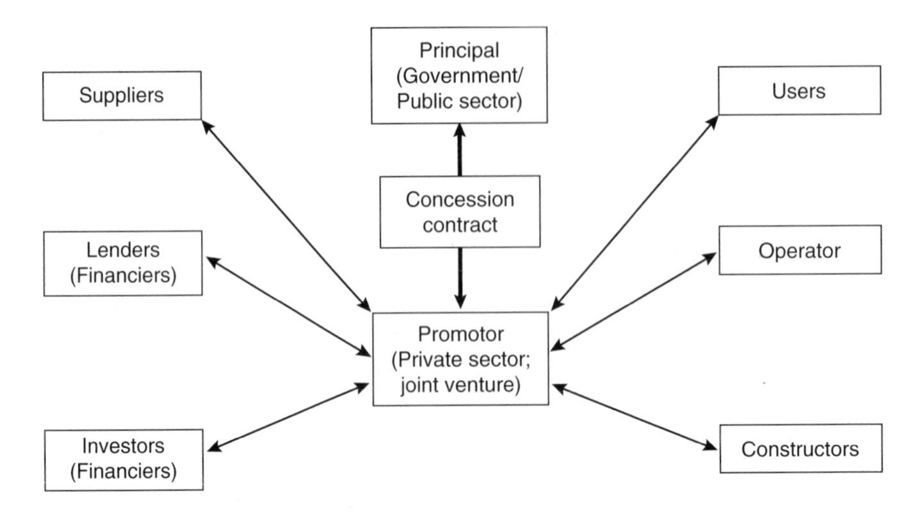

Fig. 7.1 Typical arrangements for concessions. After Merna and Smith (1994).

A typical arrangement for a concession project is shown in Figure 7.1.

The attraction for the Government is that the provision of the capital project, commonly known as the infrastructure, is financed by the private sector firms involved in provision of the project; those firms earn profit from revenues on operating the project for the prescribed period of the concession granted by the government. A further supposed advantage for the public sector is the overall lower price of the project due to the greater efficiency of the private sector. Finally, the cash flow stream for paying for the project means that actual users pay for it over a longer period and later than the capital sum would be paid by the government under conventional arrangements; however, such cash flow considerations will be taken into account by the private sector in determining the prices generating the revenue stream.

Thus, to be viable, a concession project must generate a robust stream of revenue, and one which is reasonably predictable, to enable the promoter to cover all the costs involved and generate sufficient profit. Naturally, the calculations must be done on a discounted basis and will incorporate allowances for the risks. A further means of risk reduction is for the project to be designed using well-known technology to meet the specification provided by the principal.

Concession arrangements may be contract-led or market-led. For contract-led concessions, contracts are set up at the outset for the sale of the output of the completed facility, often on a take-or-pay basis; for such arrangements there must be a tangible output, such as in power generation projects. For market-led concessions, the revenue stream is determined by use of the facility; a sum is charged per use, such as toll roads and bridges. Merna and Smith (1994) demonstrate that contract-led concessions are less risky due to the removal of the otherwise dominant risk of the revenue stream.

Commonly, promoters of concessions are joint venture companies including a major constructor, who will carry out the construction work. Often, such joint venture companies are single project joint ventures.

When the technology involved with a concession project is changing quickly and/or the operating life of the project is quite short, the principal may require the promoter to retain ownership of the facility, thereby making it a BOO (Build-Own-Operate) project. Such arrangements accord with the doctrine of privatization.

Concession projects, due to their often large size, are likely to be financed by a mix of loans and equity. Banks, despite becoming more accustomed to lending on a no-recourse basis (such a loan is secured against the project revenue stream rather than against fixed assets of the promoter - the promoter may have no significant assets other than those of the facility), they remain reluctant to lend for more than ten years. Thus, the common 30-year concession periods mean that promoters issue bonds. However, difficulties in bond financing are associated with project cost over-runs (need to obtain more finance), the pattern of finance requirements (the costs build up over a considerable design and construction period, hence, the timing of the bond issue is difficult) and cost under-runs (which means that more debt than necessary must be serviced).

Project principals, especially the Government, are concerned that not only does the facility provide a good service over its life (both while subject to the concession and after any transfer back to the principal for continued operation) but also that the promoter does not make large profits at the expense of the public. As borrowing rates for the private sector are significantly above those of the public sector, efficiency gains on concession projects must, at minimum, compensate in order to give the principal (and, hence, the public) value for money and allow that promoter reasonable profit. Waites (1996) depicts the problem: see Figure 7.2.

7.3 Capital structures

The capital structure of a firm is governed by many variables: the field of operations, the finance available and the sources of finance, the perceptions and views of managers and investors, and (perhaps of most obvious importance) the costs of capital from the available sources.

Assuming comparable managerial expertise among firms of any certain type in a particular field of operation, the predominant factor determining the actual capital structure of a particular firm is the cost of capital. In the construction industry, capital structures will vary: a property company will have a very large proportion of fixed assets; a precast concrete

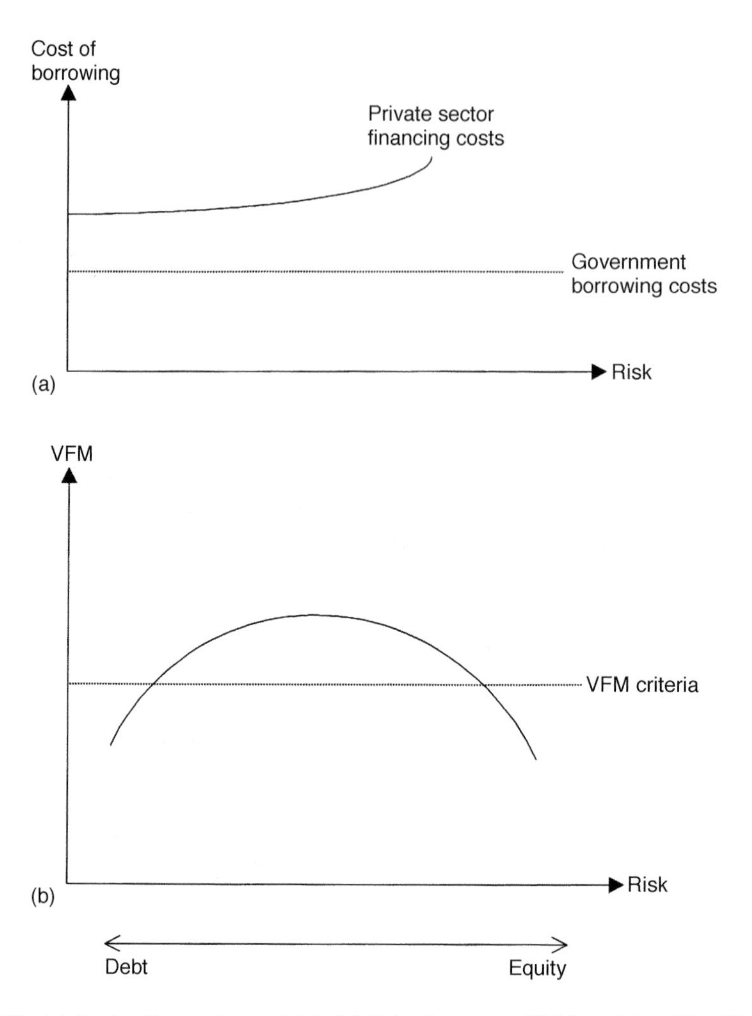

Fig. 7.2 (a) Costs of borrowing and risk. (b) Value for money (VFM) and risk. After Waites 'Risk and capital structure of successful deals' *PFI Intelligence Bulletin* (November 1996).

manufacturer will have considerable fixed assets but also a considerable stock of materials and work-in-progress; a main contractor will have fixed assets of a head office, yard, some plant but a large amount of work-in-progress; while a jobbing builder may have almost no fixed assets, a negligible stock of materials but a reasonable amount of work-in-progress (often including a significant amount of completed work for which no payment has been received due to inadequate credit control). Thus, each type of firm will need a different capital structure to suit its operational requirements.

Having considered the common sources of capital and their applicability to the various types of firm already, the cost of capital is now considered as a primary determinant of capital structure.

7.3.1 *Cost of capital*

Some major influences

The cost of capital to a firm cannot be divorced from the cost of capital to the investors, which can be viewed in an opportunity cost manner as the return, net of tax, which investors could earn in the next best investment (or in alternative investments) after making due allowance for any variations in the risks involved. Risk is a very important element in the consideration of the cost of capital.

It is vital to determine what factors are of importance to a potential investor's decision. Many of the major factors have been mentioned but may be summarized as purchase price, return (both capital growth and revenue), inflation, taxation, liquidity, and security or risk. It is possible to combine these factors in order that yield is the prime consideration.

Yield considers the return in relation to the purchase price. However, the return which should be considered is not simply the monetary interest or dividend and growth in the capital invested but should include the appropriate allowances for tax, inflation and risk. In fact, any rate of interest has three components: inflation, risk, and time-preference.

Inflation. Inflation (the opposite of deflation) is a persistent raising of the price level in an economy relative to money incomes. If money incomes are held constant, their purchasing power is eroded progressively, i.e. real incomes are reduced. Inflation has been a common feature of all developed economies in recent years.

Inflation represents a major problem for building firms; not only must they consider how to deal with the effects of inflation in the valuation of fixed assets, work-in-progress, stocks, etc., for their accounts (and accountants are not completely certain as to how to deal with inflation) but must allow for inflation in any firm price tenders. Incorrect predictions of inflation may mean the firm makes a loss instead of a profit.

Taxation. Taxation is influential upon the cost of capital by two means: first, through the system of corporation tax and, second, through income tax levied on individuals in respect of both earned and, more importantly in this context, unearned income. In outline, corporation tax is levied on the profits of companies after the deduction of interest payments by those companies but before any profits are available for distribution.

Investment allowances in respect of capital expenditures are available for set-off against a company's corporation tax liability. The allowances vary, depending upon the type of capital expenditure (e.g. plant, which attracts allowances of 25 per cent per annum on a reducing balance basis from the date of purchase; industrial buildings, which may be written off at 4 per cent per year on a straight line basis). One further factor complicating the allowances is the location of the capital expenditure: if it is within a type of development area, the allowances will be more generous.

Risk. Risk may be viewed as being the opposite of security, higher risks being less secure. In this context, risk also embodies uncertainty. Naturally, as the security of any investment proposition is reduced, the financial incentive necessary to procure investors increases; thus, the cost of capital is inversely proportional to the security offered. Hence, investment in government securities carries a relatively low return as such an investment is generally regarded as being almost risk-free. As has already been noted, a bank will consider three forms of risk when evaluating a possible loan: the risk of total default, the risk of inability to repay the loan in the pattern and at the time(s) the bank requires, and the risk of inability to pay if the bank recalls the loan.

Comparison of risk and time-preference. It is now reasonable to postulate that for any group of possible investments of a similar nature (as inflation may have differential effects) over a given time period, the factor which dictates the returns the investments must earn to appear equally attractive to a potential investor (i.e. the investor is indifferent between the alternatives) is risk. Likewise, for a spectrum of investments of equal risk, the factor dictating the return the investments must provide to render the potential investor indifferent between them is the period of the investments (due to time-preference). Although a time-preference rate is very difficult to determine, the social time-preference rate (the time-preference rate of society) has been estimated to be possibly as low as 3 per cent (Seeley 1996). Thus, even allowing for a considerable increase in the time-preference rate and assuming that inflation affects all parts of the economy in a uniform manner (it is unlikely that actual inflation influences would produce a significant divergence from the results obtained by using this assumption), it is risk which is the prime distinguishing factor between a spectrum of possible investments.

Gearing. Gearing is the ratio of (fixed interest) debt finance to equity finance (owners' stake in the firm). Thus, gearing indicates the security of a fixed interest investment: the higher the gearing, the lower the security offered and so, the higher the necessary recompense required by the debt investors. High gearing also increases the risk of the ordinary shareholders, as greater profits must be earned by the company to enable them to receive a dividend. Comparing two companies, one with high gearing and the other with low gearing, which earn the same rate of return on their total capital, the company with high gearing will be able to pay higher dividends per share, provided the rate of return earned is in excess of the rate of interest payable on its loan capital.

Normally, the cost to the firm of the various forms of capital will encourage it to raise debt finance, especially due to the advantageous tax position of the interest payments being offset against the firm's tax liability. However, as the amount of debt finance increases (the firm becomes more highly geared) so does the amount of risk and, hence, the cost of capital also rises, thereby effectively limiting the proportion and amount of debt finance which the firm may obtain. Further, the variability of earnings of a firm is influential on the gearing it may adopt. In investment appraisal, the risk associated with an investment is measured by the variability of its return, hence firms with volatile levels of earnings are perceived to be higher risk - that not only necessitates those firms providing higher yields on investments made in them but also acts as a limit on the proportion of fixed interest finance such firms can obtain. As returns in the construction industry tend to be volatile, firms tend to be relatively low geared.

7.3.2 *Cost of individual types of capital*

It has been demonstrated that, to a significant extent, the capital structure of a firm is dictated by the type of firm and the nature of its activities. However, there is still scope within these parameters for variation in capital structures, a major determinant of the capital structure adopted being the cost of capital.

If a firm behaves as a rational economic individual, it will seek to minimize its cost of capital. In doing so, it will select capital from various sources. Hence, it is of value to consider the costs of individual forms of capital and, then, how the aggregate cost of capital may be minimized.

Equity capital: new issues

In evaluating any new issue of shares a company must consider the yield on existing shares, the expenses associated with the new issue and the discount it must offer on the new issue to make it sufficiently attractive for take-up. The company must also ensure that the existing shareholders are not disadvantaged due to the new issue. Some existing shareholders will purchase new issue shares and the remainder of the new issue will, therefore, be purchased by new shareholders. Thus:

$$\text{Cost of capital} = y_1 c + y_2 k + e$$

where y_1 = existing yield on the company's shares
y_2 = projected yield after new issue
c = proportion of new issue taken up by existing shareholders
k = proportion of new issue taken up by new shareholders
e = issue expenses
(derived from Merrett and Sykes 1973).

The firm must pay dividends sufficient to provide the shareholders with their required yield on the investment, so from the firm's viewpoint the yield is the ratio of the dividend to the capital provided by the shares.

Equity capital: rights issues

$$\text{Cost of capital} = y_2 \frac{k_n}{K} + y_1 \left(\frac{k_0 - s}{K} \right) + e$$

where e, y_1 and y_2 are as previous example
k_n = capital from new shareholders
k_0 = capital from existing shareholders
s = proceeds from sale of rights and
K = total new capital
(derived from Merrett and Sykes 1973).

Retained earnings

On the face of things, retained earnings have no cost to the firm. However, they do constitute an increased stake in the firm by the shareholders as they represent profits which have not been distributed as dividends. Thus, retained earnings have an opportunity cost. Merrett and Sykes (1973) argue that a firm employing retained earnings to finance its activities is able to accept a return on its investments which is lower than the return on shareholders' alternative investments by the difference between the tax deducted at source on distribution (Advance Corporation Tax (ACT) - from 6 April 1999, the UK Government abolished the ACT requirement so dividends are now paid gross) and the rate of personal taxation of the shareholders. This presumes that the shareholders are taxed on their income at an average rate in excess of standard rate. This analysis also ignores any capital gains tax payable on the increased market value of shares arising from the retained earnings.

Depreciation

Depreciation provisions may be considered in a similar manner to retained earnings: they have an opportunity cost and represent an increased stake in the firm by its shareholders. However, a distribution of depreciation provisions would produce a capital reduction, probably requiring outstanding debts to be repaid due to the depletion of the capital base, the security against which the debt was obtained. This indicates a proportional combination between the cost of debt repaid and the cost of retained earnings to calculate the cost of capital in the form of depreciation provisions.

Debentures

Usually debentures are issued at a discount. They are subject to a prescribed rate of interest per year until they reach maturity, at which time, the loan is repaid by the company.

Hay and Morris (1979) consider the cost of debenture finance to the company to be i in the equation:

$$B_0 = \sum_{t=1}^{t=n} \frac{K - bB_N}{(1+i)^t} + \frac{B_N}{(1+i)^n}$$

where B_0 = issue price
B_N = terminal value
b = nominal interest rate on the debenture
n = life of the debenture
t = time elapsed from issue

It must be noted that interest paid on debentures is allowed as a cost of the firm's operations prior to the computation of any corporation tax liability. This effectively reduces the cost of this source of capital.

Trade creditors

Merrett and Sykes (1973) argue that only discounts allowed or foregone should be used to compute the cost of capital from trade creditors. They consider it preferable to treat payment delays, the true credit element, as adjustments to cash flow and their reasoning has been followed in this analysis.

Discounts for prompt payment are very common in construction in respect of almost all types of subcontractors' and suppliers' accounts. However, due to the ravages of inflation there has been a marked tendency for domestic suppliers particularly to tighten their credit control, thus domestic suppliers and subcontractors should each be considered on an individual basis (possibly leading to a weighted average calculation for practical uses). It is usual for only the discount to be forfeited by late payment and not for interest to be charged on the outstanding account. The cost of foregoing discount is obtained from:

$$\left[1 + \frac{d}{100 - d}\right]^{(52/n)} - 1 = R \text{ per cent per annum}$$

where d = per cent discount offered
n = period for payment in weeks
R = rate of interest per annum as cost of foregoing the discount

The annual equivalent rate of interest obtained must be reduced by a factor of $(1 - T)$, where T is the rate of tax which the firm pays as the loss of discount adds to the firm's costs which are, of course, tax deductable.

7.3.3 *Marginal cost of capital*

The marginal cost of capital is of use to a firm in deciding whether or not to undertake a project or to expand. It considers the cost of the requisite capital from the possible sources and is, obviously, likely to produce a cost in excess of the firm's average cost of capital due, *inter alia,* to the additional risks likely to be involved.

7.3.4 *Weighted average cost of capital*

The weighted average cost of capital for a firm is of use in two major areas: in consideration of the firm's position and in evaluation of proposed changes necessitating a change in the firm's capital. Thus, a weighted average technique may be used in a quasi-marginal way to evaluate a proposed investment project, such as the construction of a new building.

The weighted average cost of capital is obtained by multiplying the net of tax cost of capital (usually expressed as a percentage per annum) for each source of capital by the proportion of total capital (for the firm or project) from that source and summing these products to obtain a single percentage.

7.3.5 *Optimum capital structure*

The optimum capital structure is that which provides the greatest benefit to the firm. This may be considered to be providing the greatest possible return to the owner's investments in the firm. Naturally, this is a somewhat complex goal: the requirements and views of the owners will vary (especially for a public company), short-term and long-term objectives and possibilities may conflict and so, given these and other problems of optimizing, it is likely that satisficing will be employed to determine the capital structure of a firm. (A 'satisficing' situation is one of compromise: it is acceptable in respect of each individual criterion, but is suboptimal. See Chapter 8.)

The optimum capital structure would be such that, within the physical parameters, the marginal cost of capital employed by the firm will be equal from all available sources. This minimizes the cost of capital to the firm and so, in consequence, should maximize the owners' returns.

Satisficing, in the context of the long-period minimum return which must be earned on the owners' investment (normal profit), should permit greater flexibility in the selection of the capital structure. In practice, the selection of a firm's capital structure will be subject to the objectives of the firm, notably the main objective which managers are believed to pursue - growth - as well as the return on investment considerations; further, the availability of capital in the market, given restrictions of time, etc. in searching for capital, is important. Thus, it is useful to reconsider the primary objectives of firms in the modern market economies in which management has become so divorced from ownership that Baumol (1959) concluded that firms operate to maximize growth of turnover subject to a minimum profit constraint. Hutton (1996) argues that, due to pressures from investors (mostly institutions) expressed via Stock Exchange activities, firms are forced to behave in ways to provide short-term returns, especially growth in dividends. Further, despite advances in information technology, search patterns for capital remain subject to bounded rationality. Hence, within limits, the capital structures of firms are becoming increasingly oriented to facilitation of expansion and provision of short-term returns.

Summary

The construction industry contains a large number of firms, varying in size from single-person enterprises to multinational public companies. Contracting comprises all forms of business enterprise but the construction professions are usually single-person enterprises or, more commonly, partnerships. The cooperative movement is evident in the industry but is still in its infancy and on a small scale.

Numerous sources of capital are available, both internal and external to the firms. Limitations on the sources exist due to the form of business enterprise and are especially evident in respect of long-term capital provisions. Special provisions have been introduced to assist the small firms and the evidence available indicates that the construction industry benefits directly from these provisions as well as indirectly via the general stimulation of fixed capital investment which such schemes promote. Short-term capital provision is particularly important to contractors. The Stock Exchange provides not only a market place for trading in securities but also acts as a scrutineer to safeguard the investing public. Sale and lease-back arrangements are important in property development due to their influence on developers' capital structures and their facilitation of investment in specialist activities.

The capital structure of any firm is related to the form of the enterprise, its objectives and the cost of capital. The cost of capital is subject to and governed by many variables which often operate independently of each other. The firm must consider these influences and their effects on the cost of the individual types of capital to determine the most suitable capital structure. The marginal cost of capital and the weighted average cost of capital will provide an indication of the optimum capital structure from the cost of capital view. The optimum capital structure is unlikely to be achieved, the concept of satisficing being more commonly employed due to the many divergent considerations and the common requirement for a degree of flexibility.

Questions

1. What forms of business unit are to be found in the UK construction industry? Describe and discuss the activities and rules of each.
2. Why is it rare for a contracting firm to be in the form of a partnership?
3. With what requirements must a firm comply to become a company? Why may such compliance be considered essential?
4. Describe the categories into which the capital of a firm is usually classified and discuss the role of time in the classification.
5. Discuss the reasons why companies are in an advantageous position over other types of business unit for the raising of capital.
6. Discuss the role of the commercial banks in the provision of capital for the construction industry.
7. Discuss the factors which determine the cost of capital for a firm.
8. How may a firm determine the most suitable capital structure to adopt?

References and bibliography

Baumol, W.J. (1959) *Business Behaviour, Value and Growth*, Macmillan.

Beecham, A. and Cunningham, N.J. (1970) *Economics of Industrial Organisation*, Pitman.

Bromwich, M. (1976) *The Economics of Capital Budgeting*, Penguin.

Buckle, M. and Thompson, J.L. (1992) *The United Kingdom Financial System in Transition: Theory and Practice*, Manchester University Press.

Buckley, P.J. and Casson, M. (1988) 'A Theory of Cooperation in International Business' from Contractor, F.J. and Lorange, P. (Editors) *Cooperative Strategies in International Business*, Lexington Books.

Dietrich, M. (1994) *Transaction Cost Economics and Beyond*, Routledge.

Fellows, R.F. (1984) 'A Study of Cost Escalation in the Building Industry', *Proceedings CIB W-65 Symposium*, Waterloo, Canada, July (pp. 927–928).

Grimmond, J. (1980) 'Co-ops offer a third way', *The Guardian*, 13 October.

Hay, D.A. and Morris, D.J. (1979) *Industrial Economics – Theory and Evidence*, Oxford University Press.

Hutton, W. (1996) *The State We're In* (Second Edition), Vintage.

James, P.S. (1979) *Introduction to English Law* (Tenth Edition), Butterworths.

Kay, J.A. and King, M.A. (1980) *The British Tax System* (Second Edition), Oxford University Press.

Latham, Sir M. (1994) *Constructing the Team* (The Latham Report), HMSO.

Lipsey, R.G. (1979) *An Introduction to Positive Economics* (Fifth Edition), Weidenfeld & Nicolson.

Merna, A. (1993) *Investigation of the agreement and risks for build-own-operate-transfer projects*, PhD thesis (unpublished), Victoria University of Manchester.

Merna, A. and Smith, N.J. (1994) 'Concession contracts for power generation', *Engineering, Construction and Architectural Management*, Volume 1 Number 1 (pp. 17–27).

Merrett, A.J. and Sykes, A. (1973) *The Finance and Analysis of Capital Projects* (Second Edition), Longman.

Midgley, K. and Burns, R.G. (1972) *Business Finance and the Capital Market*, Macmillan.

Morris, D.J. (ed) (1979) *The Economic System in the UK* (Second Edition), Oxford University Press.

Rayner, M. (1978) *National and Local Taxation*, Macmillan.

Seeley, I.H. (1996) *Building Economics* (Fourth Edition), Macmillan.

Sizer, J. (1989) *An Insight into Management Accounting* (Third Edition), Penguin.

Waites, C. (1996) 'Risk and capital structure of successful deals', *PFI Intelligence Bulletin*, November.

Williamson, O.E. (1985) *The Economic Institutions of Capitalism*, The Free Press.

8 Budgeting

Budgeting is concerned with two quite distinct areas: costs and revenues. These two aspects are drawn together by often quite sophisticated systems of performance monitoring and control, systems without which budgets are of little value. A budget is an estimate of the costs and incomes to be generated if a proposed project is undertaken. Budgets are predictions and, therefore, are subject to accuracy constraints in respect of the techniques employed, information available, expertise of personnel, and so on. As no budget is completely accurate, it is important that the inherent variability is acknowledged when monitoring and implementing control. Error should, of course, be minimized but it may be uneconomic to increase the sophistication of the budgeting system significantly (and, hence, its cost) once an acceptable level of accuracy has been achieved.

Fitness for purpose is a concept which should be applied as equally to budgeting as it should be to building design. The budget should provide the requisite information at an acceptable level of accuracy and for a reasonable cost to enable management to make decisions as well as to monitor and control the project.

Budgeting, coupled with monitoring and control, is important to everyone concerned with construction projects; the client requires a prediction of the total cost of the project and the associated fees and charges. Usually, designers bid for their aspects of the work or ensure that their expenditures are maintained within a predetermined professional fee; almost invariably, contractors will tender for the construction work and monitor their activities for profitability control. It is important to realize that such budgets are not merely predictions of a total cost or price but are time related and will indicate the pattern of accruals. It is in this time relationship that much of the importance of budgeting lies.

Updating of budgets is an important function. Budgets are usually based upon limited information and so require updating and amending as more information becomes available and as circumstances affecting the budgets change. An up-to-date budget is essential for effective monitoring and control.

Survival

In order for any business unit to survive, economic theory dictates that, in the long period, it must earn at least normal profit. How can a firm be certain of achieving this minimum profitability? Why do some firms dissolve during slump periods while others appear to prosper? Further complications in analysing the performance of organizations arise due to the increasing diversity and complexity of their operating criteria and constraints: the growth criterion, image considerations, penetration into new markets, etc. Threat of takeover, in an increasingly international/global arena, must be taken into account in the operational and financial management of firms.

There are very many questions regarding the survival and prosperity of firms.

This chapter discusses some general principles and techniques to aid success in the construction industry.

8.1 Costs

In the construction industry, almost all budgeting techniques are based upon cost. However, in any sale three basic concepts are present: cost, price and value. The cost to the purchaser is the seller's price, with value acting as arbiter between the parties. Cost is what must be given (or foregone) to obtain something. Price is what is received in return for giving up something. Value is a measure of the utility of the item(s). Thus, the price the seller asks for the good or service is the cost of that good or service to the purchaser (usually expressed as a sum of money). The value of the good or service is subjective but will be related to the exchange price such that it may be concluded that the value to the purchaser of the commodity at least equals the sum of money given up (the opposite applies to the seller).

Commonly, the terms 'cost' and 'price' are used interchangeably, which sometimes results in ambiguity. In this text the above definitions of the terms apply.

Several classifications of costs are in common usage. Fixed, variable, and semi-variable is a basic Economics classification but, in the context of the construction industry, the classification of costs as direct costs and indirect costs is more widely used and understood. (Indeed, it is upon such a classification that most tenders are prepared by contractors and subcontractors.) The addition of direct costs to indirect costs gives the total costs.

Direct costs are those which vary proportionally with output, whereas indirect costs are independent of the level of output (at least, in the short period). Broadly, the definitions of the prime costs of labour, materials and goods, and plant contained in the *Definition of Prime Cost of Daywork carried out under a Building Contract* (RICS 1975) are descriptions of the components of direct cost, and definitions of incidental costs, overheads and profit describing indirect costs. (Care should be taken not to apply the definitions so given absolutely as certain items defined as incidentals – e.g. tool allowances – are more correctly direct costs.)

8.1.1 Contractors' costs

A contracting organization will incur a great variety of costs which may be broadly classified as shown in Table 8.1.

Table 8.1 Classification of contractors' costs

Cost centre	*Cost classification*	
Site operative labour	V	D
Materials	V	D
Subcontractors	V	D
Plant:		
Hired	V	D
Owned:		
Depreciation, obsolescence	F	I
Wear, maintenance, etc.	V	D
Line management:		
Walking gangers, etc.	V	D
Forepersons	SV	I
Site agent, etc.	SV-F	I
Head office services to site (QS engineering, buying, etc.)	F	I
Supplementary departments (estimating, marketing, accounts, etc.)	F	I
Equipment:		
Leased	SV	I
Purchased:		
Depreciation, etc.	F	I
Maintenance, etc.	V	D
Head office and other premises	F	I
Capital provision:		
Operating capital charges	V	D
Other charges (some postponable)	F	I

Key: V, variable; SV, semi-variable; F, fixed; D, direct; I, indirect.

Site operative labour

The cost of site operative labour is considerably in excess of the basic wage rates specified in the NWRA due to several factors, those of major cost significance being:

- the bonus system – the level of bonus often being significantly related to prevailing supply and demand conditions

- plus rates – tool allowance, travel allowances, lodging allowances, etc.
- employer's statutory contributions – National Insurance, etc.
- severance payments – related to period of employment
- non-recoverable overtime payments
- non-productive time payments
- sick pay
- holiday pay
- clothing and safety provision – jacket, hard hat, etc.

Generally, the cost of site operative labour will be significantly higher in some areas than in others due to such factors as construction workload, alternative employment and conditions offered by those alternatives, labour supply, unionization, and construction wage expectations. Thus, areas such as London and Liverpool experience high labour costs while labour is cheaper in areas such as Devon.

However, it is not merely cost of labour which is important but its productivity. In this context such aspects as a good bonus scheme and conducive and safe conditions of work will promote labour productivity and hence be cost reducing.

On a 'typical' building project, using labour employed directly by the contractor, the labour content will approximate to 30 per cent of the project cost. However, there is an increasing, and international, tendency for main contractors to sublet the vast majority of site operations, thereby reducing the significance of the directly employed labour cost category whilst commensurately increasing the significance of the subcontractors cost category. Indeed, in several countries, including the UK, it is quite unusual for the main contractor to undertake any of the construction work using directly employed operatives – a few operatives are employed directly on projects to clean up and undertake similar, general labouring tasks only.

Materials

Materials usually account for between one-third and one-half of the cost of a building project. Smaller firms, whose size precludes the use of bulk, centralized purchasing, will have a proportionally higher cost of materials as they cannot obtain the discounts and credit facilities enjoyed by larger firms. However, particularly for the smaller projects which such firms execute, it is common for them to receive a 'mobilization payment' at the commencement of a project for the purchase of materials.

The second important aspect of materials cost is concerned with measurement and estimating. If a traditional standard method of measurement (SMM) based bill of quantitites (BQ) is used, materials are measured net. Thus, allowances for laps (and other unmeasured requirements) must be added (the amounts of laps being stated in the preambles section of the BQ) and costed accordingly in the estimating process.

Third, waste of materials is a significant source of unrecovered cost (and hence loss). Estimators, traditionally, included quite nominal allowances ($2\frac{1}{2}$ per cent–5 per cent) in their materials' estimates for waste: off-cuts, loss, theft, damage, misuse – the list of sources of waste is extensive. However, the BRE has demonstrated on several occasions (Skoyles 1974, 1976, 1978, 1981) that waste of building materials on site is greatly in excess of such allowances. Skoyles (1981) found that:

> ...the overall loss of principal materials is about 100 per cent more than is usually allowed for in estimating.... This figure is, of course, highly variable and is much lower on some sites....

Due to increasing awareness of the actual material waste incurred, the allowances included by estimators have tended to increase so that the allowances used now reflect the levels of materials waste likely to be incurred more accurately (5 per cent–15 per cent). It is interesting to postulate the possible consequences of an estimate including only one-half of the requisite materials wastage allowance, especially during a period of slump, when small profit margins prevail.

If it is assumed that the total materials content cost is one-third of the project cost according to the estimate, waste allowance being an average of 6 per cent, actual waste is thus 12 per cent and profit addition on cost is 2 per cent. Thus:

$$\frac{\text{Project cost}}{3} = 1.06 \times \text{materials basic estimate } (m)$$

$$\frac{100\%}{3} = 1.06m$$

therefore

$$m = \frac{100\%}{3.18} = 31.45\%$$

But actual waste of materials is 12 per cent $\times\ m$; thus, the 'excess' waste not included in the project estimate (and hence excluded from the project price) is:

$$6\% \times m = 1.89\%$$

Contract sum (project price) is:

$$100\% + (2\% \times 100\%) = 102\%$$

The anticipated profit margin is 2 per cent. But cost increase due to 'excess' materials waste is 1.89 per cent. Therefore, achieved profit margin is:

$2\% - 1.89\% = 0.11$

The anticipated profit is 'almost completely eroded' due to insufficient allowance in the estimate for waste of materials – the real explanation is, of course, not that anticipated profit has been eroded but that only 0.11% profit was actually included in the price due to the inadequate allowance for waste of materials in the estimate. However, recognition of the additional costs of materials waste has two consequences: increased managerial attention to reduction of such waste and increase in the level of cost estimates. Recognition of such higher cost does not, of itself, produce higher prices but does give management more accurate information on which pricing decisions can be based.

Subcontractors

Traditionally, subcontractors have been used in building to carry out specialist operations but, particularly during the post-1945 period, the use of subcontractors has increased progressively, notably in the basic building trades which were the province of contractors' directly employed operatives. The growth of subcontracting may be attributed to several major causes:

- successive governments' use of the construction industry as an economic regulator
- higher costs of labour, including 'employers' contributions'
- greater employee protection and the trend away from casual labour
- higher earnings of operatives in subcontracting organizations
- ability of main contractors to pass on responsibilities to subcontractors (defective work, delays, etc.)
- greater flexibility for main contractors
- volatility of demand

Today, some sites are run with almost a total absence of main contractors' directly employed labour. Under management contracting systems, the almost complete absence of any management contractors' operatives is usual and expected but the situation is quite commonplace across the spectrum of contract forms. Thus, in operating practice, many projects are executed as management contracts, irrespective of the 'formal' procurement approach employed.

While it may be seen as being cheaper for a main contractor to sub-let work as the responsibilities for performance, notably including responsibility for defects, may be passed on to the subcontractor – e.g. the main contractor does not have to organize the details of work execution nor have to bear the costs of non-productive time (unless due to default by the main contractor), does not have to keep expensive operatives (craft and specialist) in employment when there is no work for them in order to ensure their availability for future projects, and so on, there are certain cost-increasing disadvantages caused by subcontracting. When subcontracting, the contractor has less control in terms of standard of workmanship, output and performance generally; coordination is more complex and so more highly skilled (and, hence, more expensive) management will be required; the reputation of the contractor is, to some degree, in the hands of the subcontractors and the subcontractors themselves aim to make a profit. Thus, the increased use of subcontracting has shifted both risk and responsibility from the main contractor to the subcontractors. A power shift has occurred to some degree too – during tendering, many packages of work are bid by subcontractors, the main contractor assembles the selected bids to form the base cost for the tender. Uher (1990) noted that such a process gives power to major subcontractors, whose bids will form a significant proportion of the total cost of the project, to submit differential bids to tendering main contractors and thereby, potentially, determining which contractor will submit the lowest tender and so be awarded the work. However, such power will be highly constrained, especially during recessionary periods, during which competition for work amongst all firms will be very keen and main contractors may hold 'Dutch auctions' amongst subcontractors (and suppliers) once the main contractor has been awarded the work.

For several reasons, subcontracting may be advantageous: it promotes division of labour and specialization which should lead to greater productivity; it helps the retention of specialists who, under a subcontracting system, may do work for a variety of main contractors and so keep their skills available in the industry (it is rather rare for an operative who has left the construction industry to return to it); and it allows main contractors to concentrate upon their own specialisms of organization and project administration.

Clients' representatives often view subcontracting with a good degree of scepticism. This is probably because the chain of communication and control is longer and more complex, thereby making it more difficult for them to protect their clients' interests. However, it is apparent that, under the process of nomination (and derivatives – such as naming), such subcontractors have much closer contact with, and exhibit greater allegiance to, the clients' representatives than the contractor although their contract for work is with the latter. This demonstrates that firms show allegiance with the party they perceive as being the true source of their sales.

Thus, subcontracting is often viewed by main contractors as a cost-saving exercise which also permits greater flexibility than using directly employed labour. The flexibility occurs in the main contractors' ability to call subcontractors on to the site only when work is available for them and to require them to comply with periods for their work in accordance with the construction programme. Delays and variations frequently produce programme changes and, while in slump periods a main contractor may be in a strong position relative to the subcontractors, during a boom the positions are often reversed and the subcontractors will require continuous site working or will charge extras for leaving and returning to site and for delays, disruptions, etc.; further, they may be reluctant to execute complex or unusual items of work which will be time-consuming and hence detract from their earning capacity (which is greater on long runs of straightforward work). It is also notoriously difficult to persuade subcontractors to return to site to complete work when their skills are in great demand. Some less scrupulous subcontractors are very reluctant to return to a completed project to remedy defects.

It may be concluded, therefore, that subcontracting is not necessarily a cost-reducing process but, if current industrial trends have a sound basis, there are strong indications that this is the case at present.

Plant

Construction plant is usually hired by the contractor but may be owned by the contractor directly, as is common for smaller, frequently-used items of general plant. In the case of a large contractor it is usual for a plant hire subsidiary to own all the plant and to hire it out to the construction divisions. Apart from the advantages to the company of having its own equipment specialists, it also permits the company to benefit from discounts on cross-hiring from other plant companies when very specialist plant, not owned by the contractor's organization, is required.

The hire of large, driven plant usually includes a driver; the hire charge covers normal plant usage costs, driver's wages, fuel, etc., but not such items as breakdowns or damage attributable to the main contractor (e.g. a puncture caused by rubbish being left on an untidy site). Major plant will also have 'on' and 'off' charges as lump sums for bringing the plant to the site and taking it away. The normal hire charge will be a certain amount per hour or per week. Scaffolding will be subject to not only 'on and erect' and 'dismantle and remove' charges (which will be relatively large sums) but to hire charges for the scaffold's being on site and further, lump sum charges for its periodic adaptation as the work proceeds; clearly, it is economic to not only hire the scaffolding for the minimum time but to minimize the adaptations – commensurate with efficient working. The use of mobile scaffold towers and mobile mechanical towers has become common in internal installations for buildings (e.g. false ceilings) and has largely, obviated the need to provide close boarded decking.

Although the contractor must pay a greater amount for the use of plant if it is hired rather than owned by the contractor, it must be remembered that the contractor also saves by hiring plant as the costs of plant being idle between projects, of obsolescence, of major servicing, etc., are borne by the plant hire company. For owned plant such costs are included as overheads (although there will, of course, be an inclusion in the hire charges for those costs as well as an addition for the profit of the plant hire company). Perhaps the most significant aspect, however, is that hiring is a revenue charge only but, if plant is owned, a capital investment is required for the purchase. Thus, hiring plant allows the contractor to operate upon a smaller capital base which, other things being equal, should produce a greater return on the capital employed.

One significant costing problem in connection with hired plant occurs where plant is not required on site for a continuous period. In such a situation the decision whether to keep the plant on site for the entire period or to off-hire it when it is programmed to be idle should be based upon an examination of the hire charges for the programmed idle period against the additional on-off charges. It is obviously good programming practice to require plant to be on site for one period only and to be fully utilized within that period. The issue is particular to more specialist items of plant and requires careful attention in programming the project operations to secure economy; further, even if such economy is achieved via programming, it is essential that the information is conveyed to site in a form which is useful to the site management, including as an aid to decisions over any concurrently competing demands for use of the plant. (This analysis is similar to that for subcontracted work.)

Small tools are usually owned by the contractor. Costing systems vary between firms, from those who do not cost small tools to a particular site (their cost being part of overheads) to those who operate a complex system of cost to the site and credits for returns (such a system creates problems of second-hand valuations). A common system is to charge small tools to a site at their new price (even if second-hand) and give no credits for returns (but to require the physical return of still useful items).

It is apparent that the amount of plant used on a project increases as the size of the project increases. This trend is likely to continue and, indeed, will probably be enhanced by the tendency for construction to become an on-site assembly process and a more mechanized process. Thus, plant costs are likely to increase as a proportion of total project costs.

Line management

Line management includes all those who supervise the production work but do not actually execute the work themselves. In the context of a construction contractor, the line management will comprise walking gangers, forepersons, site agents, and contracts managers (the titles allocated to those functions vary between firms).

Walking gangers are employed to supervise the work of several gangs, often the labouring gangs, on a site. Their conditions of employment are governed by the rules appertaining to operatives and so this type of management may be considered to be a variable cost.

Forepersons are often employed on a more permanent basis and now are usually salaried. As they are frequently tradesmen who have been promoted to a supervisory role, one foreperson will generally oversee the working of one (or several related) trade on a site. This intermediary level of management is a semi-variable cost to the firm.

Site agents, including subagents, are salaried staff and, in most instances, are fixed costs, but in smaller organizations, and in more traditional firms, all staff who are site-based are wage-earning and so, due to the different conditions of employment applicable, they may be regarded as forming part of the semi-variable costs. There is an increasing trend for site management to be salaried and have conditions of employment similar to head office staff. Thus the trend is away from semi-variable towards fixed costs.

Hillebrandt (1974) considers that as site management may be hired and fired with ease and with little expense it is a variable cost. The argument also hinges on the assertion that site management costs increase in a directly proportional way to the output of the organization. Such a situation may occur due to either more site management being employed to cope with an increase in the contractor's workload or better quality (more able), and hence more expensive, site management being employed or some combination of these two.

Hillebrandt also considers that there is a lack of highly skilled construction managers, that their recruitment is more expensive and that firms will wish to retain their services, indicating that they are a fixed cost. This argument is probably most applicable to the contracts management who are usually salaried and employed as head office based staff. It is probably due to a lack of skilled managers at all levels in construction that the trend has been towards employing such personnel on a salaried, permanent basis.

The total cost of line management is small in relation to a contract sum or a contractor's annual turnover. The importance of management lies in its influence over the site production activities through its organizational abilities and so, despite its being itself of quite minor cost significance, it is of major importance to the profitability of the firm.

Overheads

Generally, overheads are considered to be those costs which must be borne even if there is no output from the firm. They are fixed costs. In a contracting firm, overheads will comprise costs such as head office rent, estimating department, accounts department, company cars. It would be exceedingly time-consuming and expensive to record accurately, and then cost, the overheads to each individual project undertaken by the firm.

Overheads are, therefore, *apportioned* among the firm's activities and are *absorbed* by each activity in its costing system.

Head office services to site. Although the personnel involved work largely on site, normally they are employed and costed as head office staff. The discussion under 'line management' in respect of the supply, expertise, cost significance, etc., of managers, applies to these personnel also.

Supplementary departments. These head office staff constitute fixed costs, although, as with any staff, the more junior personnel constitute semi-variable costs. These departments are charged to projects as part of general overheads.

It is noteworthy that the estimating department (and in some ways marketing) is unusual as most of the work undertaken produces no return for the costs incurred, i.e. the majority of tenders submitted do not result in the firm being awarded the contract. Thus, if on average the firm is successful in one of every six tenders submitted, the one-sixth of successful tenders must bear the cost of the five-sixths of unsuccessful tenders. Hence, if estimating costs can be reduced while maintaining the level of accuracy of estimating, the firm's profitability will be enhanced or, if the savings are reflected in slightly lower tender prices, the firm will win more contracts (other things remaining constant).

A pilot investigation into tendering costs carried out at Brunel University by Harding (1980) indicated that the cost of submitting a tender for a 'typical' building project of £1 million to £3 million contract sum is approximately 0.8 per cent of the contract sum, but with quite wide variation about this figure.

Sir Maurice Laing stated in 1978 that 'Tendering throughout the industry costs one quarter of one per cent or less of all work. . . .' However, the Economist Intelligence Unit (EIU 1978) stated that '. . . it appears likely that average estimating costs for building projects fall below 1 per cent of turnover'.

It appears, therefore, that the larger firms undertaking high-value, large projects enjoy economies of scale in tendering. It must be noted that success rates have a major influence over tendering costs in the context of cost as a proportion of turnover. Anecdotal evidence indicates that the cost of tendering for design and build projects is about three times the cost of tendering for 'traditional' projects; a result is that design and build contractors tend not to be willing to tender if more than three firms are tendering for a project.

Equipment. Equipment such as office machinery, computers, cars, etc., may be purchased outright by a firm and so constitute a capital lock-up in a similar manner to plant purchase. Increasingly popular alternatives to outright purchase are leasing or renting the equipment (usually cars are either purchased outright or leased). Equipment ownership remains with the leasing company (lessor).

If equipment is leased, the lessor always owns the equipment. The lessee (trading firm) agrees to make payments to have the use of the equipment for a prescribed period (usually in terms of years) and to take out a maintenance contract on the equipment. For computers, annual maintenance contracts may exceed 10 per cent of the purchase price and so, are unlikely to be worthwhile for any but large installations – PCs, etc., become outdated extremely rapidly and so, with increasing capacity and speed of operation of such computers and the increasing array of software, much of which operates only on up-to-date hardware, periodic replacement may be the most economic option.

Equipment may also be rented. Under such an arrangement the trading firm pays only for the period for which the equipment is required (and rented) and does not bear the cost of repairs (these forming part of the rental charge). Renting is thus the most flexible, but also the most expensive, option.

Renting is possible but usually on quite minor equipment only. For major items of equipment, purchasing or leasing are often the only alternatives. Leasing has advantages for cash flow by not tying up considerable amounts of capital and leasing arrangements often permit the lease period to be extended (a popular option during a recession).

Head office and other premises. These are fixed costs and are charged to projects as part of general overheads. The costs of premises contain two major elements: the initial capital purchase of the lease or freehold, and the annual expenses associated with the premises. However, as for other fixed assets, it is only the annual costs of those assets which should be allocated for absorption – the depreciation of the capital sum. As many buildings increase in realizable value over time (actually, it is the land on which the buildings stand which appreciates, while the building itself depreciates, yielding a net appreciation of the combined land and building), and may be revalued periodically in the firm's accounts, any charges to overheads will be in respect of the depreciation of the building itself for freehold premises. In the case of leasehold premises, similar considerations apply but, for shorter-term leases, the appropriate charge is any reduction in the realizable value of the lease from year to year. Some of the annual expenses associated with premises are postponable (e.g. maintenance), while others are, to a certain degree, variable in that they are dependent upon the occupancy and use of the premises (e.g. heating, lighting).

For some time construction economists have been concerned with costs-in-use in attempting to predict the life cycle costs of buildings. In these evaluations much attention has often been paid to the time intervals between works of maintenance, due not only to the maintenance costs but also the costs of disruption occasioned by the work. However, it would now appear that these issues are of less significance and more attention must be devoted to running cost economies, particularly related to heating costs. Flanagan and Norman (1983) demonstrated that, for many buildings (and for individual elements) the capital cost of provision is dwarfed by the replacement and maintenance costs when the discounting techniques of life cycle costing are employed.

Capital provision. The costs of capital have been discussed in Chapter 7. It is important that the capital itself is not added to the other costs of the firm's operations as that would be double counting (the real capital being physical items used in the production processes). The costs of the capital provision (costs of finance), however, are part of the total costs of the firm's operations and must be included in the costing processes. The inclusion could be effected by calculating and adding the cost of working capital provision together with the interest on longer-term debt (as parts of overheads) and return to the owners (as the profit addition).

8.1.2 *Cost patterns*

Cost patterns are useful to indicate how costs vary over different levels of output, and between different types of project. Figure 8.1 shows a simple long-period cost pattern where variable cost increases linearly with output but the fixed cost increases are staged.

However, the various categories of fixed cost will have stages of cost

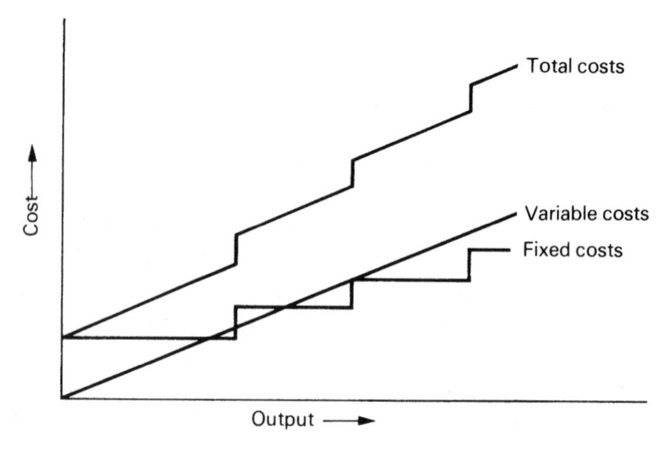

Fig. 8.1 Simple long-period cost pattern.

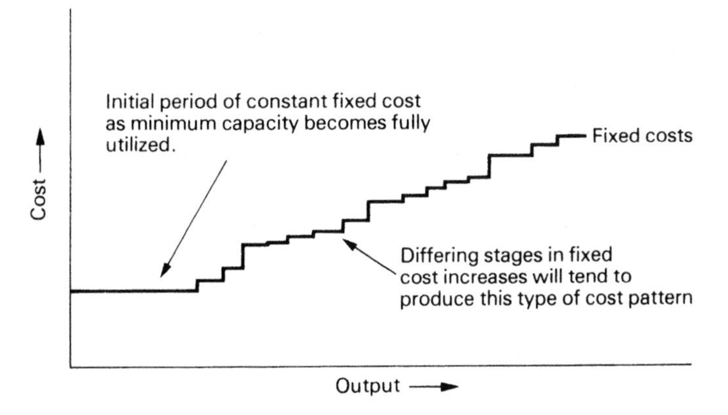

Fig. 8.2 Changes in fixed costs with output.

increase of different size, occurring at different times, thereby tending to change the pattern of fixed costs to one which is a stepped upward slope, as shown in Figure 8.2.

It is probable that certain staged fixed cost increases, such as the purchase of an additional office, will be of such a size as to exert a major influence over the cost pattern. It is likely that these major stages of size alteration will be possible over a range of outputs and will comprise not just one major fixed cost increase but a number of associated increases, e.g. a firm's workload increases and thus the firm requires more office staff, then merits larger office premises and, again, further staff. Such a situation is illustrated in Figure 8.3 and shows the levels of activity at which changes in scale occur.

Figure 8.4 shows the influence of the changes in variable costs due to changes in efficiencies of operation (economies of bulk purchasing, inefficiencies of crowded workplaces, etc.). These are particularly notable as the firm grows from zero output and over each range of output where changes in the scale of production may occur.

8.1.3 Systems of costing

The system of costing used depends primarily upon the nature of the production system, but the aim of any costing system is cost control which is achieved by recording the costs actually incurred, comparing those costs with a predetermined standard, and taking any necessary corrective action. The construction industry largely produces major, one-off products, each of which contains many common and repetitive operations that are usually executed under slightly varying conditions. Also, the price of the product is, almost invariably, agreed prior to its actual production (speculative housing and similar developments being notable exceptions in most cases).

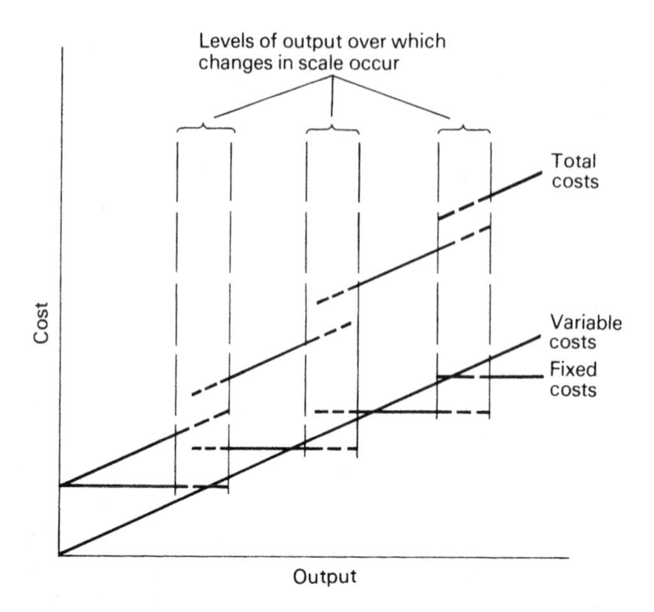

Fig. 8.3 Stages during which changes in fixed costs may occur.

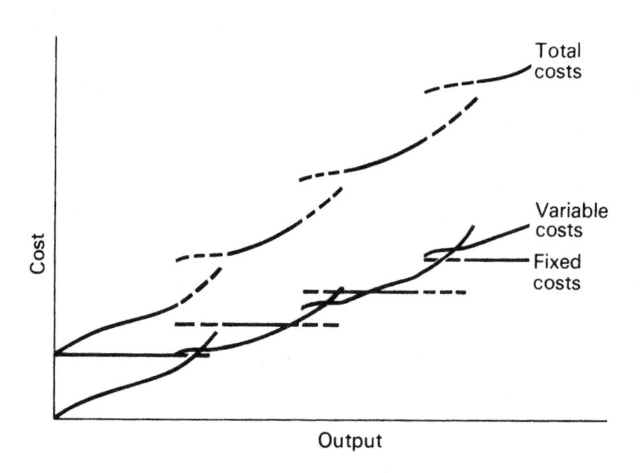

Fig. 8.4 Long-period cost pattern, combining fixed costs' change stages with variable costs.

For a typical construction project the usual model price, both of individual items and of the entire project (the contract sum), is assumed to be based upon a cost prediction (the contractor's estimate) and is agreed between the client and contractor prior to the work being commenced.

Overall historic costing

This is the simplest costing system whereby the overall cost of a completed project is compared with the revenue generated to determine whether the project achieved the required level of profit. The results of such an analysis may be used to predict the cost of proposed projects, after making adjustments for such variables as general cost escalation (fluctuations), specification changes (quality), quantity changes, cost effects of different locations, technological changes, etc. However, it must be remembered that the cost to the client is the price of the 'successful' contractor and that the final cost to the client is the *final* contract sum with additions for the numerous fees paid to consultants: design, legal, etc.

A great deal of data are required for accuracy in this exercise. Such data are provided from several sources, the primary source being the Building Cost Information Service (BCIS) of the RICS. BCIS provides data of projects' initial contract sums to assist its member subscribers (usually consultant quantity surveyors) to predict costs of proposed buildings. Such predictions are, almost invariably, of the probable initial contract sum for the proposed project and are given as 'single figure' forecasts. Fortunately, increasing awareness of the variability of such forecasts is leading some practices to include measures of variability in the forecasts (e.g. range with most likely cost) thereby giving far more appropriate information to clients and empowering the clients to participate more realistically in the cost planning processes for the proposed projects. Cost planning techniques are detailed in many texts such as Seeley (1996) and the variabilities inherent in the resultant forecasts, *usually, of at least 5 per cent even if made at the stage of preparing bills of quantities*, are examined in Bennett (1982), Morrison (1984) and Ashworth and Skitmore (1983).

By comparing the total cost of a project with the total revenue received, a contractor is able to calculate the profit derived from a project. However, attention should be given to the overheads which the project has absorbed, settlements of claims (with the client and with subcontractors, etc.) to determine the level and sources of profit. Such overall analysis has only elementary use, more detail is necessary for real control. Overall historic costing records may be useful in considering the level of tenders for similar projects in the future.

Periodic historic costing

This is a similar process to the overall historic costing described above, but is carried out for each project at predetermined intervals (often monthly). This enables the contractor to determine if the project is operating within its budgeted cost and to take action, when necessary, to get the project back 'on course' (see also 8.5.1).

Job and unit costing

This system is most suited to production in which a unit of output may be readily identified throughout the production process so that all costs attributable to the production of that unit may be allocated to it. Overheads are allocated on a predetermined basis; in construction, this will usually be as a percentage of cost.

The system should be designed to facilitate the comparison of the estimated cost with the actual cost achieved on site and such comparison should provide useful feedback to the estimating department. Potential problems arise under this costing system in respect of overhead allocation and miscellaneous non-productive times (due, for example, to bad weather, delays in normal activities, etc.). Overheads are thus best treated as a separate allocation. Non-productive times should also be recorded separately to provide programming information.

Standard costing

This is the costing system most extensively used in the construction industry. It is a system whereby costs are predetermined and subsequently compared with actual costs achieved to facilitate control. The predetermined costs are standard costs and are obtained through the application of work study and extensive cost recording. In the construction industry the labour constants used in estimating are of the nature of standard costs (note also production bonus target times).

Comparison of the cost of actual production with the standard cost gives the variance. Variances will usually be calculated for each cost centre of labour, plant, materials, and overheads for each operation or set of operations.

Naturally, due to the variety of products produced by the construction industry, the application of standard costing is rather complex. The more extensive use of computers in connection with cost recording (particularly for bonus calculations) and with estimating (to determine labour constants) has led to this system of costing being used more widely.

It is usual for cost variances to be determined on a weekly basis as costing is usually carried out in conjunction with the system of bonus payments - the measure for labour production bonus calculation being used to determine the cost of the direct labour expended upon the project. Difficulties arise when comparing the bonus measure with the BQ measure (the basis of valuation and hence revenue) as the measurement methods are not always the same. However, these difficulties may be overcome, enabling this system to be used for rapid response project cost control.

The measure for production bonus purposes is of an operational nature whereas the measure for valuations is usually based upon a prescribed standard method of measurement (e.g. SMM7), which is more geared to quantity surveying than production requirements. Attempts have been made to bridge this gap, such as the operational bills developed and promoted by BRE some years ago but as yet, no universally satisfactory solution has been found.

8.2 Profit

Profit, in its basic forms of normal and supernormal profit, has been defined in Chapter 7. It is also useful to consider the accountant's concepts of profit: gross profit is total sales revenue minus production and sales expenses; net profit is gross profit minus depreciation and interest on loans; profit after tax is net profit minus tax payable upon that profit and, thus, represents the earnings available as a surplus, which may be used as a source of capital or may be distributed among the owners of the firm.

Neoclassical economics regards profit as the return rightly accruing to the entrepreneur (the owner(s) of a business) for enterprise and use of funds. Marxist economics, however, regards profit as a surplus earned by the labour employed and appropriated by the entrepreneur, asserting that the single source of profits is the labour element of production in converting raw materials into consumables. A basic belief is that everyone is entitled to share equally in the resources available. Therefore, quirks of birth, ability, inheritance, etc., should be ignored and the fruits of human labour, in various forms and in various activities, should be shared equally by society.

In a capitalist society, profitability dictates which organizations survive and prosper and which do not; need is subordinate to effective demand. Thus, most economies are mixed to ensure the supply of essential goods and services which a free market economy would either ignore or supply inadequately to fulfill the needs of society.

In measuring and analysing profit, it is advisable to consider the total amount of profit, the profit per period of time, profit related to types of work (i.e. sources of profit) and profit related to the investment required to produce it (yield). Such analyses assist in determination of which areas of activity the business should pursue, given the levels of risk perceived and the growth and turnover requirements.

8.3 Financial policy

The basic assumption of economic theory, that a firm operates in such a way as to maximize its profits, has come under considerable scrutiny in recent times. It has been acknowledged that firms have other operating criteria which render the profit-maximizing assumption inadequate. Although the objectives of each individual firm are peculiar to that organization, a generally applicable spectrum of objectives would include profitability, growth, continuity of existence, market share, turnover, size, image, and influence.

Further parameters may be brought to bear on a firm's operations. Lack of managerial ability may preclude a firm's growth beyond a certain size. The size of a firm may be restricted by the owner wishing to retain control or to retain the type of organization. The firm, or industry, may be regarded as a 'way of life' and so a firm will remain in business despite earning very low profits (notable among small, personally controlled construction firms). Many more parameters exist.

Firms, therefore, operate under somewhat conflicting objectives (increased turnover, for example, is by no means necessarily synonymous with increased profits although this is a quite widely held belief among construction managers). In such a situation the operational pattern is one of compromise: acceptable in respect of each individual criterion, but suboptimal. This behaviour was described by H A Simon (1960) as 'satisficing'. A firm will thus require a given profit level in the long period but this is unlikely to be the maximum achievable, being mitigated by the firm's other objectives. (This concept is perhaps especially relevant to the consideration of dividends.)

It is likely to be the objective of the firm's continuing existence which predominates during slump periods. In such times, interest rates and unemployment tend to be high, work scarce and, hence, profits low. During these periods some construction firms 'buy' work (that is, they bid for work at a price which is less than the direct costs of carrying out the work plus overhead costs, etc. and normal profit; in fact, in extreme cases the price may be less than the costs of executing the work). Buying work may be justified for several reasons, such as: it keeps resources employed and therefore avoids redundancies; it keeps the name of the firm at the attention of potential clients and consultants; and it ensures the continued existence of the firm in readiness for an improvement in the market.

Theoretically, as long as the firm covers its variable costs as a minimum, it should continue in business in the short-period, because any revenue earned in excess of the variable costs is a contribution to fixed costs which the firm must pay whether it produces or not. In essence, a firm may act rationally to buy a project provided it prices that project to cover the marginal costs of executing that project as a minimum. Buying work is a short-term expedient only. There are considerable dangers in even a short-term policy of this nature as it can very rapidly lead to a downward spiral, ending in bankruptcy. A firm losing money on one project may be tempted to finance that project from one which is profitable; it may then be forced to take on more projects at an overall loss (due to the economic situation in a recession and the necessity that the firm must obtain more work) but with the pricing so manipulated that the operations executed early in the projects are profitable at the expense of subsequent, loss-making operations ('front-end loading'). This situation is quite obviously rapidly progressive, is very difficult to reverse and is a common cause of construction liquidations.

Although buying work may be seen as occasionally essential to survival, it is important that the pitfalls are recognized and that progress is monitored vigorously. Buying work is also occasionally employed in more buoyant periods as a policy in exceptional circumstances, the most common of which are: to obtain a small project for an influential client (usually a client who requires a large amount of construction work), or to obtain a particular project of great prestige.

Particularly in the period since 1945, it seems that the construction industry has been used as a regulator for the UK economy. A positive aspect was that government could use its demand for construction to maintain a fairly even aggregate level of activity if orders from the private sector declined; however, more commonly government uses its demand power to effect macroeconomic policy via the Keynesian multiplier but subject to the lag effects in the accelerator principle. A further aspect is that, since the widespread move to privatization of public sector activities, the proportion of demand for construction work exercised by the UK public sector has declined to below 35 per cent; this has restricted the potential influence of government on the industry as a whole but may magnify the effects of policy instruments sectorally, notably for civil engineering where the public sector is by far the major source of demand. Parry-Lewis (1965) found that, throughout modern times construction had been subject to sizeable cyclical fluctuations in workload. The influence of government action has tended to make the workload of the industry less predictable and has, therefore, acted to deter investment, mechanization, research and development, training, employment, in fact, most of the requirements for a modern and efficient industry. Thus, although lead times for initiating or curtailing work can be quite long, the difficulties in moving construction operations between sectors of the industry tend to dominate - most resources are not very mobile occupationally.

Government appears to regard construction as a homogeneous entity when, in reality, this is far from the case. Civil engineering contractors, for example, cannot and do not switch from major motorway construction to building advanced factories. There is a set of distinct sub-industries within the construction industry. Nor is there a pool of good construction labour of all categories available at the end of a recession. Once a person leaves the construction industry it is unlikely that that individual will return, despite higher wages brought about by labour shortages at an upturn (other working conditions are also very relevant). Thus, the labour shortage in construction is a very common feature at the start of an industrial recovery, for not only has much labour permanently left the industry during the recession but there have been few new entrants and little training has been done.

Likewise, fluctuations in demand for construction output promote conservatism in the industry. New, untried and unproven techniques, plant, etc. are ignored and investment is minimized. Research and development and education and training are all too frequently the first budgets to be cut by firms when a recession begins.

Firms do respond to fluctuations in workload and shifts in demand. Larger firms tended to diversify their activities horizontally and vertically: horizontally to give a greater spectrum of activities at about the same stage of production (e.g. a contractor setting up a rehabilitation division), and vertically to diversify activities over the stages of production (e.g. by a contractor setting up a brick production division or a property development division). Such diversification, which occurs during boom periods, permits the firm to divert resources more easily to the profitable areas of activity in the event of changes in the pattern of demand and provides a broader base for continuing existence since, although all sectors of the industry may be subject to recession, some sectors will be more adversely affected than others. However, in recent times the trend has seen several instances of reversal with firms retrenching to what they regard as 'core business', evidently perceiving specialization as a risk-reducing mechanism at a time when the construction operations are carried out by subcontractors rather than using directly employed resources.

Small firms are often more adaptable and can switch from speculative 'new-build' to extensions to renovation work with only minor difficulties. In fact, small firms often prosper (relatively) in recessions because there is an expansion in small renovation and extension work; firms and householders tend to 'stay put' and 'make do', and will renovate or extend existing premises in preference to moving to new premises. In recessions, preventative maintenance is usually regarded as a postponable cost.

In recessions, it is usual for medium-size firms to suffer most. Such firms are 'squeezed' as large firms 'trade down' the scale of work (having more financial resources and stability, through diversity of activities, to bid low) and small firms 'trade up' (are able to bid low in seeking larger projects than they would usually by having lower overheads).

During recessionary periods the need for good credit control is paramount. Interest rates are high, credit periods are reduced, and finance is expensive and in short supply. It is vital that debtors pay promptly and that the periods of credit allowed by the firm do not exceed those allowed to the firm. Inadequate credit control can easily turn a potential profit into a loss.

Thus, although effective cost monitoring and control should be employed by any firm, the importance of this action is greatly magnified during recessions. It is essential that firms formulate policies for their operations but it is also vital that the limitations and implications of those policies are fully appreciated and taken into account in budgeting.

8.4 Revenues

Under a budgeting system, the complement of cost prediction and control is revenue prediction and control. In the construction industry, revenue predictions are vital to enable a contractor to match expenditure and income. However, although several techniques are available, no prediction method or formula is universally applicable.

Generally, revenue prediction may be divided into two subsections: pre-contract and post-contract. In the pre-contract situation, the prediction is likely to be of a less detailed nature, while in the post-contract period, the prediction will be based upon information appertaining to each particular project.

Most revenue prediction techniques, due to the methods normally used to price construction projects, are based upon cost predictions. The cost control techniques already discussed in this chapter are usually applied to projects on a micro (or operation) level, whereas the predictions used for revenue purposes are usually applied on a macro (or project) level.

8.4.1 Predictions of revenue and cost

Pre-contract predictions are frequently of a universal nature, based upon a typical project model: commonly mathematical models derived from regression analyses of previous projects are employed. Post-contract predictions should be related to each individual project, either by modelling projects by individual types or, even more specifically, by determining the cash inflow and outflow patterns from the particular project's estimate, priced BQ and construction programme. As such, pre-contract predictions tend to provide a guide to revenue and cost expectation without necessarily making allowances for the peculiarities of the particular project in question. The situation is justifiable as there is no certainty that the firm will carry out the project and therefore a mere indication of the costs and revenues which the project will generate is sufficient.

Some predictive techniques, discussed above in the context of costs, may be applied to revenue predictions but these also tend to be of a general nature.

A method that has been developed and found to be useful in practice is the S-curve analysis. The technique is fully described by Cooke and Jepson (1979), and in application to health service projects in Hudson (1978). Cooke and Jepson utilize a model of a typical building project in which the pattern of value accrual is based upon the cost accrual over the project duration (precontract costs are recovered as part of overheads).

In the typical project, the cost accrual assumes the following pattern:

1. During the first third of project duration, the cost accumulates in a parabolic pattern to achieve one-quarter of costs incurred at one-third project duration.
2. During the second third of the project duration, the cost accumulates in a linear fashion such that at two-thirds project duration, the accumulated costs total three-quarters of project total costs.

3. During the final third of project duration, the cost accumulation is a mirror image of the first third duration, to achieve 100 per cent cost at physical completion.

The value of the project also accumulates in the same pattern but naturally exceeds the cost accumulation by the mark-up applied for contractor's profit. Obviously, it is equally possible to construct the cost pattern from the value pattern as it is to construct the value pattern from the cost pattern. In its simplest form, only the value (or cost), mark-up and duration are required to carry out this projection.

It is quite a simple task to program a computer to output cumulative project costs and values based upon this pattern of accumulation. Using the typical parabola equations, the S-curve is given by:

$$y = \frac{9x^2}{4}; \qquad 0 \leq x \leq \frac{1}{3}$$

$$y = \frac{3x}{2} - \frac{1}{4}; \qquad \frac{1}{3} \leq x \leq \frac{2}{3}$$

$$y = \frac{9x}{2} - \frac{9x^2}{4} - \frac{5}{4}; \qquad \frac{2}{3} \leq x \leq 1$$

where x is the cumulative proportion of project duration ($0 \leq x \leq 1$) and y is the cumulative proportion of project budget cost or value ($0 \leq y \leq 1$). (The equation developed by DHSS, in Hudson (1978), is slightly at variance with the above but was devised specifically for DHSS projects.)

The predictions may also be executed manually, either by using the equations given above or by purely graphical means, the middle section (a third time, a quarter cost to two-thirds time, three quarters cost) being a straight line and the parabolic 'lead-in' and 'lead-out' sections being drawn freehand. From the basic cost S-curve, the value S-curve and the actual receipts graph (delayed, stepped and adjusted – to reflect retention per cent withheld and release pattern – all in accordance with the terms of the contract to be used) may be constructed. A further sophistication is to apply a delay factor to the payments made (usually weighted average). Thus, the resultant graphs and tables give an adequately accurate cash flow forecast for each individual project (Table 8.2).

The information so obtained may be further utilized to produce a breakdown of the finance requirements for the project, both long-term finance and short-term finance. Such information is of value to a firm in deciding whether to tender for a project and to assist control of its capital both for individual projects and, by arithmetic summation of the individual project requirements together with an appropriate 'contingency' allowance, in aggregate for the firm.

The analysis described above is for a typical project. Each type of project and each firm will have its own individual peculiarities and organizations may develop prediction techniques on the S-curve basis that are adapted to suit their project types, methods of working, and organization. Although the S-curve prediction is reasonably accurate, it is still perhaps best suited for use as a preliminary predictor at a relatively early stage.

A somewhat more refined, and hopefully more accurate approach, is to cost and price the project programme on the basis of large cost centres at tender stage and more precisely, perhaps by operations, at the construction stage (as different levels of accuracy are required at these two stages). These exercises require more information and are more time-consuming than S-curve predictions, but have the advantage of being based upon the individual project data. Most quantity surveyors currently carry out this form of analysis to recover the preliminaries of a project.

S-curve example

S-curve analysis may use either cost or value as its basis. In this example, cost is used. (*Note*: If the value is, say, cost + 10 per cent, then cost is 10/11 × value, *not* 90 per cent × value.)

Estimate for a firm price project of one year's duration = £100,000
Overheads = 10% × estimate
Profit = 5% × cost
Certificates are monthly (always calendar months)
Income is received two weeks after certification
Weighted average payments delay by contractor = 1 month (see Table 8.3, p. 214)
Retention = 5%; half released at practical completion, half released at end of defects liability period
Defects liability period (DLP) = 6 months
Assume: 1 month = 4 weeks; no holidays
Cost of project = £100,000 × 1.1 = £110,000
Value of project = £110,000 × 1.05 = £115,500

The budgets produced from S-curve and programme analyses provide the overall framework within which the firm can monitor and control its activities. The monitoring and controlling functions are reliant upon information from the site as both cost and value data.

The cost data are usually provided via the accounts department for materials, wages, and salaries and via the quantity surveyors in relation to subcontractors. Overheads are treated as a separate cost, usually as a percentage addition to the total direct costs. Commonly, cost surveyors provide data on the cost of site operations – those data are useful in effecting cost control at the operational level through comparison with the costs of the operations predicted in the estimate.

Table 8.2 Cash flow calculations

(A) *Month No.*	(B) *Cum proportion time elapsed* (%)	(C) *Cum proportion project completed* (%)	(D) *Cum cost* (£)	(E) *Cost per period* (£)	(F) *Cash outflow per period* (£)	(G) *Cum value* (£)	(H) *Cum retention* (£)	(I) *Cum net value* (£)
½	4.2	0.4	440	440				
1	8.3	1.6	1,760	1,320		1,848	92	1,756
1½	12.5	3.5	3,850	2,090	440			
2	16.7	6.3	6,930	3.080	1,320	7,277	364	6,913
2½	20.8	9.7	10,670	3,740	2,090			
3	25.0	14.1	15,510	4,840	3,080	16,286	814	15,472
3½	29.2	19.2	21,120	5,610	3,740			
4	33.3	25.0	27,500	6,380	4,840	28,875	1,444	27,431
4½	37.5	31.25	34,375	6,875	5,610			
5	41.7	37.5	41,250	6,875	6,380	43,313	2,166	41,147
5½	45.8	43.75	48,125	6,875	6,875			
6	50	50.0	55,000	6,875	6,875	57,750	2,888	54,862
6½	54.2	56.25	61,875	6,875	6,875			
7	58.3	62.5	68,750	6,875	6,875	72,187	3,609	68,578
7½	62.5	68.75	75,625	6,875	6,875			
8	66.7	75.0	82,500	6,875	6,875	86,625	4,331	82,294
8½	70.8	80.8	88,880	6,380	6,875			
9	75.0	85.9	94,490	5,610	6,875	99,214	4,961	94,253
9½	79.2	90.3	99,330	4,840	6,380			
10	83.3	93.7	103,070	3,740	5,610	108,223	5,411	102,812
10½	87.5	96.5	106,150	3,080	4,840			
11	91.7	98.4	108,240	2,090	3,740	113,652	5,683	107,969
11½	95.8	99.6	109,560	1,320	3,080			
12	100	100	110,000	440	2,090	115,500	2,887	112,613
12½					1,320			
13					440			
18								115,500
18½								

Notes: Column B is calculated on a straight line basis. Column C is calculated from the typical S-curve equations for cost on value accrual. Column F is as column E but with a delay (derived from weighted average payments delay) applied. Column K is as column J but with the delay on receipts of cash applied. Column N (no payments delay) is column M minus

Value data are usually obtained from the firm's quantity surveyors. As this information is normally prepared on a monthly basis for valuation and certification purposes, the costs and values are also normally reviewed by the firm at the same time intervals.

Firms differ in the ways in which costs and values are kept under review. This may be carried out in detail every month or in outline each month but with a detailed review at longer intervals, e.g. every three months (Fig. 8.7). The time intervals should be selected to suit the firm's requirements and are governed by such considerations as the value of the project, the rate of physical progress, the project value as a proportion of

(J) *Net value per month* (£)	(K) *Cash inflow per month* (£)	(L) *Cum cash outflow* (£)	(M) *Cum cash inflow* (£)	(N) *Net cash flow* (£)	(P) *Max cash flow* (£)	(R) *Net cash flow* (£)	(S) *Max cash flow* (£)
				−440	−440		
1,756				−1,760	−1,760		
	1,756	440	1,756	−2,090	−3,850	+1,316	−440
5,157		1,760					
	5,157	3,850	6,913	−3,757	−8,914	+3,063	−2,094
8,559		6,930					
	8,559	10,670	15,472	−5,648	−14,207	+4,802	−3,757
11,959		15,510					
	11,959	21,120	27,431	−6,944	−18,903	+6,293	−5,648
13,716		27,500					
	13,716	34,375	41,147	−6,978	−20,694	+6,882	−6,944
13,715		41,250					
	13,715	48,125	54,862	−7,013	−20,728	+6,737	−6,978
13,716		55,000					
	13,716	61,875	68,578	−7,047	−20,763	+6,703	−7,013
13,716		68,750					
	13,716	75,625	82,294	−6,586	−20,302	+6,669	−7,047
11,959		82,500					
	11,959	88,880	94,253	−5,077	−17,036	+5,373	−6,586
8,559		94,490					
	8,559	99,330	102,812	−3,338	−11,897	+3,482	−5,077
5,157		103,070					
	5,157	106,150	107,969	−1,591	−6,748	+1,819	−3,338
4,644		108,240			−7,188		
	4,644	109,560	112,613	+2,613	+2,613	+3,053	−1,591
		110,000				+2,613	+2,613
2,887							
	2,887		115,500	+5,500	+5,500	+5,500	+5,500

column D. Column P (no payments delay) is column M for the previous month minus column D for the current month. It shows the maximum capital required immediately prior to cash inflows being received. Columns R and S (with payments delay) correspond to columns N and P, above, but are calculated using column L in place of column D.

the firm's turnover, and familiarity with the methods of working. In all instances, and regardless of the system used, it must be remembered that control can be exercised only over events which will occur in the future so, while *analysis* of historic costs and values on the project are vital in providing input data to the control process, the focus must be on prediction of future performance for control to occur. Usually, the assumption is that the performance is *deterministic* (that the future will replicate the past) hence immediate past performance is regarded as the best indicator of future performance and so the best guide to any control required.

As short-term finance is cheaper than long-term finance, a firm should

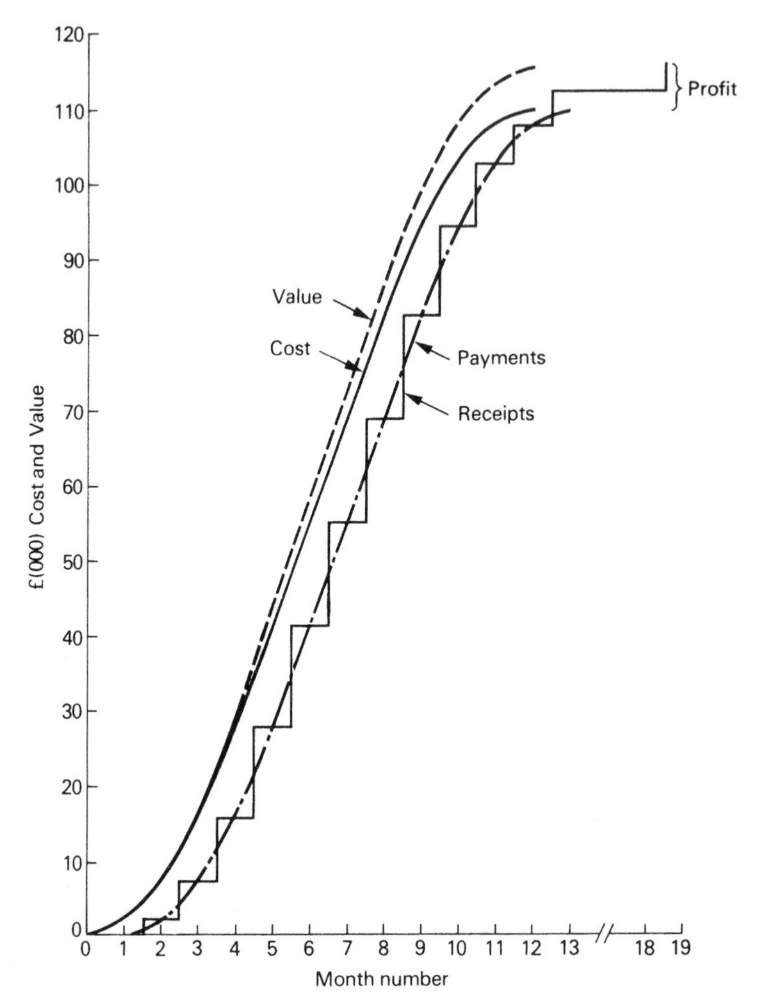

Fig. 8.5 S-curve analyses. *Note:* the payments curve may be constructed with a lead-in delay accumulation from the origin. It is shown here as a simple delay on the cost curve throughout its length.

keep its long-term finance to the minimum. At any time, the maximum cash flows required by projects (in aggregate) minus the funding available on a short-term basis (including any overdraft facility not taken up) indicates the long-term finance required by the firm. (Naturally, however, some overdraft facility will usually be maintained as a contingency.)

8.4.2 Forecasting financial requirements

Cash flow models for projects may be employed in aggregation to forecast the financial requirements of the firm. Depending on the models which are employed, the reliability of the forecasts will vary; clearly, it is useful

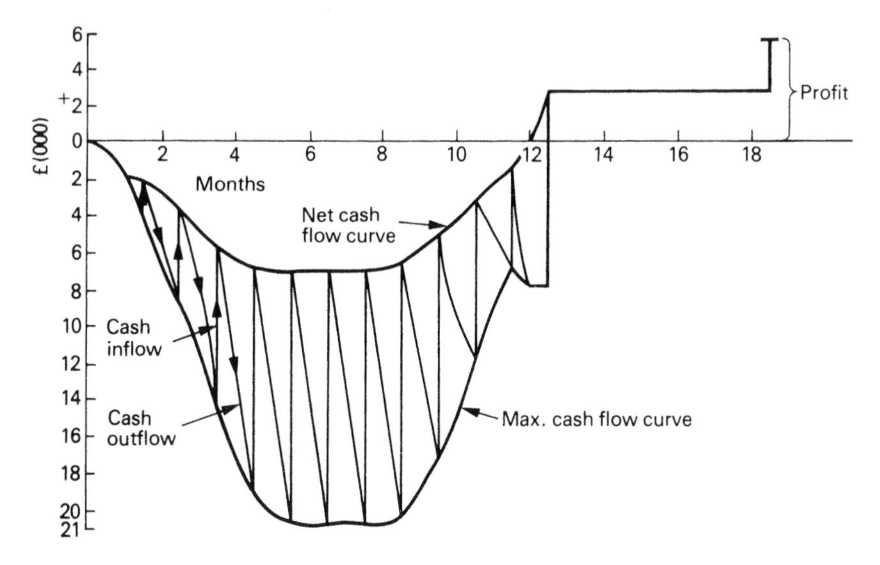

Fig. 8.6 Cash flow curves.
Note:
1. Net and maximum cash flows may be determined directly from the relevant S-curves.
2. The area between the abscissa and the net cash flow curve, whilst the latter is below the abscissa, shows the long-term finance required for the project.
3. The area between the maximum cash flow curve, while below the abscissa, and the lower of the abscissa and the net cash flow curve, shows the short-term finance required for the project.
4. The project is self-financing (entirely) only when both net and maximum cash flow curves are above the abscissa.
5. In the example quoted, payments delay (of one month weighted average) by the contractor obviates the requirement for long-term capital for the project.
6. In neither instance shown in the example does the project become entirely self-financing until the payment (inflow) following practical completion is received (month $12\frac{1}{2}$).
7. The payments delay by the contractor also greatly reduces the project's short-term finance requirements.

to monitor the forecasts to determine their accuracy, together with the nature (random or systematic – bias) and size of errors. The resultant forecasts of financial requirements may be reviewed for 'smoothing' – in a similar way to resource smoothing in project programming – because significant fluctuations in financial requirements are likely to be cost increasing.

As financial markets classify finance provision according to the period involved and the risks, long period finance and/or high risk activities attract high costs of finance capital. Further, suppliers in the finance market specialize in the types of finance they provide and so knowledge of the finance market and details of the finance required are essential to secure a 'good deal'.

Portfolio analysis shows that aggregate risk is not necessarily the

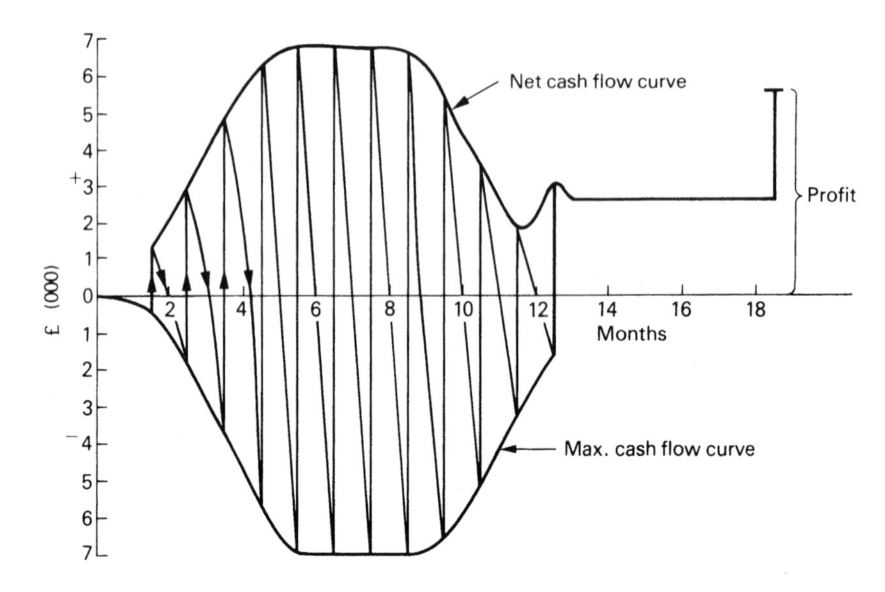

Fig. 8.7 Cash flow curves with payments delay applied. *See also Notes to Fig. 8.6.*

arithmetic sum of the individual component (project) risks: a properly diversified portfolio of projects results in reduced overall risk for the firm which should yield a lower cost of finance. Thus, especially for high risk, international projects, it is likely that the individual project risk will significantly exceed the corporate risk – so, corporate finance will, in such cases, be less expensive than project finance. (A similar argument may be applicable to the provision of 'country finance' in international aid, etc. such as advances by the World Bank instead of 'project finance'.) Further methods of reducing risk 'exposure' on projects are through joint venturing arrangements and by securing an export credit guarantee from the Government (for international projects).

Aggregation of the forecast portfolio of projects' predicted cash flows and examination against sources of finance may be shown in graphical form as in Figure 8.8. The aggregation does not show inherent variability, which could be incorporated by 'square root of the sum of the squares' addition of components' individual variabilities to give envelopes of financial requirements. Maintaining the profile of maximum total available finance above the expected total capital requirement by use of the firm's negotiated overdraft limit provides a financial 'cushion' to enable the firm to cope with unanticipated, often sudden, calls for finance.

Note in Figure 8.8 that overdraft is treated as the final source of funding and so is additional to other sources. The overdraft limit is negotiated periodically (six-monthly?); the amount necessary and the amount actually used are determined by the profile of total capital required.

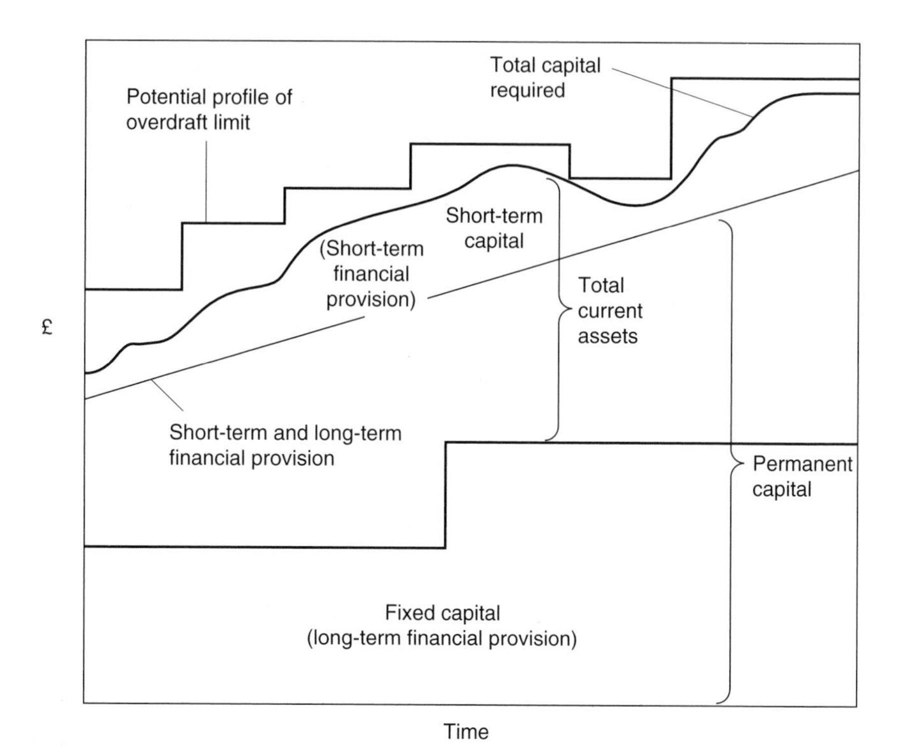

Fig. 8.8 Finance capital aggregate requirements.

8.5 Monitoring and control

8.5.1 Cost value reconciliations

While cost control systems, such as standard costs and variance analyses, are used extensively at site level for cost control of the individual project (often on a weekly basis), cost/value reconciliation is the form of budgetary control most commonly used by upper (usually head office) management. Cost/value reconciliations are usually prepared by the firm's quantity surveyors and are statements of the total costs and values in respect of each project, usually at the date on which the interim (normally monthly, based on progress of the site work) valuation was carried out. It should be noted that, although progress payments are the most common basis of interim payments, they may be based on project *stages* of completion as agreed between the contractor and the client.

The value section comprises the gross valuation (as agreed and checked against the associated interim payment certificate issued by the architect) plus statements of the retention held, the amounts any claims should actually yield, the value of materials on site, the amount of fluctuations, the value of variations included in the valuation and the amounts certified for nominated subcontractors. Adjustment should be made to reduce the value in respect of any claims against the contractor from the client.

The cost section comprises materials and wages paid plus any necessary accruals for items delivered but for which payment has not been made and for time worked for which wages have not been paid. The materials cost should be net of discounts and the wages bill should be the cost to the firm of using that labour. Hired plant, owned plant and small tools can be costed from plant returns and invoices with any necessary accruals. Subcontractors may be costed from their total account paid to the relevant date and adjusted for discounts, claims, and contracharges. Normally, nominated subcontractors (and, often, nominated suppliers) are listed in detail separately, as particular contractual conditions appertain to their accounts. Amounts of values - payments due - to nominated subcontractors will be listed in a schedule attached to the interim certificate; such sums must be discharged by the main contractor in accordance with the contractual terms (both main and nominated subcontract) and so, only contracharge amounts permitted to be made should be included as adjustments of those sums.

Overheads are usually added as a prescribed percentage of cost and the profit or loss, both to date and for the month, are clearly indicated.

It is of obvious importance that all relevant costs and values are included, making accruals, adjustments and estimates where necessary. As this document is a management control tool, its accuracy is of the essence. Top management should inform the relevant people of the profit pattern the project should produce, as this may easily have been manipulated by project pricing techniques.

The cost/value reconciliations will be used to monitor profitability performance against some predetermined expectation (S-curve analysis, priced programme, etc.). For such monitoring to be meaningful, management requires notification of the accuracy of the information and techniques being employed. Thus, a knowledge of any error inherent in estimating, programming, overhead allocations, and so on, as well as any error in the budgeting, costing, and valuing systems is required. Any distortion of the information - such as not declaring early profits to 'feed' them in at later stages of the project in order that a more even profit-earning profile is reported - may lead to unrequired and incorrect actions being instigated.

It would be useful if, for profit monitoring, a technique were employed similar to the statistical technique used for quality control. The expected profit level should be plotted for the duration of the project, with the calculated error limits plotted on each side. Upper and lower control limits would then also be established and the achieved profits would be superimposed as they occurred. This system shows the profits expected together with possible inherent variances and the actual achievements.

Trends can be identified easily and corrective action taken at an early stage. The system will also show whether the variance is due to error in the prediction system employed or to performance requiring corrective action.

8.5.2 Variances

Whatever system of cost prediction is used, it is highly probable that actual costs incurred in the production process will differ from those predicted, i.e. cost variances will occur. As there are very many causes of variances, it is desirable to implement a monitoring system which permits the existence, sizes and causes of variances to be detected at an early stage in order that corrective measures can be implemented.

In the construction industry, the cost/value reconciliations will show if a contractor is performing as predicted by regularly examining costs, value, and profit which may then be compared with a prediction. If a variance occurs which cannot be adequately explained (e.g. claims outstanding), more detailed investigation will be necessary to establish the cause(s) of the variance. On site, the production data, often used for bonus calculations, are commonly compared with the *costed* BQ to detect labour variances. Materials and plant cost control may be executed in a similar manner. (Note: Variances arising due to price increases must be examined in the context of the fluctuations recovery provisions of the contract.)

Labour variances

The predicted cost of labour for an operation (or set of operations, if more convenient to aggregate) may be obtained from the estimators' data – the estimate for that particular project. The cost will have been calculated by multiplying an estimated time for the execution of the work (the labour 'constant', adjusted as necessary to suit the particular project to produce the *standard hours*) by the applicable wage rate (*standard wage rate*). The site will then obtain data of the time taken to perform the operation(s) and the labour cost of that work (*actual hours* × *actual wage rate* = labour cost). These data permit the source(s) of any variance to be established. It is clear (see Figure 8.9) that labour cost variances arise due to actual hours differing from standard hours or/and actual wage rate differing from standard wage rate. The wage rates are functions of labour and management negotiations which occur at national level and may be modified at local level. The time to execute work is a function of innumerable variables, such as weather, managerial abilities, health, and payments and incentives, etc.

In examining an efficiency variance it is unlikely that the true causes will be established absolutely and evaluated accurately but it is necessary to determine if the variance is due to an incorrect allocation of standard

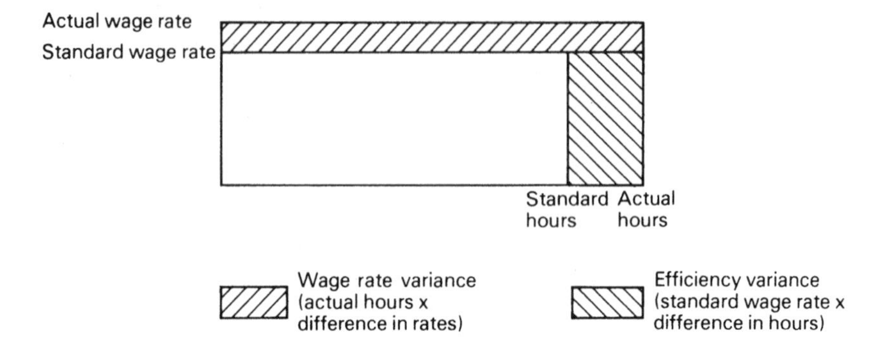

Fig. 8.9 Labour variances. After Bierman (1963).

hours or the actual performance (work study should be used for this purpose). If the actual performance has caused the variance, it should be investigated to establish if it is attributable to managerial organization or to operative performance. The results of all investigations should be fed back into the firm's data banks for future use by estimators, work study engineers, and managers.

Materials variances

Materials price variances may be largely eliminated by efficient purchasing. The estimators should obtain quotations for the supply of the requisite materials for the project (at least, the major materials supplies) on the same fluctuations recovery terms as those applicable to the main contract. (This should also apply to the quotations obtained from subcontractors.) The quotations should remain open for acceptance for a sufficient period to allow them to form the basis of an order should the main contractor be awarded the contract. (*Note:* If a firm price quotation and order are used upon a fluctuations contract, this will probably result in a positive material price variance for the main contractor.)

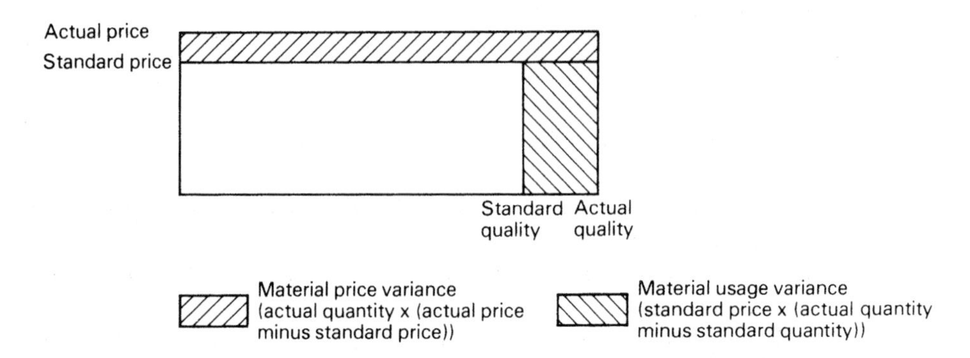

Fig. 8.10 Materials variances. After Bierman (1963).

Materials usage variances arise for many reasons and should be investigated in a manner analogous to that employed for the investigation of labour efficiency variances. For instance, if the material usage variance on timber joists is high, is it due to pilferage, damage, or larger off-cuts due to incorrect sizes having been purchased or due to management inadequacies where joists are used in the wrong location? Attention should be given to the allowances included for waste of materials and the levels of waste achieved on the project. Again, feedback is essential for avoidance of similar problems in the future.

Plant variances

These are governed by somewhat different factors depending upon whether the plant is owned or hired. In the latter case, minimum hire periods are relevant.

The variance may be split into price and usage (or efficiency) components. In any situation the operation of the plant is governed by two factors: the plant and the operator. Thus, any efficiency variance analysis must consider the choice of plant for the operation and the abilities of the operator. Clearly, the operation of plant may lead to materials and other variances also (e.g. too large a bucket on an excavator for digging a trench).

Overhead variances

Overhead variances may be usefully considered as variable overhead variances and fixed overhead variances. In the construction industry overheads are calculated as a budget for the period and are allocated to projects on the basis of expected total project value for the period or expected total direct labour cost for the period.

Variable overhead variances occur in similar ways to labour variances and may be analysed as:

- budget variance = (actual variable overhead) – [(actual hours) × (variable overhead rate)]
- efficiency variance = (variable overhead rate) × [(actual hours) (standard hours)]

Fixed overhead variances may be analysed into three components:

- budget variance = (actual fixed overhead costs) – (budgeted fixed overhead costs)
- idle capacity variance = (fixed overhead rate) × [(budgeted hours) – (actual hours)]
- efficiency variance = (fixed overhead rate) × [(actual hours) – (standard hours)]

The idle capacity variance highlights the problem of excess capacity and is caused by the actual time worked being different from the budgeted hours. Graphical representations of variable and fixed overhead variances are illustrated in Figure 8.11.

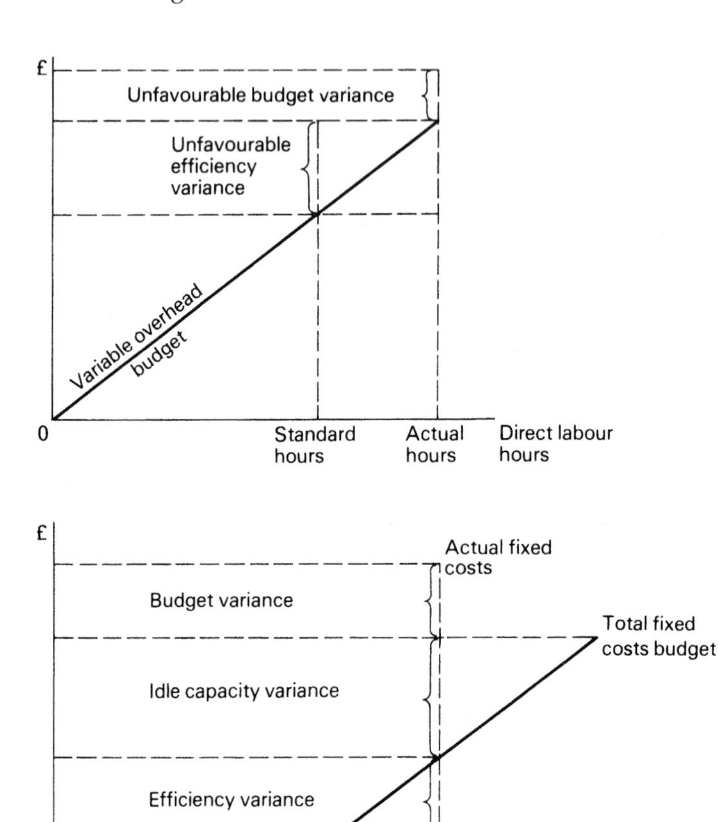

Fig. 8.11 Overhead variances. After Bierman (1963).

It is unusual for overheads to be allocated in an exact manner to the individual projects that a contractor undertakes (consultants usually allocate overheads in a similar way to contractors). Thus, a firm will examine overhead allocations, absorptions, and variances periodically (often annually) and on an aggregate basis (i.e. for the whole firm). It is on the basis of such investigations that adjustments to future budgets are made to take account of variances.

8.5.3 *Pricing manipulations*

It is usual to assume that any construction project is priced in such a way that the price of each item comprises the cost of that item plus the relevant share of overheads and profit (and any other) additions. This is not always the case. It has been shown earlier (see S-curve) that a typical building project does not become self-financing until it is near completion and in addition requires the contractor to provide considerable amounts of capital as both working capital and long-term capital. As a cost is associated with the provision of capital (see Chapter 7), contractors often seek to minimize the amount of their own capital used in the execution of a project and to make the project self-financing at as early a stage as possible.

Thus, the capital requirements of the contractor may be reduced by a site set-up payment (as often happens in international contracting: a mobilization payment, possibly around 15 per cent of the initial contract sum and set-off against interim payments), by a reduced retention deduction and earlier retention releases, by more frequent payments to the contractor, by subcontracting more work, and so on. It is also possible for the contractor to manipulate the item prices in a project to achieve these ends. The technique is called front-end loading and occurs where items of work which the contractor expects to be executed early in the project have prices which contain a disproportionately large content of overheads and profit and items of work to be executed in the later stages of the project have their prices reduced accordingly. The situation is of further benefit to the contractor due to the time value of money.

Front-end loading manipulations are limited in the extent to which they may be carried out by the consultant quantity surveyor (PQS) in checking priced BQs prior to the contract for the works being awarded.

Imprecision of the techniques commonly employed to determine the value of work to be included in interim progress payments gives scope for the contractor to have limited 'negotiation' over the amounts to be included. Whilst such negotiations may favour the contractor's cash flow, the total sums become 'self-corrected' by the final account, at the latest. Such cash flow advantages act to reduce the costs of finance for the contractor and hence the total cost.

As variations are an expected occurrence on any building project (for extent and value prediction techniques see Bromilow 1970, 1971, 1974), contractors are occasionally tempted to price low those items which they anticipate will be omitted or reduced in extent and to price high those items they expect to be increased. Again the checking role of the PQS acts as a restriction upon such manipulations.

It is arguable whether the contractor should be required to provide much capital for the execution of a construction project or whether the capital should be provided by the client. Construction is considered to be a high-risk industry and so, returns on investments in construction should be higher than returns on comparable investments elsewhere. Around 35 per cent of UK construction work is executed for the public sector, a low-risk sector of the economy. Most private sector clients of contractors will be engaged in activities which are less risky than construction; therefore, clients should enjoy a lower cost of capital than contractors. Thus, if clients provided the capital for construction projects (given sufficient safeguards which really already exist), the prices charged for construction should be reduced. The alternative is, of course, that if such a system of financing construction projects were implemented, it would result not in price reductions through savings being passed back to client organizations but in contractors maintaining price levels and thereby earning greater profits. A manifestation of the argument has been the trend to reduce the retention amounts on building projects.

8.5.4 *Inflation*

Inflation, in application to the prices of construction projects, is covered by the inclusion of a fluctuations' recovery provision. (In respect of the JCT 98 contract, see Fellows and Fenn (2001) for a discussion of the provisions.) Fluctuations may be non-recoverable or recoverable in whole or in part. If full recovery of fluctuations is permitted, adherence to the prescribed system is necessary to ensure recovery. Full fluctuations are recovered in most cases either by application of the NEDO formulae or by application of a manual system. Usually, firm price arrangements are employed on all but very large projects of over two years' duration; when full fluctuations apply, it is normal for the formulae system to be employed as it is far cheaper and easier to operate and gives less of a shortfall in recovery than the manual system.

The manual system permits recovery in respect of all but minor items. The normal partial fluctuations system (firm price arrangements) permits recovery of tax and similar government imposed fluctuations only and a fixed-price contract permits no fluctuation recovery at all.

It is evident, therefore, that the price the contractor submits for the project should include an adequate allowance for fluctuations which will be incurred by the contractor but which the contract states to be non-recoverable. This non-recoverable element must therefore be predicted for inclusion in the tender sum. Several methods of prediction exist. One useful method analyses cost increases to date and then extrapolates from them three predictions: an average, a high prediction, and a low prediction. It is for the contractor to assess the most likely prediction of cost increases in each individual case, whether on an overall project basis for each tender or trade by trade for each tender. The tender programme will be of value in this analysis to indicate the timing of operations. A slight variation might be to apply escalation indices of building costs to the S-curve cost prediction and thence the derived price curve, etc.

Failure to make sufficient allowance for non-recoverable fluctuations will lead to a reduction in project profit. Non-recoverable fluctuations must include any anticipated shortfall in recovery as well as the contractual non-recoverable elements; allowance must be made for the costs of operating the system of recovery also, commonly via overheads. Inclusion of an excessive allowance for non-recoverable fluctuations may mean that a project is awarded to a competitor.

8.5.5 Periods of credit

Periods of credit are of importance in determining capital requirements. It has already been shown that a prudent firm will attempt to give shorter periods of credit to its debtors than it enjoys from its creditors.

The period of credit a contracting firm must give to the client is usually prescribed in the contract. Under JCT '98, the JCT Standard Form of Building Contract 1998 edition, normally certification is monthly and the amount is determined by progress of the work (an alternative is 'stage payments'). Payment to the contractor must be made within two weeks of the certificate date. Thus, the contractor receives the first payment one month and two weeks after starting work on the site, and at monthly intervals thereafter.

Assuming the value of each month's work is spread evenly throughout the month, the period of credit given by the contractor for the sums paid monthly is one half of one month plus two weeks. (The average value of the month's work occurs half-way through the month.) The retention provision means that the contractor gives credit in respect of the accumulation of retention held by the client; this should be the subject of a credit assessment and should be combined with the analysis of the monthly payments for evaluation.

Various credit periods are enjoyed by the contractor. Table 8.3 indicates the usual periods, the proportion of total cost of each cost centre and the resultant weighted average payments delay (which can be incorporated into an S-curve and other budgets). Retention on subcontractors may also be evaluated, although they have been ignored in Table 8.3 and are usually of minor significance.

Once the credit facilities offered and enjoyed have been assessed, the relevant sums of money should be applied to them to ensure that the credit offered is, at minimum, no less than the credit enjoyed.

Table 8.3 Weighted average payments delay

Cost Centre	*Payment interval*	*Average delay (weeks)*	*Project cost proportion (%)*	*Weighted average delay (weeks)*
Direct labour*	weekly	$\frac{5}{6}$	10	0.08
Salaried labour, etc.	monthly	2	5	0.11
Domestic suppliers	as JCT nom.	$6\frac{1}{2}$	20	1.30
Nominated suppliers	as JCT	$6\frac{1}{2}$†	15	0.98
Labour-only subcontractors	weekly	$\frac{1}{2}$	10	0.05
Domestic subcontractors	monthly	2	14	0.30
Nominated subcontractors	as JCT	$4\frac{3}{4}$‡	20	0.95
Plant hire	monthly	2	5	0.11
				3.88

3.88 weeks weighted average payments delay = 0.9 months.

* Direct labour:

Assume that a production bonus scheme is in operation and that pay = $\frac{2}{3}$ basic plus $\frac{1}{3}$ bonus; bonus payments lagging behind basic by one week.

$$\left.\begin{array}{ll}\text{Basic pay} = \frac{1}{2}\text{ week delay} \times \frac{2}{3} = \frac{2}{6} \\ \text{Bonus} = \frac{3}{2}\text{ week delay} \times \frac{2}{3} = \frac{3}{6}\end{array}\right\} \frac{5}{6}$$

Salaried labour, etc.

Assume this includes other overhead payments

† $\frac{1}{2}$ month + 30 days = $6\frac{1}{2}$ weeks

‡ $\frac{1}{2}$ month + 17 days = $4\frac{3}{4}$ weeks

8.6 Clients

Clients, especially industrial or commercial organizations, will use budgeting techniques in respect of their own activities which, although often similar in their underlying principles to those used in the construction industry, will be adapted to the clients' requirements. Thus, for their financial forecasting and control purposes, clients are usually reliant upon information supplied by the design team. (Large industrial, commercial, and governmental clients who undertake a good deal of construction work often have 'in-house' construction experts and sometimes in-house design teams.)

The information required is dependent upon the stage that the project has reached. During the initial stages (see RIBA 1973 and Figure 8.12) – inception and feasibility – the employer will be primarily concerned with the overall cost of the project (the timing of cash outflows is also receiving greater attention, even at the early stages of a project, due to the prevailing high interest rates and increasing use of investment appraisals which employ discounting).

Once the decision to proceed with the project has been taken, the amount of detailed cost information required by the client will increase as the design progresses and the construction period approaches. The PQS, or cost consultant, will prepare a cost plan for the project which will become more refined and precise as the design evolves. The cost planning

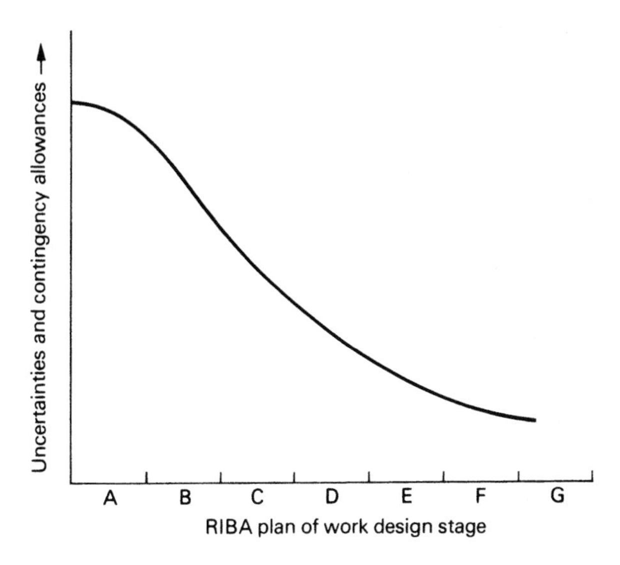

Stages: A inception; B feasibility; C outline proposals; D scheme design; E detail design; F production information; G bills of quantities.

Fig. 8.12 Reduction of necessary contingency allowances as design progresses.

activities are an integral part of the design process. The cost plan will be used to provide the employer with quite detailed cost information regarding the cost consequences of design requirements and a pattern of anticipated cash outflows during the construction and completion stages. Only after tenders have been obtained, scrutinized, and the contract awarded can the cost plan be adjusted to reflect the actual contract sum and its distribution within the work sections – hence becoming the *cost analysis* of the accepted tender. JCT 98 normally requires that the contractor should provide a master programme for the project (a requirement which reflects previous good practice) and this will also considerably aid accurate cash flow forecasting for clients.

As a project progresses through the construction stage, numerous alterations will be made to the cost analysis to keep it up-to-date. The alterations necessary will be caused by occurrences such as changes in client's requirements, design amendments, other variations, fluctuations, delays and other claims. A useful system to ensure that the client is aware of cost developments during a contract is to provide a statement at monthly intervals (i.e. a statement will accompany each interim certificate) detailing the expected payments for each of the following three months, the expected total of the payments for the three months following those, and the anticipated final account sum. Such a statement will keep the client aware of the total commitment to the project and of the liquidity requirements of that commitment. (The liquidity aspect is important, not merely to ensure that payments can be made when due but also to permit the employer to maximize his earnings from his available capital as liquid investments attract a lower return than illiquid investments.)

8.7 Consultants

Consultants – the architect, PQS, structural engineers, and services engineers – are usually employed under standard contracts (conditions of engagement/appointment, produced by the appropriate professional institution) with the client. Consultants charge a fee for their services which is determined via competitive bidding (historically, such consultants were paid fees on scales prescribed by the appropriate professional institutions). Their activities are primarily design-oriented and highly labour-intensive (although the situation is changing with increased use of computers). Thus, their budgeting activities, analogous to those of contractors, are mainly concerned with keeping their costs within a predetermined limit to secure the anticipated profit margin. Their budgeting, therefore, is frequently in the form of time allocations of personnel for the performance of prescribed tasks.

It may be argued that although the fees charged by consultants are a reasonably small proportion of a construction project's total cost (about 10 per cent on large building projects), the significance of the consultants lies in their influence over the construction cost element and the costs of using the completed building. Good design can significantly reduce both construction and building-user costs.

Cost planning plays its most vital role in the achievement of good design, although, as demonstrated by Kelly (1982), approximately 80 per cent of the cost of a project has been committed by the time 20 per cent of the design period has elapsed; hence, to be of significant influence, input must be provided early in the design phase. Further, Lera (1982) noted in respect of the initial design:

> ...the tradition persists whereby the architect prepares a sketch plan from which the other consultants work. Frequently alone, and often in a matter of hours, the architect arranges spaces in a structure using predominantly aesthetic criteria.

The cost plan will initially establish the overall cost of the proposed building and, once the cost has been agreed by the client, will progress to consider all cost aspects of the design, as it evolves, in increasing levels of detail. Cost planning permits the cost consequences of each design decision to be fully evaluated to ensure that a good solution to the client's requirements (contained in the project brief) is obtained by the provision of a building in which the costs are correctly balanced. However, the variability of the data and techniques of prediction employed should be not only borne in mind but made explicit in order that the resultant decisions are better informed.

Naturally, at the early stages when a comparative exercise is commonly carried out to establish the project cost, several contingency sums should be included. These should be reduced or may even be completely eliminated as the design evolves and construction progresses. Perhaps one important aspect, which to a certain extent is psychological, is that the client will invariably remember, and base everything upon, the initial project cost projection; therefore, reasonable savings will be welcomed but even small cost escalations will be viewed with abhorrence. Consequently, it is important that each change in the project cost is recorded and justified. However, such a belief about client behaviour is likely to have repercussions on the cost predictions given and, potentially, on the realized cost of the project. The belief will encourage cost forecasts given to the client to be 'pessimistic' – at the high end of the range of variability. A corollary is that, in order that savings will not be 'excessive' (because setting aside too much liquid funding for the project has an opportunity cost for the client), design modifications may be introduced which enhance the building but also increase its cost to an out-turn cost so that it is close to the prediction. If the variability in cost forecasts is made express (rather than their following the 'tradition' of being single-figure forecasts), then much of such issues of the cost predictions becoming self-fulfilling prophesies will be avoided.

8.8 Investment appraisal

Investment appraisal is most widely considered in the context of the decision of whether to purchase, or acquire by some other means, capital equipment, usually in the form of plant or buildings. In the context of construction, investment appraisal is used by clients (and consultants) in their evaluation of proposed construction works whether as new work, refurbishment or rehabilitation (normally, repair and maintenance work is paid from the revenue budget). Contractors use investment appraisal techniques to assess their own capital investments (plant, equipment, buildings, etc.) and in their assessments of potential projects, i.e. it is part of the bidding process. In the context of bidding (by competitive tender or an alternative method) investment appraisal is used to evaluate the envisaged capital requirements of the projects.

In an investment appraisal, only the incremental expenditures and receipts directly attributable to the project under scrutiny should be included (the marginal costs and revenues of the project); sunk costs (i.e. those which have been incurred prior to the decision) should be ignored as they are irrelevant to decisions about the future.

8.8.1 Payback method

The payback method is very simple. It calculates the period required for the incremental net cash inflows generated by the investment to amount in total to the initial incremental capital outlay. It is not a discounting method but is nonetheless popular, probably due to its simplicity.

As it is not a discounting method, its applicability is limited to short-duration investments, due to the resultant inadequacies in ignoring the time-value of money. The period over which this method may be considered reasonable decreases with each rise in the interest rate. The method also fails to take account of any cash inflows which the project generates beyond the break-even point (after the payback period). Nor does it take into account the pattern of the payments stream prior to the break-even point. Its value lies in its simplicity and the ease of its use as an initial filter (commonly, a small number of years to achieve payback) for investment proposals; only investment possibilities which pass the initial payback filter are considered in greater detail using more sophisticated discounting appraisal techniques.

8.8.2 Discounted cash flow (DCF) methods

Discounting is a technique which takes the time-value of money into account and may be considered to operate as a reversal (or perhaps more accurately a reciprocal) of compound interest. The technique facilitates comparisons of projects with different investment requirements and with varying cash flow patterns. It is also possible to compare projects with different lifespans but additional care is required to select the most appropriate method to employ.

One major problem with any DCF method is the determination of the rate of interest to use in the evaluation. A nominal or market rate of interest comprises three component elements – time (or liquidity) preference, risk and inflation. The components are combined in either an additive or a multiplicative model:

$$(1 + i_n) = (1 + i_t) + (1 + i_r) + (1 + i_i) \quad \text{or}$$

$$(1 + i_n) = (1 + i_i)(1 + i_r)(1 + i_i)$$

Where: i_n is the nominal or market rate of interest
i_t is the time preference or liquidity preference rate of interest
i_r is the component for risk
i_i is the rate of inflation

In practice, isolation of the components of the market rate of interest may require some approximations to be made. Individual investigations may be necessary to determine the time preference rate for a particular investor. Seeley (1996) suggested that a social time preference rate approximates to 3 per cent. There are many measures of inflation: the most usual measure for an economy is the consumer price index but it is advisable to select a measure more particular to the type of investment being considered. Short-period investments in government stocks are virtually risk-free and so the risk component in a rate of interest can be calculated by comparison or given the market, time preference and inflation rates.

Normally, a rate of interest increases with the duration of an investment. However, on occasions, short-period rates of interest may be very volatile and so be above the rates for longer periods – a reverse yield gap.

Market guidance to appropriate rates of interest include:

- commercial bank base rates and trends therein
- interest rates on government securities
- stock market indicators, e.g. P/E ratios
- interest rates prevailing in the commercial sector for similar ventures

Basic calculation formulae

1. *Amount of £1* (compound interest): the amount received at the end of a period in which £1 is invested at the start of the period at a given rate of compound interest.
   $A = (1 + i)^n$ where A is the amount of £1, i is the rate of interest, and n is the number of years. For example, £1 invested for 3 years at 10 per cent gives:

Year 1	$A_1 = 1.00 + 0.10 = 1.10$
Year 2	$A_2 = 1.10 + 0.11 = 1.21$
Year 3	$A_3 = 1.21 + 0.12 = 1.33$

   Now using $A = (1 + i)^n$, the answer for year 3 above is:

   $$A_3 = (1 + 0.10)^3 = \mathbf{1.33}$$

2. *Present value (PV) of £1* (reciprocal of compound interest): the present value of receiving £1 at the end of a given number of years with a certain rate of compound interest prevailing.

   It has been shown above that if an investment made today for a period is subject to compound interest, the investor will, at the end of the period, receive a sum greater than the original sum invested. Considering the PV, the sum received by that investor at the end of the period is worth £1 to him today. Thus

   $$PV = \frac{1}{A} = \frac{1}{(1 + i)^n}$$

3. *Amount of £1 per annum (APA)*: the amount of £1 per annum is the amount which will accumulate if £1 is invested each year of a period at a prescribed rate of interest (investments are assumed to be carried out at the end of each year), e.g. endowment policies.

$$\text{APA} = \frac{\text{A} - 1}{i} = \frac{(1+i)^n - 1}{i}$$

4. *Annual sinking fund (ASF)* (reciprocal of amount of £1 per annum): the amount which must be invested annually to produce £1 at the end of a given period, where the amount invested is subject to compound interest at a given rate (again, end of year investments are assumed).

$$\text{ASF} = \frac{i}{\text{A} - 1} = \frac{i}{(1+i)^n - 1}$$

5. *Present value of £1 per annum or years' purchase (YP)* (reciprocal of rate of compound interest): the capital value of an investment: the present value of receiving an annual amount of £1 for a given number of years at a given rate of interest – usually applied to the income from rental property.

$$\text{YP} = \frac{1}{i} \text{ (for long periods)}$$

For example, an investor will purchase a freehold costing £10,000 only if it will yield him at least 10 per cent per annum.

$$\text{i.e. } £10{,}000 \times \frac{10}{100} = £1{,}000$$

Consider the situation in reverse: A freehold yields net income of £1,000 per annum. If an investor requires 10 per cent return on any investment, how much should he pay for the freehold?

$$£1{,}000 \times \frac{100}{10} = £10{,}000; \quad \frac{100}{10} = 10 \text{ YP}$$

However, if the property were leasehold, the capital invested to purchase the lease must also be considered as this may be regarded as a 'wasting asset' and so should be redeemed at the expiry of the lease. The ASF is used to provide the capital redemption:

$$\text{YP} = \frac{1}{i + \text{ASF}}$$

where i is expressed as a decimal and ASF is the figure derived from the table for the period at the given rate.

Thus, YP calculates the capital value, the ASF being to provide for the redemption of the original capital outlay at the expiry of the lease.

Interest and ASF rates often will be different, interest usually being the greater.

Note: As YP is PV of £1 per annum, then

$$YP = APA \times PV = \frac{A-1}{i} \times \frac{1}{A}$$

$$= \frac{(1+i)^n - 1}{i} \times \frac{1}{(1+i)^n}$$

$$= \frac{(1+i)^n - 1}{i(1+i)^n} = \frac{1}{i}\left[1 - \frac{1}{(1+i)^n}\right]$$

which, for long periods, approximates to:

$$YP = \frac{1}{i}$$

6. *Effects of tax:* the rate of tax (RT) will affect the calculation of all the above. This is accounted for by a modification to the rate of interest (RI) applied to obtain the effective rate of interest (ERI). Thus:

$$ERI = \frac{1 - RT}{1} \times RI$$

For example, if RI = 5% and tax is 40%, then

$$ERI = \frac{100 - 40}{100} \times \frac{5}{100} = \frac{300}{100} = 3\%$$

The sinking fund is different. The net ASF is significant, i.e. matured fund at the end of the period is paid to the investor free of tax as the tax authorities regard the ASF deposits as capital deposits and as such are made after payment of tax. This means the ASF is taxed *before* it is paid.

For example:

Investor's income	£1,000 per annum (net before tax)
Tax at 50%	500
Income net of tax	500 per annum
ASF to redeem capital = (say)	100 per annum

Thus: Net of tax income to provide ASF payment = £100 per annum

Gross of tax income to provide for ASF = £100 × 2 = £200 per annum

Thus, ASF (net) to obtain gross of tax, ASF payments must be multiplied by

$$\frac{£1}{£1 - RT}$$

Thus, YP (dual rate), with allowance for tax on the ASF, is

$$YP = \frac{1}{i + ASF\frac{1}{100 - T}}$$

where T = % tax

Net present value (NPV) method

The NPV method is a discounting method in which a predetermined rate of interest is used to discount the incremental net cash flows generated by an investment throughout its entire life to (usually) the present-day value. The evaluation includes inflows and outflows, the capital expended on the investment and any inflow from selling the investment (scrap value for plant, etc.) or outflow occasioned by disposal of the investment at the end of its life.

For a single project analysis, the investment should be undertaken if the NPV is positive. Where a spectrum of investment projects are assessed, those with a positive NPV should be undertaken or, if the capital available for investment is limited (i.e. capital rationing applies), those projects with the highest positive NPV should be undertaken in descending order until the capital available is fully expended.

Thus, an NPV calculation assumes the form:

NPV = (NPV of incremental net cash inflows)
± (NPV of terminal cash inflow/outflow)
− (Initial incremental capital investment)

The NPV method is quite flexible and can include the effects of taxation (with the tax cash outflows occurring the year after the liability is generated), multiple rates of discounting and reinvestment possibilities as well as incorporating possible different rates of inflation applying to different projects or to different elements of a single project. NPV requires both the cash flows and the discount rate(s) to be predicted. The decision rule of the NPV method pays no attention to the life of the investment; thus, for comparison of investments of different lives, it makes no acknowledgement of a shorter-life investment releasing capital earlier for new investment projects and so, careful attention to the constituents of the analysis is advisable.

Internal rate of return (IRR) method

The internal rate of return is that rate of interest which, when applied to discounting the net incremental cash flows of an investment, produces a net present value equal to the capital sum expended on the investment. Thus, the NPV so determined minus the original investment equals zero. In an IRR calculation, the cash flows predicted (these are usually net of tax although for greater accuracy tax payments may be incorporated as they are envisaged to occur and discounted separately - this separate tax treatment is not worthwhile unless large sums are involved and predictions may be carried out with precision) and the rate of interest are the determinants of the project's viability.

Provided the IRR is at least as great as the rate of interest the firm must pay on the capital to be invested in the project (usually the marginal cost of capital), the project is a reasonable investment. Where a spectrum of possible projects is under examination, that (or those) with the largest positive IRR, hence the largest positive differential between the IRR and interest to be paid on the capital, should be selected.

The actual calculation commonly employs trial and error as an interpolation to find the IRR. Initially, a reasonable rate of interest is selected, discounting calculations are made, and the original capital sum is then deducted. If the result is a positive sum, the rate of interest selected was too low, and vice versa. Discounting tables have rates of interest in stages, often of one or one-half per cent; thus to determine the exact IRR, interpolation between a close high result and a close low result is employed, normally assuming a straight line, proportional relationship to exist.

Although IRR has the advantage that it is understood by business people easily, it has the disadvantages that it produces only the average return that the project should earn over its envisaged life and, depending on the cash flow profile, may produce two different IRR values!

Annual equivalent (AE) method

The annual equivalent method is derived from the NPV method. Its value lies in its usefulness for comparison of investments with different life periods and the readiness with which it is comprehended by management.

This method considers the sum of equivalent annual expenditures and receipts appertaining to a project over its life: the initial, annual, periodic and life-end cash flows. To obtain annual equivalents, the capital sum is multiplied by the rate of interest expressed as a decimal; annual sums require no amendment as they are already annual; and periodic and life-end sums are discounted to NPVs and then multiplied by the rate of interest expressed as a decimal. If a sinking fund is required for capital maintenance (as with the purchase of a lease), the sinking fund factor (calculated or from tables) is multiplied by the relevant capital sum.

Annual equivalents are used to evaluate proposals such as whether to purchase or lease a building in which the capital sums, cash flows and lengths of the investment periods will be vastly different.

8.8.3 *Return on capital employed*

The capital employed is the capital used in the business but excludes current liabilities except bank loans and overdrafts. The capital employed in relation to a particular project should be the capital required for that project, the incremental capital.

Provided that the return generated by the project exceeds the required return on the capital employed, the project is viable. It should be noted that discounting is commonly not used in this analysis.

8.8.4 *Break-even analysis*

Break-even analysis is a sensitivity analysis technique to aid decision-making where a spectrum of investments is being considered. The investment appraisal methods described above may be supplemented by break-even analysis.

Graphical presentation is useful as management can easily appreciate the information shown. The two analyses most commonly employed consider the rate(s) of interest at which two (or more) projects have zero NPV (i.e. the IRR of the projects). The alternative, which is of particular value in the consideration of running costs of buildings, examines the NPV of each investment at the end of each selected period (usually each year or five years) of its life, the rate of interest being prescribed. This alternative is of great use in the evaluation, both total and as comparisons, of various heating systems for buildings (these systems constituting an increasingly large proportion of a building's running costs due to fuel cost increases). It is common for a system with a low capital outlay to have a high running cost and vice versa. Break-even analysis will indicate which system is cheapest for a given time period as it is likely that the system with low installation but high running costs will become the most expensive after some time in use. It is important that suitably short intervals for the analysis are selected to obtain an appropriate picture of the project – notably the likely step-pattern of the graph: intervals which are large may give too much smoothing which, by not revealing sufficient detail, may hamper analysis and decisions.

8.8.5 *Sensitivity analysis*

Sensitivity analysis is a final stage in investment appraisal. It examines the effects of changes in the variables. As it is usually quite easy to determine which variables are of greatest influence over the outcome of the appraisal, only those major variables need be considered. Each of these variables is altered incrementally in turn and the effects on the outcomes of the appraisals are noted. Only if two investments present quite close outcomes in the original analysis is there likely to be a significant change in the results.

If the results of the original appraisals are well distinguished (in value, etc.) and if each variable is only of limited significance to the outcome of the analysis, then sensitivity analysis may not be worthwhile. If the original results are close and/or if a few variables exert major influences over the results, sensitivity analysis should be employed.

Although sensitivity analysis is usually applied to investment decisions, it could be of considerable use in tendering where, for instance, the effect on the tender sum of a change in the bonus payments to direct labour could be evaluated (via the effects on hourly rates of pay, and the labour content of the estimate).

8.8.6 Capital rationing

Normally, organizations do not possess and cannot acquire enough capital to enable them to take up every investment opportunity available; other resources will limit the scope of such activities too. Hence investment decisions are made under situations of capital rationing and so, supplements to the decision rules applicable to the individual investment appraisal methods must be employed. The situation may be complicated further by the pattern of availability of capital over time - that will necessitate examination of the expected cash flow patterns of potential investments both individually and in aggregate.

For the array of viable investments (those which have passed initial 'filters') where not enough capital is available for all to be undertaken, the basic selection is likely to follow the hierarchy of projects in descending order of benefit, e.g. select the project with the greatest NPV first. However, it may be more appropriate to take up projects which accord with the performance criteria of the organization, notably yield on investment: hence investments offering the highest returns on capital employed should be selected until the available capital is exhausted.

While financial considerations are essential in investment appraisal, it is also important to recognize the involvement of risk and non-financial operational requirements. Hence factors including utilization of capacity, possible postponability and effect on the investment portfolio should be considered to provide a fairly comprehensive, utility-based evaluation. Those investment opportunities offering the greatest expected utility, subject to the capital rationing would be adopted. (Expected utility is the aggregate of utility factors multiplied by the probabilities of their realizations.)

8.8.7 Developer's budget

At the initial stages of a development proposal a developer will often be faced with the problem of how much to pay for a site. The relevant sum is usually determined by a residual method which utilizes the planning parameters, forecast construction cost, fees, finance costs, and the estimated value of the completed development. This technique is of merit as it implicitly recognizes the effects on the value of land due to developments carried out or to be carried out (the development potential, or hope value) upon it. The following simple example illustrates the technique.

A vacant plot, frontage 40 metres and depth 30 metres has planning permission for an office building with the parameters: may cover up to two-thirds of the site, with the building plot ratio = 3. A valuer considers that the proposed building should produce a gross income of £110 per m^2 usable floor area. Landlord's outgoings will be £20,000 per annum. Construction costs for the building will be £400.00 per m^2 and site works will cost £90,000. Circulation space is 15 per cent of gross floor area. Construction period is two years.

Floor area of building:
Assume road to front of site of overall width 10 m
Site area = 40 × 30 = 1,200 m^2
Area for planning purposes includes half of the road width:
40 × 35 = 1,400

Planning area	1,400
Plot ratio	× 3
	4,200

Usable floor area to generate income: 4,200 × 85% = 3,570 m^2

	(£)	(£)
Value:		
Annual rental income (3,570 × £110)		392,700
Less: Landlord's expenses		20,000
		372,700
Years' purchase in perpetuity for offices, say 7%		× 14.3
Gross development value (GDV)		5,329,610
Deduct costs:		
Cost of building 4,200 × £400	1,680,000	
Cost of site works	90,000	
	1,770,000	
Professional fees at 10%	177,000	
	1,947,000	
Finance for construction, compounded at 15% p.a. (1,947,000/2 at 15% p.a.)	313,954	
Legal, agents, etc., fees at $2\frac{1}{2}$%	133,240	
Developer's profit at 20% × GDV	799,442	3,193,636
Value of site plus site finance		2,135,974
Cost of site finance, compounded at 15% p.a. for two years = 2,135,974 × (0.3225/1.3225)		520,871
Site value		£1,615,103

The developer should pay to £1,615,103 (usually this would be rounded to £1,615,000) for the site.

Costs:	(£)
Construction, etc.	1,947,000
Finance for construction, etc.	313,954
Legal, etc.	133,240
Site	1,615,103
Finance for site	520,871
GDV minus developer's profit	£4,530,168

A net return of £372,700 p.a. on £4,530,168 = 8.2%

This return should be compared with the prevailing returns on similar investments and on alternative investments, as well as being examined in the context of the developer's requirements, to determine the viability of the project.

The developer's profit included in the above calculation is the profit the developer requires for undertaking the project as the developer. The building contractor's profit on the construction work will constitute part of the construction and site works costs to the developer.

Summary

Survival is probably the primary objective of any business enterprise. Particularly during periods of economic recession, construction firms are exceedingly conscious of the problems of survival and seek to predict, monitor, and control costs and revenues with a diligence far surpassing that employed during more buoyant times.

Predicted costs are the normal basis for price calculation and may be classified in several ways but, whatever classification is used, the cost elements remain, each being worthy of separate attention. Patterns of cost may be determined and employed for predictions. Costs should be monitored by a suitable system, tuned to the requirements of the firm or the individual project.

Profit is often seen as being the primary objective of economic activity. This view is open to question, especially with the divorcing of ownership and management which suggests that growth is regarded as a major corporate objective. 'Satisficing' may be used to evaluate the profit goal in the context of all the objectives of a firm. Indeed, profit may be regarded as an unjustified appropriation of the returns due to the factor of production labour. Some firms may dispense with profit altogether in the short period to facilitate their survival for a profitable future, but such a policy is fraught with problems.

Revenues are calculated on the basis of the predicted total cost of the work plus a profit mark-up. It is possible to predict both revenues and costs for a project at an early stage and, subsequently, to monitor the performance but it is important that inherent sources and sizes of variability be recognized. Naturally, predictions should be updated to take the latest available information into account.

Both costs and revenues must be monitored and controlled. This may be achieved by analysis of variances so that the required corrective action is instigated. To foster effective control, monitoring must be accurate and timely. Pricing manipulations may distort expected cost and revenue patterns and inflation may erode profits. Credit control is of great importance. The period of credit offered to customers should be no greater than that enjoyed by the firm.

Clients also require cost information regarding construction projects. Total and periodic costs are both important, the former indicating their total commitment and the latter, the cash flow requirements of the project.

Consultants usually obtain commissions through competitive bidding and operate under conditions of engagement prescribed by the appropriate professional institution. Thus, while the revenue is predetermined, cost control is essential to ensure adequate profit.

Investment appraisal is used to determine whether a proposed project is worthwhile or to select the most suitable project from several options. Many techniques are available, the more suitable of which for long-life investments include the time-value of money in the evaluation. In situations where several projects produce close results, sensitivity analyses should be employed to determine the responses to changes in the basic data, thereby providing more information upon which a decision may be based. Most investment decisions are subject to capital rationing and so, selection on the basis of yield, advisedly using utility analysis also, is advocated.

Questions

1. Discuss the various costs incurred by a building contractor in the context of the firm's ability to survive during a recession.
2. Why is subcontracting of major importance in the building industry?
3. What systems of costing are appropriate for firms engaged in the building industry? Outline the main features of the systems.
4. 'If a firm acts rationally, it will endeavour to maximize its profits.' Discuss.
5. Why is it important for a client to receive a budget for a building project at an early stage of the design process? Discuss an appropriate budgeting technique and any necessary modifications as the project progresses to completion.
6. Why is it usually necessary for a contractor to monitor the profitability

of each project as it progresses? How may such monitoring be *effectively* achieved?

7. Discuss the importance of credit control in the building industry.
8. What factors are considered to determine the most appropriate investment appraisal technique to use? Why is the use of discounted cash flow techniques fraught with problems?

References and bibliography

Ashworth, A. and Skitmore, R.M. (1983) *Accuracy in Estimating*, Occasional Paper No 27, Chartered Institute of Building.

Bierman, H. Jr (1963) *Financial and Managerial Accounting – An introduction*, Macmillan.

Bierman, H. Jr and Swieringa, R.J. (1987) *Financial Accounting: an introduction*, Dryden Press.

Bromilow, F.J. (1970) 'The nature and extent of variations to building contracts', *The Building Economist*, 9 (pp. 93–104).

Bromilow, F.J. (1971) 'Building contract cost performance', *The Building Economist*, 9 (pp. 126–38).

Bromilow, F.J. (1974) 'Measurement and scheduling of construction time and cost performance in the building industry', *Chartered Builder*, 10 (pp. 57–65).

Cooke, B. and Jepson, W.B. (1979) *Cost and Financial Control for Construction Firms*, Macmillan.

Drebin, A.R. and Bierman, H. Jr (1978) *Managerial Accounting: an introduction*, Saunders.

Economist Intelligence Unit (1978) *Public Ownership in the Construction Industries*, Economist Intelligence Unit.

Fellows, R.F. (1984) 'A Study of Cost Escalation in the Building Industry', *Proceedings CIB W-65 Symposium*, Waterloo, Canada, July (pp. 927–928).

Fellows, R.F. and Fenn, P. (2001) *JCT Standard Form of Building Contract 1998 Edition*, Palgrave.

Flanagan, R. and Norman, G. (1983) *Life Cycle Costing for Construction*, RICS.

Freedman, R. (1961) *Marx on Economics*, Penguin.

Harding, J. (1980) 'Tendering in the Construction Industry', Final Year Project, Department of Building Technology, Brunel University.

Hillebrandt, P.M. (2000) *Economic Theory and the Construction Industry*, Third Edition, Macmillan.

Hillebrandt, P.M. and Cannon, J. (1990) *The Modern Construction Firm*, Palgrave.

Hillebrandt, P.M. Cannon, J. and Lansley, P. (1995) *The Construction Company in and out of Recession*, Palgrave.

Hudson, K.W. (1978) 'DHSS expenditure forecasting model', *Quantity Surveying Quarterly 5*, Number 2, Spring.

Kelly, J.R. (1982) 'Value Analysis in Early Building Design', from Brandon, P.S. (editor), *Building Cost Techniques: New Directions*, E & F N Spon (pp. 115–125).

Laing, Sir Maurice (1978) 'Cool response to public ownership', *Building*, 24 March.

Lera, S. (1982) 'At the Point of Decision', *Building*, 28 May (pp. 47–48).

Morrison, N. (1984) The accuracy of quantity surveyors' cost estimating, *Construction Management and Economics*, 2, pp. 57–75.

Parry-Lewis, J. (1965) *Building Cycles and Britain's Growth*, Macmillan.

RIBA (1998) *Plan of Work*, from RIBA (2000) *The Architect's Plan of Work*, RIBA Publications.

RICS (1975) *Definition of Prime Cost of Daywork Carried Out under a Building Contract*, Royal Institution of Chartered Surveyors and National Federation of Building Trades Employers.

Seeley, I.H. (1996) *Building Economics* (Fourth Edition), Macmillan.

Simon, H.A. (1960) *Administrative Behaviour* (Second Edition), Macmillan.

Skoyles, E.R. (1974) 'Wastage of building materials on site', *BRE CP 44174*, HMSO.

Skoyles, E.R. (1976) 'Materials wastage – a misuse of resources', *BRE CP 67176*, HMSO.

Skoyles, E.R. (1978) 'Site accounting for waste of materials', *BRE CP 5/78*, HMSO.

Skoyles, E.R. (1981) 'Waste of building materials', *BRE Digest Number 247*, HMSO.

Turin, D.A. (editor) (1975) *Aspects of the Economics of Construction*, Godwin.

Uher, T.E. (1990) 'The Variability of Subcontractors' Bids', *Proceedings, CIB 90: Building Economics and Construction Management* 6, University of Technology, Sydney (pp. 576–586).

9 Financial Performance

Performance is concerned with the achievement of objectives. The objects of a company are contained within the Memorandum of Association (see Chapter 7) and concern such aspects as the activities of the company and its sphere of operation. Other types of organization will have objects of a similar nature. However, any firm will also have unwritten objectives, objectives which must be determined from examination of the firm's performance and discussion with those involved: the owners and the managers.

The unspecified objectives will include profitability, growth, continuity of existence, market share, turnover, size, image and influence. The achievement of these objectives is affected by a vast number of both endogenous and exogenous variables.

To permit a firm to work towards the achievement of its objectives, decisions must be made. The decisions prescribe what the firm will do and the method of execution. Decisions are made in a great variety of ways from pure intuition to quasi-scientific analysis. Decisions use data and information derived from the past to forecast the future. Hence the outcomes are subject to both risks and uncertainties.

The outcomes of the decisions taken by a firm invariably have financial consequences either directly or indirectly. Direct consequences may be monitored as the action progresses but that is not so with indirect consequences. The indirect financial consequences are often separated from the decision(s) by a time lag. Thus these effects may be evaluated only in a global fashion by examination of the firm's performance as detailed in the accounts.

The accounts contain a wealth of information about the firm but their interpretation is not always straightforward. Comparisons with competitors and analysis of trends are as important as the information contained within each set of accounts. The information contained in accounts, and associated documents, provides much useful information for actual and potential investors.

9.1 Value management

Value Management (VM) concerns determining what is required – usually in a hierarchy of functions and classification of functions – and structuring, organizing and controlling the processes involved to provide the functions to the greatest degree possible within a set of (identified) constraints.

The terms 'cost', 'price' and 'value' are often used interchangeably and hence their meanings become confused. Cost is what must be given, or foregone, to obtain a good or service: it is what the purchaser must pay and is usually expressed as a sum of money. Price is what the seller receives in exchange for the provision of the good or service to the buyer. Hence the purchaser's cost is the seller's price: that sum of money is the *exchange value* of the good or service. However, both buyer and seller will have their own, subjective valuation of the good or service which has been the subject of the transaction – these are the *use values* of the buyer and seller. It is the conversion of the *use values* into money equivalents which determines whether a transaction will occur. Provided the maximum money equivalent of the buyer at least equals the minimum money equivalent of the seller, a transaction can occur. The market price at which the transaction occurs is the *exchange value* of the good or service.

Whilst VM in practice is focused on exchange value and cost considerations, it is important to remember how exchange value is determined. Further, in amalgamation with cost considerations, the objective of VM should be maximization of the relationship of use value to cost, subject to the constraints of the client – notably capital rationing – and the necessity to secure appropriate benefits for the other project participants in order to ensure their adequate commitment to the project and, hence, good performance.

The initial tasks of managing value are to determine what functions are necessary, to quantify the functions in a hierarchy of (relative) importance, to identify the constraints and then to determine how to secure maximization of performance measured against the functions required. The problems then concern effecting the action and controlling the processes of 'production', i.e. a transfer from 'value management' to 'production management'.

Value management, sometimes called 'value engineering', has been defined by Venkataramanan (1991) as:

> ... an organized, creative technique directed at analysing the functions of a product, service or a system with the purpose of achieving the required functions at the lowest overall cost consistent with the requirements which comprise its value...
> In actual application, VE (value engineering) consists of a series of step-by-step techniques to identify unnecessary cost and then eliminate it. To do so, it concentrates on functions and their cost ... (p. 4)

Promotion of value seeks to maximize the functional performance which is needed at minimum cost; in so doing, it is acknowledged that, in many practical instances, the performance which is designed differs from the performance which is needed, in terms of amount and/or type (by function). Consequences for value of changes in function (utility) and cost are shown in Table 9.1.

Table 9.1 Effects of changes in function (utility) and/or cost on value

Function (utility)	*Cost*	*Value*
Large increase	Increase/Constant/Decrease/Large decrease	Increase
Increase	Constant/Decrease/Large decrease	
Constant	Decrease/Large decrease	
Decrease	Large decrease	
Large increase	Large increase	Constant
Increase	Increase	
Constant	Constant	
Decrease	Decrease	
Large decrease	Large decrease	
Increase	Large increase	Decrease
Constant	Large increase/Increase	
Decrease	Large increase/Increase/Constant	
Large decrease	Large increase/Increase/Constant/Decrease	

Green and Popper (1990) and Norton (1994) suggest that a value management study has five phases:

1. Information
2. Speculation
3. Evaluation
4. Development
5. Presentation

The phases constitute the 'job plan' and are implemented once a project has been selected for study and the team of value management 'experts' has been appointed. As early decisions have the greatest consequences, to maximize benefits from a VM study, it should be undertaken as early as possible in the project cycle.

The information phase requires information and data about the intended project to be collected, notably the objectives of the main project participants. Hence, the objectives for the project, performance criteria in their various forms (i.e. functional requirements for the project) and parameters – notably finance availability in terms of total and pattern as well as time for completion of the project (and phases of it) are set. During the information phase, the performance objectives (which are variables to be maximized or minimized) should be converted into performance targets (values of the variables set for the project performance to achieve). In obtaining the information, it may be helpful to ask questions, such as:

- What is the project?
- What is it intended to do/achieve?
- What other attributes are required from the project?
- What will the project cost – in total, over time, between functions?
- What limitations apply?
- How may the project be procured?

Information about the intended construction site – topography, access, etc., and timing of project cycle activities will be important. As VM should adopt a life cycle approach (the holistic approach, including design, construction, use and disposal), information gathering should extend to incorporate the total project life cycle.

A particularly important component of the information phase is to determine and to classify the functions of the project. Functions should be classified into:

1. Primary/basic
2. Secondary
3. Unnecessary

Generally, functions should be described by using one verb and one noun (e.g. the primary function of a column is 'carry load'). Once the functions of the project, and its components, have been determined, they should be classified so that the creative phase may begin.

The creative phase involves the generation of alternative design ideas: usually this is done through 'brainstorming'. It is vital that no evaluation is carried out, as such premature evaluation may constrain the free generation of ideas, no matter how 'wild' – novel alternative solutions may prove particularly useful once evaluated; many significant advances have happened 'by chance'. Hence, the creative phase should benefit from the presence of a multidisciplinary team, preferably including people from outside the 'construction professions' so as to avoid being limited by well-known, 'traditional' solutions.

Once the creative phase has been completed – no more ideas are being generated – the evaluation phase may begin. The ideas for design solutions must be analysed against the project life cycle functions and costed; they must be examined against project constraints also. During analysis, ideas may be amended, combined, divided, expanded, etc. to improve on the initial ideas and to generate feasible solutions, through improved functioning within the constraints. Note that more cost is a good investment if the consequent increase in function is even greater – see Table 9.1.

At completion of the analysis of functional solutions, their costing and elimination of any which cannot comply with the constraints (assuming that the constraints cannot be amended to take advantage of any beneficial solutions which fail to meet the original constraints), followed by evaluation of the solutions which are feasible, must occur. The objective of the evaluation is to produce a hierarchical rank of alternative solutions on a life cycle function and cost basis. To achieve such ranking, the decision-makers must apply weightings to the criteria so that consistency is achieved.

The development phase requires the top ranked solution(s) to be 'worked-up' into detailed designs. This should follow sensitivity analysis of the top scoring solutions from the evaluation phase if the scores of several solutions are close. To avoid waste, it is important that not too many solutions are developed - if one solution is far ahead of the next in the hierarchy, only that top solution need be developed fully. As the solution is developed, its life cycle cost should be estimated, noting the variabilities involved. Throughout development, the detailed solution(s) should be checked for compliance with project requirements; during this phase, if several solutions are being developed, some may emerge as demonstrably better: concentration should be on those.

The final phase is presentation to the client and making recommendations to assist the client in deciding on the best solution. Clear presentations and arguments based on facts are essential. The outcome from this phase should be a firm decision to which all participants will be committed because it has been shown to be the most appropriate solution.

The solution must be implemented: final design detailing and construction. It is useful to monitor this process so that the assumptions made in the VM stages can be verified/modified for future studies; also, it may be possible to introduce changes to improve the value achievement of the project as it is implemented.

Once implemented, the project should be reviewed, i.e. be subject to a value analysis to check achievement of functions, costs involved and compliance with project constraints. Compliances and, especially, non-compliances should be noted carefully with reasons - it may be that conditions of implementation differed from those assumed in the study!

Throughout any VM study, unnecessary functions should be isolated and eliminated as well as unnecessary cost; frequently, unnecessary cost will be incurred only because unnecessary functions are present! Refinement and improvement of primary and secondary functions may remove unnecessary cost also, perhaps by reducing or eliminating unnecessary functions. (For example, a better quality, more efficient but more costly solution: an expensive item of plant may generate less unwanted heat in its operation and so require less heat dissipation function and cost, require less maintenance at longer intervals and require fewer repairs over its working life.)

Kelly and Male (1993) note four formal approaches to VM which are adopted widely in North America:

- charette
- 40-hour study
- value engineering audit
- contractor's change proposal

A charette is a meeting chaired by a value manager (value engineer) attended by the designers and the client's representatives who have contributed to the project brief. It is held immediately after briefing has been completed and has the objective of rationalizing the brief by functionally analysing the key elements and spaces for the project. Five major advantages are noted for a charette:

1. It is inexpensive.
2. It ensures the whole design (and client) team is briefed.
3. As it occurs early in the project, it can have very significant impact on cost and value.
4. It can be executed in a short period - usually not more than two working days.
5. It is an integrative exercise - cuts across political, professional and organizational boundaries.

A 40-hour study takes place at sketch design stage. It is carried out by a team of design professionals who otherwise are independent of the project and under the chairpersonship of the value manager. It is the most common VM form of study and follows the five phases, noted above. In the week following the execution of the 40-hour study, the report is sent to the client and the design architect, feedback is obtained and actions determined.

A value engineering audit is carried out by a VM team to ensure that the project will provide good 'value for money' (i.e. a high function per unit of currency expended).

A contractor's change proposal seeks to enhance constructability by encouraging the contractor to submit proposals for design changes which will reduce the cost of the project. (That the contractor receives a share (bonus) of any savings adopted could raise problems!) Each proposal must be investigated to ensure that value for cost, on a project life cycle basis, is not disadvantaged.

9.2 Decisions

Decisions are necessary due to the dynamism of the economy and society. A decision is a judgement: as such, it is concerned with imperfect information. If all the requisite information were available the result, or course of action, would be obvious (in fact, a decision would not be required) but this is rarely, if ever, the case, especially in connection with the construction industry where time periods are usually long and variables are numerous.

A decision is the human element in the determination of a course of action and will, therefore, be governed not only by the information available and techniques used but also by the outlook of the individual.

9.2.1 *Risk and uncertainty*

Decisions are concerned with variables which are normally classified as risks or uncertainties. Risks are unknowns, the probability of the occurrence of which can be assessed by statistical means (risks are usually insurable). Uncertainties are unknowns, the probability of the occurrence of which *cannot* be assessed (uncertainties are uninsurable). It is possible, however, for a decision-maker to assign a subjective probability to an uncertainty. As knowledge increases, in conjunction with the amount and detail of statistical data, areas of uncertainty are progressively converted to areas of risk (the evolution of weather data and forecasting techniques is a good example).

9.2.2 *Optimism and pessimism*

Optimism and pessimism are concerned with describing how a decision-maker evaluates the range of possible outcomes from the decision. Schofield (1975) considers that optimists and pessimists put very similar values upon positive outcomes, the difference in their evaluations lying in the values placed upon negative outcomes.

Thus an optimist may regard a possible loss of, say, £1,000 resulting from an investment decision to be of little consequence, whereas a pessimist would regard such a loss possibility as a very serious problem, possibly of sufficient magnitude to exclude that investment despite other possible outcomes yielding large profits.

A particular difficulty concerns the inconsistency of human behaviour. As well as individuals varying in their perspectives along the spectrum of optimism-pessimism, their perspectives also vary over time, possibly in a random fashion. Hence, it is important to identify the decision-maker and that person's decision profile for each decision situation and occasion (see Figures 9.1 and 9.2).

9.2.3 *Competition*

The type of competitive environment in which the firm operates will have considerable influence in the decision-making processes. Within the

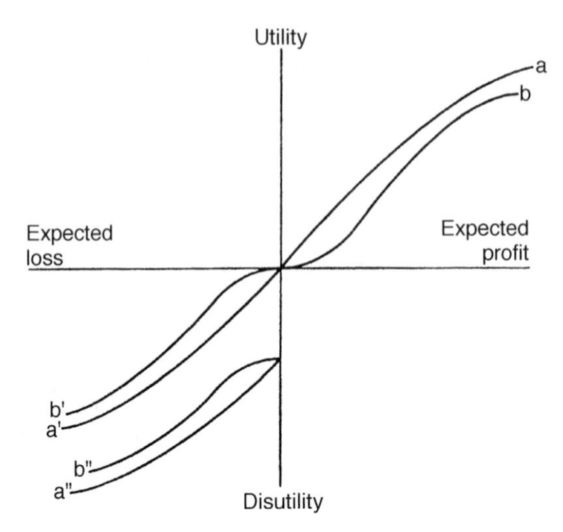

Fig. 9.1 Alternative utility profiles for optimists and pessimists. Curve a shows diminishing marginal utility (of increasing expected profit); curve b shows increasing, then constant, then diminishing marginal utility (of expected profit); curves a′ and b′ show disutility of expected losses and are 'mirror images' of curves a and b; curves a″ and b″ are as curves a′ and b′ but are displaced downwards on the utility scale to depict a minimum level of disutility of any expected loss.

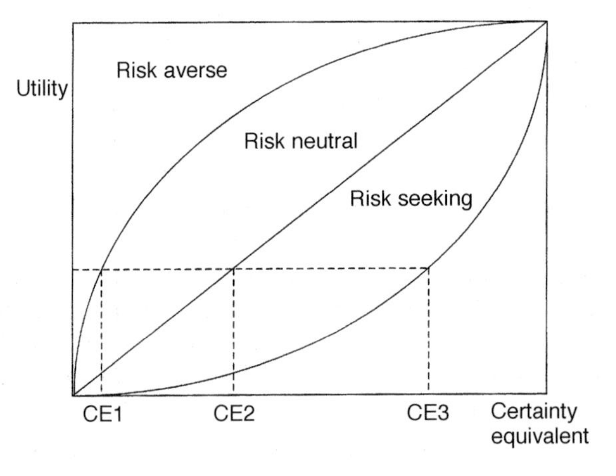

Fig. 9.2 Utility profiles relating to risk-taking. A certainty equivalent (in money) is the receipt of a sum with probability of 100 per cent. Risk attitudes are denoted as: risk averse: utility of expected value < utility of certainty equivalent; risk neutral: utility of expected value = utility of certainty equivalent; risk seeking: utility of expected value > utility of certainty equivalent.

construction industry, almost every type of competition may be found in operation. The range is from monopolistic competition (competition amongst the many, often considered as the 'real world's' equivalent to the theoretical concept of perfect competition) to near monopoly. Thus the spectrum of the various forms of imperfect competition prevails within which oligopoly (competition among the few, sometimes in the form of imperfect oligopoly where there is a dominant firm: a price leader) is of prime importance in major contracting.

On the demand side, there is an equally wide range of competition prevailing from near perfect to monopsony (a single buyer, the demand side equivalent to monopoly). The situation of monopsony is applicable to specific types of construction projects (e.g. motorways, nuclear power stations) as government bodies (or privatized organizations) constitute the only customers for these types of project in the domestic market.

Under monopolistic competition, the firm is a 'price taker': there is a market price for each job and, other things being equal, the firm should obtain a proportion of the work for which it bids governed by the number of firms submitting bids (assuming consistency of accuracy and behaviour in the bidding). In a monopoly situation, the price the monopolist puts on the work is limited by the source of the monopoly power but, perhaps more importantly, also by the existence of legislation to abolish monopolies which act against the public interest. Under monopsony, which exists for much public sector work, various forms of cost limits are imposed by government bodies (e.g. university halls of residence). However, concurrent with privatization and the objective of increasing market freedom, many of the cost limits and similar constraints have been formally abandoned.

For a large proportion of construction projects, a situation of oligopoly exists. This situation may be caused by the method of letting projects (single-stage selective tendering), by locational effects, or by sectoral effects (work type). The actions of firms operating under oligopoly can often be explained in terms of the hypothesis of qualified joint profit maximization. This hypothesis acknowledges the existence of two sets of economic forces operating concurrently: one set inducing the firms to operate in such a way as to maximize the joint profits of all the firms, the other set inducing firms to act so as to maximize the profits of the individual firm. This hypothesis may be usefully employed to explain the actions of firms, often clearly evidenced by the way in which work is allocated in a localized market wherein the local firms tacitly agree the type of project to be undertaken by each. Firms will not 'poach' each other's work and joint action is usually pursued to exclude a firm from elsewhere gaining a foothold in that market. (The reader is referred to texts on economic theory, such as those listed at the end of the chapter, for a detailed discussion of this hypothesis.)

Oligopoly is by no means limited to localized building markets. Many features of this form of competition are exhibited by national building and civil engineering contractors.

An alternative basis of competition is that of *contestable markets* (see, e.g. Button 1985) in which sunk costs of entering a market (the costs incurred to enter a market in excess of what can be recovered on exit) are what determines the level of competition in that market by constraining the number of potential new entrants - numbers and sizes of firms, etc., are held to be irrelevant.

Thus the competition of the market place provides the context in which there are spatial (as outlined above), sectoral, and time effects also - most especially, when the pattern of demand is subject to major and rapid change.

Nor can the competitive elements be excluded by obtaining contracts through negotiation rather than by competitive tender. As Hillebrandt (1985) discusses, if the contractor's price escalates too much during negotiations, the client always has the option of shelving the project or engaging another contractor; however, the more advanced the negotiations, the greater the commitment of the client and so the greater the potential for price increasing by the contractor.

Traditionally, competition is viewed as the major limiting force upon the pricing level. However, competition may also act to increase prices. The costs of submitting tenders has already been discussed in outline (see Chapter 8) and what might be considered as the ultimate situation of competition, open tendering, has received a good deal of condemnation due, *inter alia*, to its cost implications. It is generally acknowledged that the right price is not necessarily the lowest tender: the expansion of negotiated systems of letting contracts is industrial evidence of this, where the contractors obtain work on the basis of expertise, quality, performance, etc. Elements of competition other than price are becoming increasingly significant - notably attention to quality facets (through quality assurance requirements for firms to qualify to tender) and recognition of the importance of management in securing good performance (the requirement to submit CVs of the key managers to be employed on the project along with the tender and presentations by those persons of how they would run the project as a final selection element). Particularly for commercial developments, time is a vital performance criterion due to its impact on the profitability of the development - a shorter project duration gives earlier returns (sale or rental income) plus lower finance charges on the funds used for the development (which, in most cases, far exceed higher construction costs due to 'acceleration' of construction).

9.2.4 Probability and distributions

By its very nature, a decision will have at least two possible outcomes. More usually, there will be a range of possible outcomes between quite well-defined extremes. A decision-maker will, of course, be concerned with the determination of the possible outcomes but will also wish to qualify the possibilities by an assessment of how likely (or unlikely) the outcomes are. Where the possible outcomes are in the form of a continuum between extreme cases, the extremes, and several intermediate outcomes, will be examined.

Provided the extreme outcomes have been assessed correctly, it is certain that the actual outcome will lie within that range. However, it is not likely that the probability of the occurrence of each possible outcome is equal everywhere within the range, some outcomes will be more probable than others. The probabilities may be represented as a graph of probability against possible outcome, a probability density function. The normal distribution curve is the commonest example of this type of function. (However, common forecasting methods, e.g. PERT, use the beta distribution due to its flexibility.)

The shape of the probability density functions relating to the possible outcomes of a decision are of importance, particularly when considering the views of an optimist and a pessimist. In this context, the tails of the distributions will be of significance as well as the clustering of the distribution about the modal value. The greater the spread of the tails, the greater is the doubt about the outcome (it will occur within a wider range). Further, the distribution may not be symmetrical, it may be skewed towards a high or a low value or might be bimodal.

9.2.5 Probable profit contribution

Probable profit contribution (PPC) is a concept similar to that of mathematical expectation. PPC is obtained by multiplying the anticipated profit by the probability of realizing that profit. Thus, in a situation where, for example, a contractor submits bids for work, it could be used to indicate the optimum level of mark-up, or bid.

This technique is reliant upon bids for a large number of contracts being submitted; this would apply to each individual submarket as well as the overall situation for the firm. Previously submitted bids and successes, together with information about competitors' bids, are the data upon which this technique is based.

A simplified graph of the situation appertaining to a contract may be plotted, as in Figure 9.3. The distribution shown in Figure 9.3 may then be used to obtain the graph of PPC, as in Figure 9.4.

While a firm operates and submits bids in a fairly constant environment, each individual project for which a bid is submitted may be analysed, as shown in Table 9.2 (see also Figure 9.5 which shows that the cost increases with the increased number of competitors as fewer projects for which bids are submitted are obtained, thereby increasing overheads on successful bids). It should be noted that if, in the long period, the firm cannot achieve an adequate return on capital at the optimum bid level (for many projects), it will go out of business eventually.

Table 9.2 may also be represented graphically, as shown in Figure 9.6. If it is assumed that Figure 9.6 represents the situation where plant

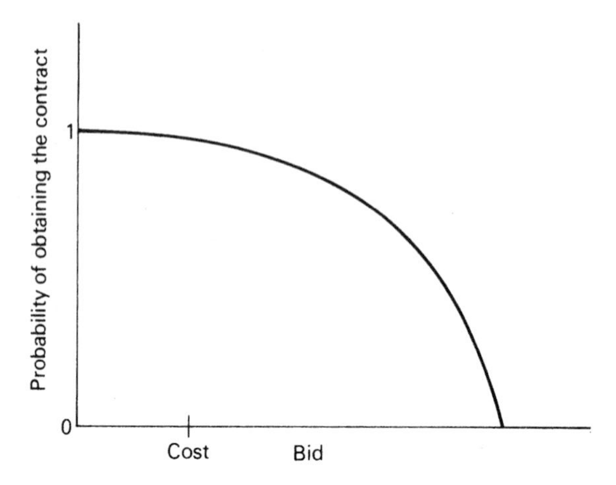

Fig. 9.3 Assessment of a bid's chance of success.

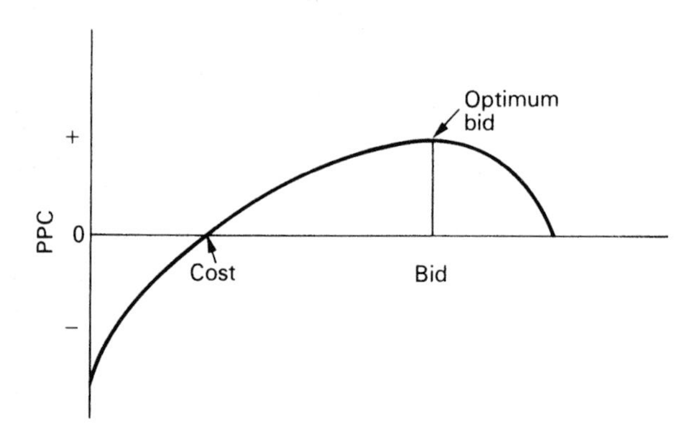

Fig. 9.4 Determination of the optimum bid.

Table 9.2 Determination of the optimum bid. The optimum bid is 12; the bid which yields the greater PPC. (Assume capital used is 50% of cost level.)

% Return on capital	*Cost*	*Bid*	*Profit*	*Probability of success*	*PPC*
−20	10	9	−1	1.0	−1
0	10	10	0	0.9	0
20	10	11	1	0.7	0.7
40	10	12	2	0.5	1.0
60	10	13	3	0.2	0.6
80	10	14	4	0.05	0.2
100	10	15	5	0.01	0.05

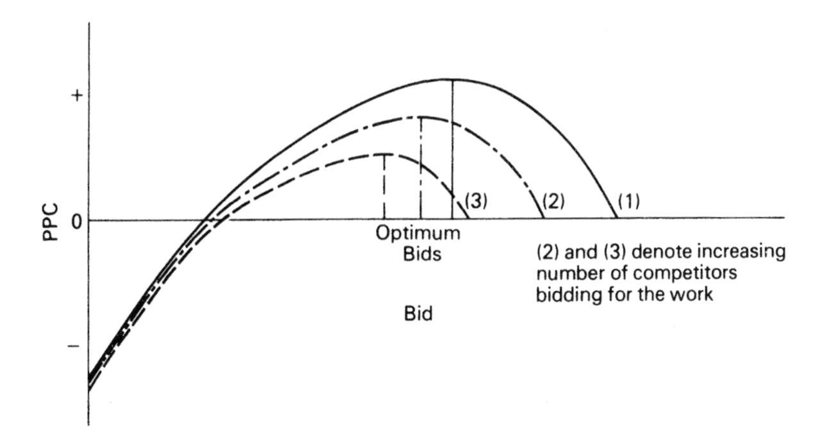

Fig. 9.5 The effects of increasing the number of competitors.

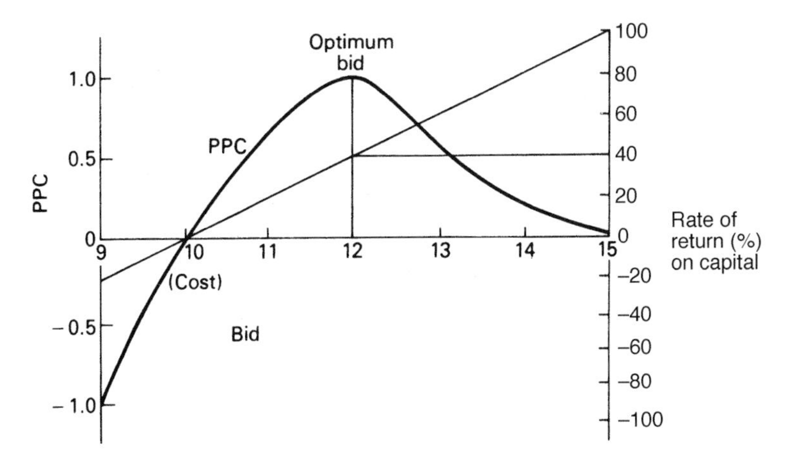

Fig. 9.6 Determination of the optimum bid and the return on capital it would yield (plant owned by the contractor).

is owned by the contractor, it is possible to contrast the situation if the plant required for the project execution were hired, as illustrated in Figure 9.7.

The hiring of plant, instead of owning it, will have several effects, often dependent upon the degree of utilization of plant if owned. However, if the assumption is made that the utilization of plant is of a similar level, whether it is owned by the contractor or by a plant hire company (this is reasonable, if viewed on the basis of a contractor having a plant hire subsidiary and the contractor's overall position is being examined; however, many firms do not have a plant hire subsidiary and so are unable to fully utilize any owned plant) the main effects of hiring plant (instead of owning it) will be:

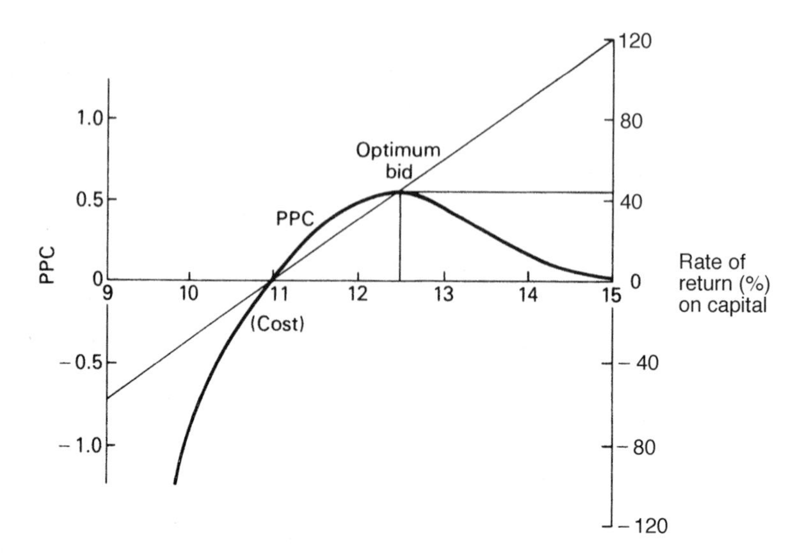

Fig. 9.7 Determination of the optimum bid and the return on capital it would yield (plant hired by the contractor).

1. The cost of the project will increase (the plant hire firm must earn a profit in its own right).
2. The PPC curve will tend to flatten in shape and the cost and optimum bid will move to the right.
3. The rate of return on capital at the optimum bid will be raised due to the decreased capital used by the contractor for the project execution.

In any situation requiring a decision where subjective probabilities are an integral part (e.g. bidding), it is useful to consider a three-point probability analysis. This analysis considers an optimistic, a pessimistic, and a most-likely probability for the particular outcome and determines the mean probability by application of formula (based on the ß-distribution, but note the analogy with the prismoidal rule in land surveying):

$$\text{Mean} = \frac{0 + 4M + P}{6}$$

where 0 is the optimistic probability, M the most likely probability, and P the pessimistic probability.

Investment decision example

An investment is available which involves a capital expenditure of £10,000. The investment has a two-year life such that at the end of year one it may produce an income of £10,000, £5,000, or £0 and at the end of year two it may produce a further income of £30,000, £7,000, or −£4,000 (a loss). The prevailing rate of interest is 10 per cent. Figure 9.8 (see also Table 9.3) shows the possibilities of the investment's cash flows over its life together with the probabilities of their occurrence. The expected NPV is of considerable significance as it expresses the probable outcome over the spectrum of possibilities.

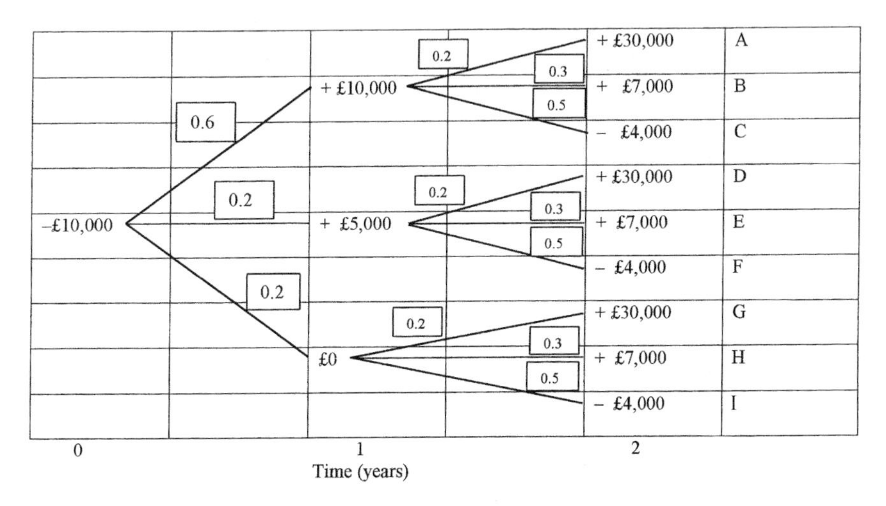

Fig. 9.8 Possible money outcomes.

Table 9.3 Calculation of monetary expectation

Path	*PV of cash flows on path (£)*	*Probability of path*	*Monetary expectation (ME) (£)*
A	+23,884	0.12	+2,886
B	+4,876	0.18	+878
C	−4,215	0.30	−1,265
D	+20,248	0.04	+810
E	+331	0.06	+20
F	−8,760	0.10	−876
G	+14,793	0.04	+592
H	−4,215	0.06	−253
I	−13,316	0.10	−1,331
Expected NPV			+£1,461

Having executed a monetary analysis of the problem, it is possible to proceed to evaluate the utility of each outcome. The assessment of utility is in itself difficult as utility is a subjective concept and can be measured only on a relative scale. By careful analysis and questioning of the decision-maker(s), it is possible to construct a scale of utility, the units of which are termed 'utiles'. A utility scale for use in this example is shown for an optimist and another for a pessimist; a graphical representation of the utilities is illustrated in Figure 9.9 (see also Table 9.4). This figure represents only one of several possible analyses of optimism and pessimism. An alternative view is identical for the optimist but the pessimist's curve is displaced downwards (kinked at the point where it becomes negative) throughout the negative portion of its length. It is likely that both positive and negative values will adhere (eventually) to the law of diminishing returns – decreasing marginal utility/dis-utility – and hence both positive and negative portions of the graph will assume an 'S' shape accordingly.

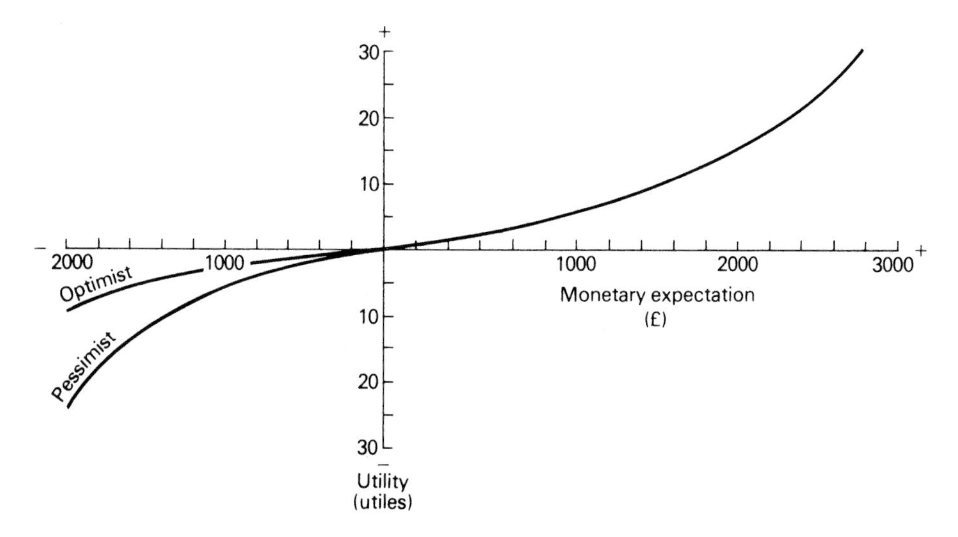

Fig. 9.9 Evaluation of monetary expectations.

The outcomes shown by Table 9.5 indicate that both the optimist and the pessimist would decide to take up the investment as the total utility values of the expected monetary outcomes of both are positive.

This type of analysis may be of value to developers in evaluating possible investments. In such a situation, the time periods would be considerably longer and the probabilities allocated to cash flows would not necessarily be independent. The technique could easily accommodate evaluation of possible construction cost escalation due to both fluctuations and variations as well as the evaluation of income flows from the completed building and/or its selling price.

The analysis is of use to a contractor for evaluation of a project's cash flow and profitability, being an extension of the evaluation techniques discussed in Chapter 8. It could also be used on a macro level for the firm to evaluate policy options, for instance, whether or not to maintain the level of the firm's resources during a period of reduced workload.

Where problems are quite complex, it may be difficult to adapt the technique described above to cope adequately. Two more comprehensive aids to decision-making, risk management and decision analysis, are considered below.

Table 9.4 Utilities of monetary expectations

Monetary expectation	*Utility of expectation*	
(£)	*Optimist*	*Pessimist*
−2,000	−8.75	−23.20
−1,500	−4.55	−11.40
−1,331	−3.65	−9.00
−1,265	−3.05	−8.30
−1,000	−2.20	−5.55
−876	−1.70	−4.50
−500	−0.65	−2.10
−253	−0.20	−0.95
0	0	0
+20	0.03	
+500	1.50	
+592	1.95	
+810	3.20	
+878	3.55	
+1,000	4.40	
+1,441	7.55	
+1,500	8.10	
+2,000	12.80	
+2,500	19.00	
+2,866	26.00	
+3,000	30.00	

Table 9.5 Evaluation of monetary expectations

		Optimist		*Pessimist*	
Monetary expectation	*Probability of path*	*Utility of outcome*	*Expectation of utility*	*Utility of outcome*	*Expectation of utility*
+£2,866	0.12	26.00	3.12	26.00	3.12
+£878	0.18	3.55	0.64	3.55	0.64
−£1,265	0.30	−3.05	−0.92	−8.30	−2.49
+£810	0.04	3.20	0.13	3.20	0.13
+£20	0.06	0.03	0.00	0.03	0.00
−£876	0.10	−1.70	−0.17	−4.50	0.45
+£592	0.04	1.95	0.08	1.95	0.08
+£253	0.06	−0.02	−0.01	−0.95	−0.06
−£1,331	0.10	−3.65	−0.37	−9.00	−0.90
			2.50		0.97

9.2.6 Bidding

Traditionally, bidding has been synonymous with price competition. The norm has been for price to be the final variable evaluated to distinguish between competing contractors and hence the primary variable determining which firm secures the project. Recently bidding has become the usual method for awarding work to consultants. However, it is recognized increasingly that such emphasis on price is somewhat misplaced and so, irrespective of the formal system adopted for bidding/tendering, post-bid evaluations of lower bidders' proposed project management personnel are carried out. Bids tend to be accompanied by both priced BQ and CVs of the staff to manage the project if awarded as required accompaniments to the basic tender. The additional evaluations are in recognition of the need to secure good performance against quality and time requirements (as well as cost) and that such performance is determined by those who are charged with managing the project primarily.

Many bidding analyses focus on price competition. Further, the models employed commonly operate on the basis that price bid comprises total (forecast) cost of the project, including allowances for risks, etc., to which an addition (mark-up) for profit is applied. Investigation of the behaviour of firms in the construction market indicates that, in practice, such a model is not very appropriate and that 'market factors' rather than forecast costs determine prices bid. Thus, from an Economics perspective, individual project bid prices are determined through short-period factors (with the lowest rational bid being forecast marginal cost), whilst long-period factors apply at the corporate level to ensure survival (i.e. cover total costs plus earn at least normal profit; following Baumol (1959) the firm will require a minimum rate of growth of turnover too).

Formal analysis of competitive bidding on construction and other projects really began with the work of Friedman (1956). Friedman proposed that, under the system of 'sealed bid auctions', which applies to construction projects for single-stage open or selective tendering, the following model could be employed to examine the likely outcome by bidders; the model assumes the normal business objective of profit maximization and that the project will be awarded to the lowest bidder.

The probability (p) of winning the project when the competition is between a known number of known competitors (by submitting the lowest, valid bid) is:

(p of 'beating' competitor A) × (p of 'beating' competitor B) × ... × (p of 'beating' competitor N)

However, Gates (1967) suggested that the model should be modified to be:

$$\frac{1}{\left[\frac{(1 - \text{p of 'beating' competitor A})}{(\text{p of 'beating' competitor A})} + \ldots + \frac{(1 - \text{p of 'beating' competitor N})}{(\text{p of 'beating' competitor N})}\right] + 1}$$

In situations where bidding occurs between a number (n) of unknown competitors, the models are:

Friedman:
(p of 'beating' the 'average' competitor)^n

Gates:

$$\frac{1}{n\left[\dfrac{1-\text{p of 'beating' the 'average' competitor)}}{\text{(p of 'beating' the 'average' competitor)}}\right]+1}$$

Willenbrock (1973) adopted a decision tree approach to consider a variety of factors contributing to the utility of bids. The ratio of bid price: cost estimate was considered against utility of winning the project at that price (not just against probable profit or the probability of winning alone). Carr (1982) followed a similar approach to examine probability of winning the project; from analysis of outputs from his mathematical model, he concluded:

> Expected value is not very sensitive to small changes in mark-up because each adjustment in mark-up is counterbalanced ... by a shift in the probability of winning. This allows for adjustment in mark-up to the level of workload, or to receive an adequate return on investment, without much change in expected value.

Although the role of market factors in determining bids is acknowledged, the practice of estimating remains common in bid production. However, given the greatly increased incidence of subcontracting, estimating activity has changed to evaluation and assembly of subcontractors' bids. Thus, as discussed by Uher (1990), power to decide main contractors' bids, and hence which contractor wins the work, has shifted significantly towards major subcontractors (who bid for packages to most of the main contractors who are bidding for the project).

If forecast costs of projects remain influential in determining bid prices (as must be the case in the long-period), analysis of accuracy of estimates is important, especially as differences in contractors' bid prices tend to be very small (as a proportion of the bid price). Fine (1975) and Harris and McCaffer (1989) provide analyses that show how estimate errors and level of competition affect the required mark-up to be applied for a firm to break even over a number of bids.

Often, there is a desire to amend the 'shape' of a bid to secure a different cash flow profile. Such shaping is limited by the possible consequences of bid evaluation by the client. Additionally, there may be a desire to change the price bid just prior to its submission. Most easily, such adjustments are made to items which command significant proportions of the total price – as demonstrated by Gray (1983), preliminaries items are regarded as suitable as a small number of items constitute about 90 per cent of the preliminaries. Bennett (1982) assessed the proportion of contract sum comprising preliminaries: for larger projects it is likely to be about 16 per cent. However, care must be exercised in manipulating prices of preliminaries items for, as they constitute 'site overheads', they are used to determine reimbursements to the contractor for loss/expense due to certain delays!

Clearly, data are crucial in bidding and bid evaluation, both by the client and by the contractor. Feedback on bidding provides names of the firms and prices bid. Thus, given those data and a firm's own forecast cost, Fine (1975) provided a means for a contractor to examine its bidding in terms of level of bids and ability to judge market movements. The objective is assumed to be maximization of profit and requires data of own predicted costs, lowest tender sum, next lowest bid where own bid won and profit earned on the array of projects considered (due to the projects won). In the first stage of analysis, own tender sums, for the total array of projects under consideration, are adjusted up and down by a range of percentage increments; for each adjustment, the total profit is calculated on the projects which would have been won at the amended bid and the maximum profit and how it is obtained are noted. In the second stage, own costs are adjusted by a series of 'across the board' percentage adjustments to yield bid figures and the profit maximizing adjustment noted. If the contractor has actually earned the maximum profit, the firm's bidding is as effective as it can be. If the first analysis yields maximum profit, the contractor's market judgement (of movements) is reasonable but is either optimistic (if a downward adjustment yields maximum profit) or is pessimistic. If the second analysis yields maximum profit, the contractor has poor market judgement.

9.2.7 *Risk management*

The management of risks concerns dealing with events, the likelihoods of the occurrence of which are variable, either of quantifiable (even, known) variability, or probability risks, or of quantifiable variability only through subjective estimation at best: uncertainties (as noted, above).

Often, risks and uncertainties are associated with negative (undesired/undesirable) outcomes exclusively which is a biased view and is termed 'downside risk' in USA. Risk and uncertain events/outcomes are subject to a distribution of possible (variable) outcomes which may be preferable to as well as worse than those used. (The view is similar to the use of mean and standard deviation – norms and variabilities.) Commonly, measures of variability are employed as measures of risk, such as the variance or, preferably, the standard deviation (due to it being on the same scale as the mean and most likely values). Use of the coefficient of variation as a measure of risk is helpful in comparative analyses due to the measure being a percentage, thereby eliminating scale effects.

Safety may be viewed as freedom from risk; from a practical perspective, freedom from unacceptable risk. Blockley (1992), whose work focuses on structural safety, identified a 'hazard' as being a set of evolving preconditions of failure. Understandably, Blockley's work is concerned with the negative aspects of risks and uncertainties but many of the concepts – notably the relationship of hazard and risk – are transferable to other facets of 'risk management' (e.g. financial). Given a 'global' view, risk concerns the consequences and chances of a hazard event, outcome multiplied by its probability, so hazard events occur in the present but risk events (consequences) take place in the future (until it happens!) Thus, neither safety nor risk (nor uncertainty) can be managed directly; they must be managed indirectly through the direct management of hazards.

Hazard concerns proneness to (probability of) failure. Individual hazards influence and allow for the incubation of other hazards and so can be cumulative; individually minor hazards can aggregate into major risks/failures.

A critical factor in risk management is the combining of the objective analysis of the risk events and the risk profile of the people involved – the decision-maker(s). People are believed to be risk averse but to vary in their degree of risk aversion, both between individuals and for the same individual over different situations and times; possible risk profiles are shown in Figures 9.11 and 9.12 (see pp. 257 and 258).

The management of risk involves four main stages:

- identification
- quantification
- allocation
- response

Risk identification concerns determination of what risks (and uncertainties) apply to the project. A systematic approach to identification is essential, analogous to SWOT analysis. Risks which are due to variables outside the control of the project should be separated from those which lie within the boundaries of the project. Those outside must be accepted and dealt with as part of the project environment, while those within the project and subject to the control (influence) of the project management may be employed as performance incentives.

Risk quantification, by definition, can apply to risks only. However, subjective assessments can be applied to uncertainties (see, e.g. Fellows 1996). Clearly, the development of ever more databases facilitates the quantification of probabilities of the occurrence of future events and hence the transformation of uncertainties into risks. (The most notable example is weather: the extensive data records and sophisticated models now in use for forecasting compared with the folklore methods of the past.) Commonly risk is quantified by measuring variability of outcome: variance, standard deviation, coefficient of variation; Monte-Carlo and Delphi techniques may be employed also. Sensitivity analysis, to determine the effect on the project outcome in terms of the important performance criteria, may be used to detect those risk variables which are likely to have the greatest impact on the project, those to which management must devote attention! A sensitivity analysis 'spider' diagram is shown in Figure 9.10.

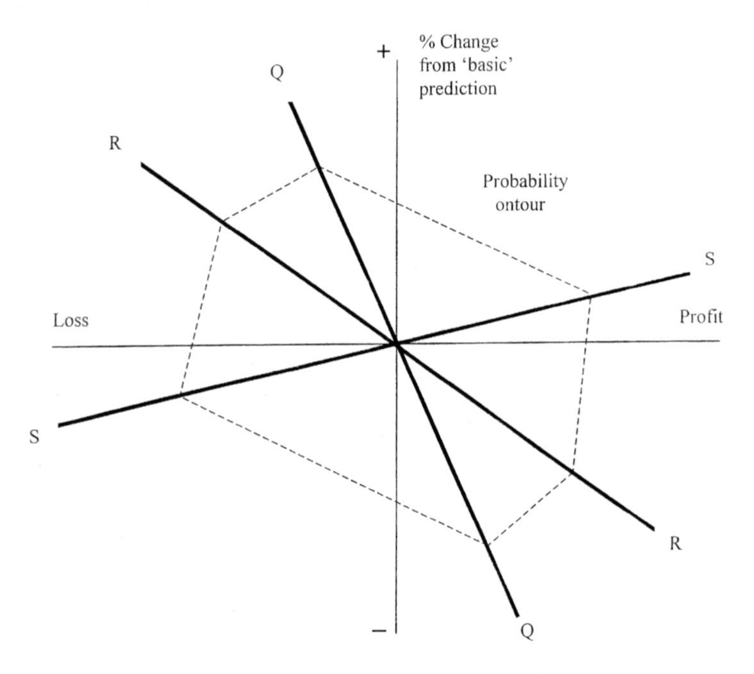

Fig. 9.10 Sensitivity analysis – 'spider' diagram with probability contour.

Risk allocation, is normally done by the contract. In situations involving negotiation, the risk allocation will be determined by the parties' trade-off between their perception of utility from the project against the risks, subject to their negotiating abilities. As, in many cases, the client and consultants determine the contract terms (bespoke contract, standard form, amended standard form): the choice of terms will be dominated by attempts to maximize client utility, mitigated by prospects of contractors' risk aversion and responses to risk allocations. A primary principle of risk allocation is that if a party can control a risk, then that risk should be borne by that party. Perry and Hoare (1992) advance six principles for the allocation of risks:

1. Allocate each risk to the party most likely to be able to control it.
2. Risks which cannot be controlled by constructors should be borne by the client.

3. A risk should not be allocated to a party who is unlikely to be able to sustain the consequences of the risk's occurring.
4. Risks should be allocated to encourage good management by the parties carrying the risks.
5. Parties not carrying a risk should be willing and motivated to assist in the management of the consequences if the risk occurs (*requires commonality of goals*).
6. If the likely impact of the risk's consequences is small, the parties tend to be indifferent over its allocation (*care is needed over accumulations*).

Risk response occurs in five main ways:

- remove
- reduce
- avoid
- transfer
- accept

Removal or *reduction* of risk is likely to involve redesign and/or significant changes to the project, e.g. relocation to a different site with better ground conditions. *Avoidance* may occur similarly or, in the extreme, may mean a constructor declining to undertake the project. *Transfer* of risk may be passing on the work to a subcontractor; insurance means that the risk is retained but the financial consequences are passed to the insurance company. *Acceptance* of a risk usually generates a price increase in compensation, dependent upon the perceived size of the risk and the risk aversion profile of the party carrying the risk.

Risk assumption

All construction projects involve risks and uncertainties of a wide variety of types (sources: people; and causes: events) and sizes (consequences, commonly assessed in financial terms) for all the participants. The risks on any project comprise basic risk (the risk associated with any such project) and specific risk (the additional risk associated with the particular project). Once risks have been identified, quantified and allocated, including any transfers, etc., compensation for the participant's residual risk-bearing can be determined. To that compensation, the following should be added:

- insurance premia for any risks insured
- compensation for subjectively-assessed uncertainties
- a contingency to reflect the level of confidence in the assessments (of uncertainties) and an intuitive provision against any uncertainties remaining; the contingency can be little more than a guess!

Tah, Thorpe and McCaffer (1994) found that contractors make financial provision for risks on projects by making additions to bid prices in one of the following ways:

- a percentage in the profit margin
- a separate percentage addition on all costs
- lump sum in the preliminaries
- a percentage in one bill, if the risk is in that bill alone

9.2.8 Decision analysis

Decision analysis is a methodology for problem-solving in that it plots all the possible decision-paths and outcomes and then enables an objective evaluation of each to be made. It is a relatively new approach, being developed primarily by Robert Schlaifer and Howard Raifa at the Harvard Business School during the 1960s.

The application of decision analysis does not itself solve a problem but, on the basis of the criteria judged to be of importance, facilitates evaluation of all the possible outcomes. Therefore it is of use in introducing a measure of objectivity into an otherwise subjective area of management. It is vital that all possible courses of action are identified and evaluated, including the 'do nothing' option which exists for every decision.

Any aid to decision-making, which presents a ranking of possibilities, especially if evaluated on some form of monetary basis, may be perceived by decision-makers as purporting to make the decision itself. This is not the case. The decision must still be made but in the light of an additional, more objective evaluation of alternatives. The utility perceptions of the decision-maker are incorporated in the identification and weightings of the criteria against which the array of possible outcomes are evaluated.

The application of decision analysis has six distinct stages which are outlined below in the order in which they are normally executed.

1. *Analysis of the problem*. Here the objective is to split the problem down into simple components such that each component may be easily managed by the decision-maker. In most cases, the problem will be complex and so a decision tree will be constructed, a type of flow chart, showing all the stages involved in the solution of the problem and all possible outcomes. It is important that each decision stage is coherent i.e. that the stages along each path through the decision tree fit together in a meaningful way with all other associated decision stages. It is useful to ensure that the decision tree is kept simple: this is encouraged by analysis of the decision paths in such a way that the nodes (sub-decision points or points of 'uncertainty') have only a small number, preferably two or three, of paths emerging from them.

2. *Description of the outcomes.* A complete description of each possible outcome is given at the right-hand end of each path through the decision tree. Once the decision-maker has identified the (performance) criteria against which the consequences (outcomes) of the decision should be assessed, it is essential that the outcomes be described fully in terms of those criteria.
3. *Assess the value of the outcomes.* It is essential that the objectives are clear. Assessment of the outcomes is based upon utility: the utility of the decision-maker at the time and in the circumstances of that particular decision. The relative desirability (utility) of each possible outcome is assessed. In the case of more complex problems where several criteria are to be considered relative to each outcome this will usually involve the use of multi-attribute utility analysis. The criteria are listed and their relative importance is evaluated, then each outcome is assessed against each criterion by using whatever data and knowledge are available to determine the most likely relative performance of each outcome against each criterion individually. Each outcome is then evaluated by summing its utility score against each criterion weighted by the relative importance of each criterion.
4. *Assessment of probabilities.* The alternative outcomes of each decision stage (the alternative paths emerging rightwards from each event node) are allocated the probability of their occurrence. These may be rather subjective assessments but past data should be used whenever possible to lend objectivity. A mix of data and intuition is usual for most probability assessments.
5. *'Fold back' the decision tree.* This may be done physically but most commonly is the process of working from right to left (i.e. commencing with the final outcomes) to determine the utility at each node, finishing the process at the start node. The expected (weighted average) utility at each event node is calculated as the folding back proceeds. Where several activity paths enter an event node from the right, the path with the highest utility is selected and the others are eliminated. The 'best' path through the decision tree is thereby determined.
6. *Sensitivity analyses (see also Chapter 8)* Sensitivity analyses are employed to identify the crucial elements of the decision. Special care is required where two or more paths are of nearly equal utilities. The analysis seeks to determine the effects of incremental changes (to the weightings of criteria, the scorings of outcomes against them or probabilities of alternatives from an event node).

Example

A simple example illustrates not only how decision analysis may be formally applied to problem-solving but also how there is a certain logic in decisions apparently made by intuition. Even if the analysis is not applied formally, the technique is of value in focusing thought upon the items of major importance in the decision.

A contractor receives a set of tender documents for a familiar type of project. The firm has a reasonable current workload and the potential client and consultants are reputable organizations. There are no peculiarities or pressures relating to the tender. The contract is to be JCT 98, private, with quantities, and the code of procedure for single-stage selective tendering applies.

The contractor has to decide what to do: first, whether to do anything, to return the documents or to submit a tender and, if to submit a tender, how to do so. It is assumed that a tender may be submitted in one of three ways:

1. by obtaining a 'cover price'
2. by preparing a tender based on an accurate estimate
3. by preparing a tender based on an approximate estimate

The approximate method is quicker and uses fewer resources. Figure 9.11 shows the decision tree derived from the above information.

The 'do nothing' option is considered inapplicable due to the role of normal commercial practice and the potential harm such a course of action might cause to the reputation of the contractor with the client, consultants and their business contacts.

Having determined the possible courses of action open to the firm, the next stage is to evaluate them. In order to carry out the evaluation, the criteria applicable to the outcome must be established and themselves evaluated. This is done by the technique of multi-attribute utility analysis, as illustrated in Figure 9.12

There is sufficient time for an accurate estimate to be prepared (see path leading to outcome E) should the approximate method produce an unacceptable result, but obviously such a situation is costly to the firm. Alternatively (path to outcome A), returning the tender documents may be perceived by the firm as unsatisfactory because it might mean exclusion from future tender lists (strictly, this should not be the case according to the code of procedure).

Thus, every argument, relating to each outcome will be considered and evaluated objectively (as far as possible) against each criterion. A scale of utility of 10 to 110 has been used for each criterion to avoid conceptual problems of outcomes with zero utility against certain criteria which could occur if a scale of, say, 0 to 100 were used. It must be remembered that, when evaluating the outcomes against each of the criteria, each criterion must be considered individually and the outcomes are assessed relative to each other against each criterion. (The numbers on the scale are for the purposes of subsequent calculation but are of value in ensuring

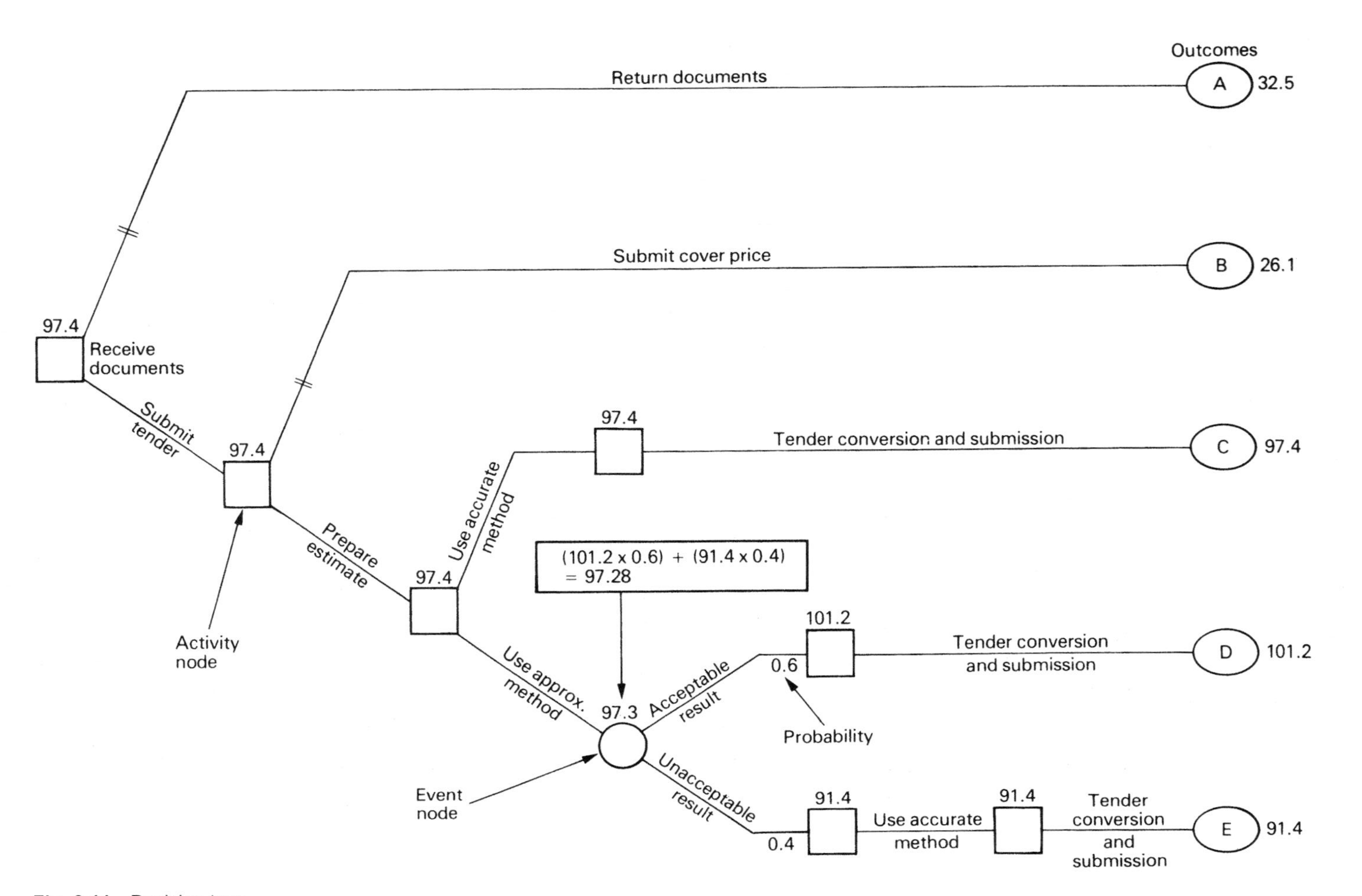

Fig. 9.11 Decision tree.

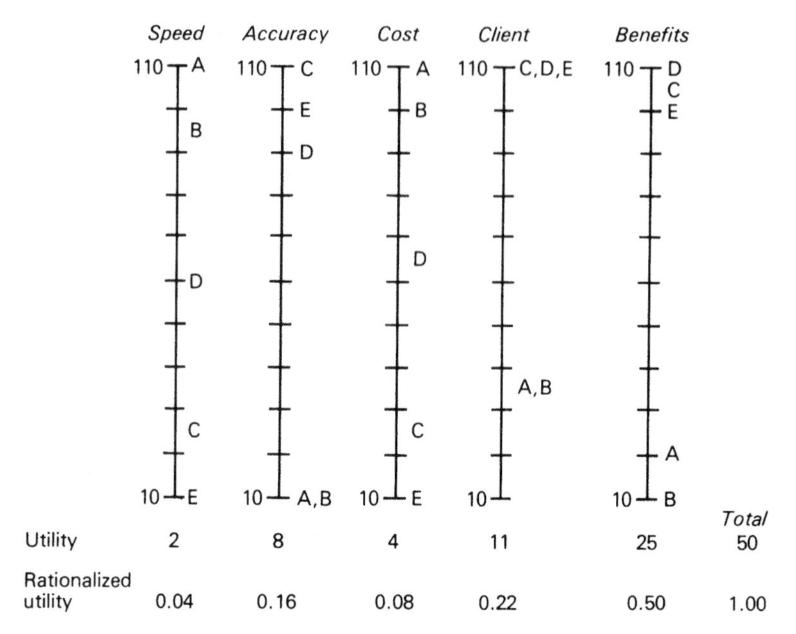

Fig. 9.12 Multi-attribute utility analysis. *Speed:* speed of obtaining a solution/tender; *accuracy:* accuracy of the solution; *cost:* cost of solution; *client:* client and consultants – attitudes, risks, etc.; *benefits:* benefits potential to the contractor – profitability, resource employment, work type, work continuity, etc.

that the relative positions of the outcomes on the scale are correct, e.g. cost criterion – A is very cheap, E is expensive, C is almost four times as expensive as D.)

Table 9.6, the outcome evaluation, is obtained from Figure 9.12 by considering each criterion in turn and calculating the utility of each outcome against each criterion by multiplying the measure on the criterion scale by the rationalized utility of the criterion. The values obtained for each outcome are then summed to obtain the total utility of each outcome, the values then being recorded against each outcome on the decision tree (as Figure 9.11.)

Where an event node occurs, the paths leading outwards from it to the right represent alternatives and so, probabilities must be allocated to each

Table 9.6 Outcome evaluation

Criterion outcome	*Speed*	*Accuracy*	*Cost*	*Client*	*Benefits*	*Total*
A	4.4	1.6	8.8	7.7	10.0	32.5
B	3.8	1.6	8.0	7.7	5.0	26.1
C	1.1	17.6	2.0	24.2	52.5	97.4
D	2.4	14.4	5.2	24.2	55.0	101.2
E	0.4	16.0	0.8	24.2	50.0	91.4

path (the sum of the probabilities of the paths from an event node must, of course, equal unity).

The decision tree is then 'folded back'. Working from right to left, the utilities at the nodes are calculated as shown. At an activity node, where more than one path enters that node from the right, a decision must be made; logically, the path with the higher (or highest) utility is selected and the utility value allocated to the node. All other paths entering the node from the right are ignored and blanking off lines are drawn across them.

Thus, the logical path for the decision is obtained (path C in Figure 9.11), being that path with the greatest utility.

It is possible that other paths will have only slightly lower utilities and so, sensitivity analyses should be employed to evaluate the effects of changes in the variables: positions on the utility scales, probabilities, rationalized utilities of the criteria and, perhaps, even the criteria themselves. If in the example the probability of the approximate method producing an acceptable result is increased to only 0.62 (from 0.6), that course becomes the logical one to follow.

The technique of multi-attribute utility analysis may be used in the evaluation of buildings, either constructed or proposed. Such evaluation introduces a significant measure of objectivity and is of great value in requiring the criteria for judgement and their relative importance to be determined.

Such an assessment has been used to evaluate housing provision such that various types of housing unit of different standards may be compared. It is also possible by use of this technique, which considers notional units of housing, to calculate total housing requirements, etc. (For a more detailed account see Hillebrandt 1985.)

However, it may be that the multi-attribute utility theory (MAUT) approach is not really appropriate to the way people actually evaluate alternatives: they may not consider and evaluate outcomes against individual, component criteria but may evaluate the outcomes against an aggregate of the criteria. Hence the evaluation technique of conjoint analysis has been developed in which alternative outcomes, with known, predetermined attributes, are rated by decision-takers. The relative importances of the component evaluation criteria are then calculated from those overall ratings, the component criteria having been identified from theory/literature, etc. and agreed with the decision-maker.

9.3 Financial reporting

Financial reporting is the process of recording and communicating the financial performance of a firm. The best-known financial reports are those which appear in sets of published accounts, particularly the profit and loss account and the balance sheet. These two reports are closely linked but of quite different natures: the profit and loss account being a statement of the firm's income and expenditure over a period (usually a year) and the resultant profit (or loss); the balance sheet being a statement of the firm's assets and liabilities at a particular instant (usually the year end).

Although a good deal of information is contained within a set of accounts and their accompanying notes, the information is frequently not in the most useful form, thereby necessitating a degree of gleaning of meaningful figures and of interpretation. Such activities cannot be carried out in isolation from the accounting principles and conventions which have considerable influence, not only on the way the accounts are presented but also upon the figures appearing within the accounts.

Analysis of financial reports and statements, useful though these may be, are also subject to quite severe time limitations for validity. Essentially, such analyses are valid only in the relatively short period; in the long period, the success of a firm is governed by such fundamentals as technology, economic conditions, tastes, and fashions: items for which there is no financial measure.

Accurate financial reporting and interpretation constitutes a useful information input to management for the decision-making function. In this context it is important not only to consider current performance but also to analyse trends in order that a time dimension may be introduced.

Essentially, a set of accounts comprises the profit and loss account and the balance sheet, however, those accounts will be supplemented by other accounts (often a requirement of legislation), the report of the directors (of a company) and the report of the auditors. In particular, the auditors' report should be examined to check that the accounts represent '*a true and fair view*' in the opinion of the auditors. Figure 9.13 illustrates the operational framework of a firm and the information required for and influences upon the profit and loss account and the balance sheet. Figure 9.14 depicts a profit and loss account for a building contractor and Figure 9.15 depicts a balance sheet. Figure 9.16 shows a manufacturing account, which shows the financial results of the company's production activities; Figure 9.17 shows the associated trading account; while Figure 9.18 is the profit and loss appropriation account and Figure 9.19 is the sources and application of funds (or cash flow) statement which shows how finance has been generated and used by the company.

The accounts report the consequences of transactions undertaken during the period from the perspective of the particular firm. There are special provisions which relate to groups of companies (parent and subsidiaries) in the form of consolidated accounts. Further special provisions concern the accounts of medium and small companies which are permitted to produce abbreviated accounts. Small and medium-size groups do not have to prepare group accounts.

If subsidiary companies are not owned 100 per cent by the parent company (whether direct or indirect – via other subsidiaries – owner-

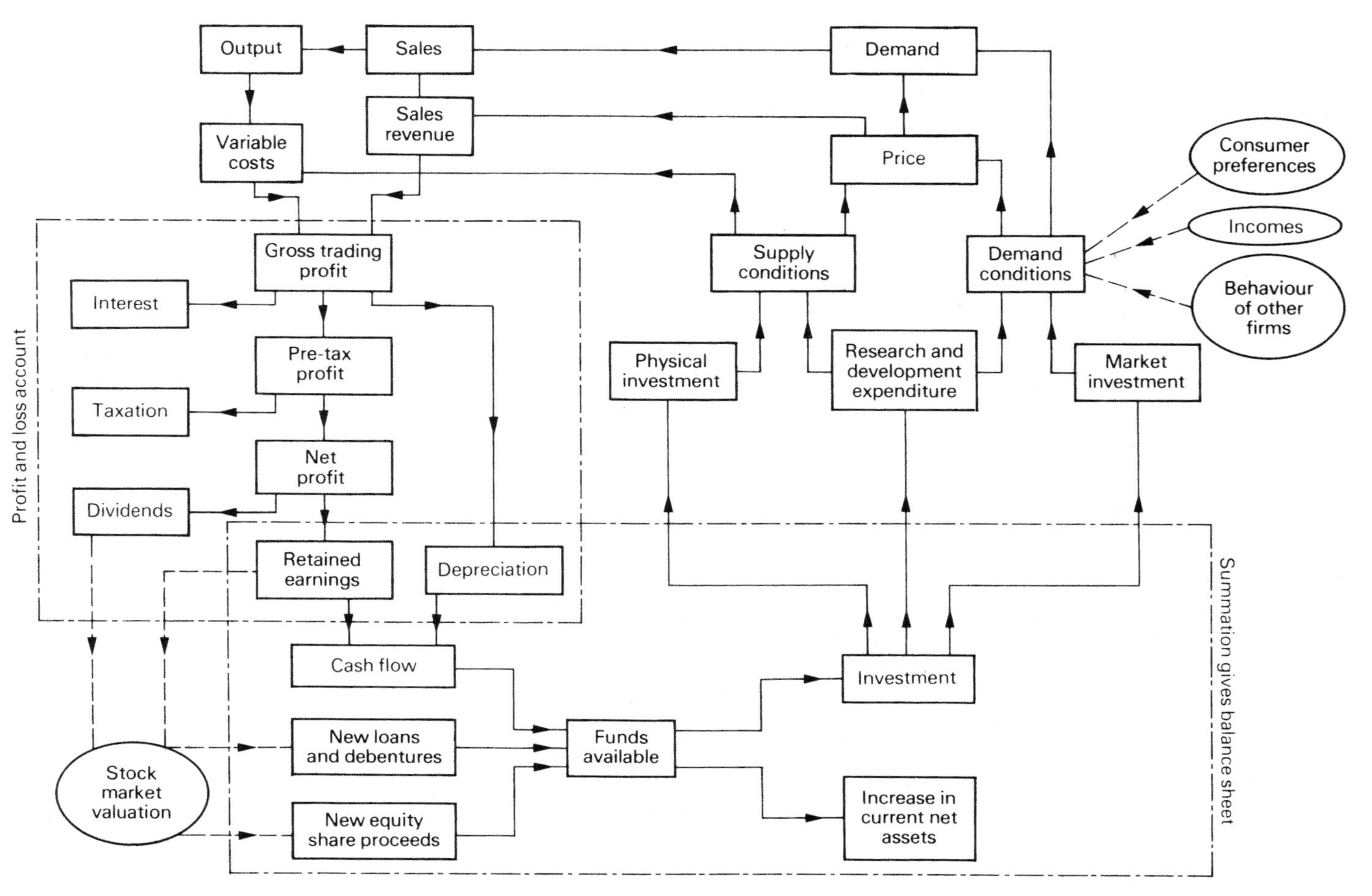

Fig. 9.13 Profit and loss account and balance sheet components. After Morris (1979).

ship), profits earned, assets owned, etc. will be reduced proportionately in the consolidated accounts to show the '*minority interests*'. Although group accounts include the balance sheet of the parent company, the other accounts of the parent are excluded. The legal definition of a *subsidiary undertaking* is quite complex but mainly concerns ownership of over 50 per cent of voting rights or the ability to control the activities of the subsidiary. A parent may hold a long term *participating interest*, normally if it holds at least 20 per cent of the voting shares in the subsidiary. Each subsidiary company will produce its own accounts but must state the name of the *ultimate parent company* in them.

Group accounts may include *associated undertakings* which are companies in which the parent has a participating interest and has a significant, rather than dominant, interest over the associate's activities.

The criteria for small and medium-size companies are amended periodically (to keep pace with changes in price levels, etc.). In the 1997/98 tax year, a medium-size company had to satisfy at least two of the following criteria:

- turnover not exceeding £11.2 million
- total assets not exceeding £5.6 million
- no more than 250 employees

The criteria relate to the production of accounts and to liabilities for corporation tax, on which small and medium companies enjoy graduated relief.

9.3.1 *Profit and loss account*

The profit and loss account (or, for some organizations, the income and expenditure account) is a statement of the incomes and expenditures of the firm which occurred during the accounting period, usually a year, and the resultant profit or loss. The Companies Acts of 1985 and 1989 prescribe two formats for profit and loss accounts. The more usual format is employed in Figure 9.14 but, in either format, the account will show:

1. gross profit from trading
2. supplementary costs and expenses
3. operating profit
4. investment income and expenses, if any
5. profit before taxation
6. taxation
7. profit after tax

It is also essential for the account (now, usually, in a separate profit and loss appropriation account) to provide information regarding the disposal

D.M. O'Lition (Contractors) plc		
Profit and loss account for the year ended 1 April 2001		
	(£000)	(£000)
Gross profit from trading account		7,250
Head office salaries and overheads	1,780	
Directors' salaries	1,100	
Audit fees	400	
Provision for bad debts	1,350	
Depreciation	200	4,830
Operating profit		2,420
Investment income		400
Profit before interest and taxation		2,820
Less: Loan interest	816	
Debenture interest	4	820
Profit before taxation		2,000
Taxation		480
Net profit to appropriation account		1,520

Fig. 9.14 Profit and loss account.

of the profit between distribution to the owners and the balance retained within the firm. Any amount retained within the firm is an addition to the owners' investment in the firm. Such retained profits (sometimes called retained earnings) are a major source of capital for the financing of future activities and for growth.

Thus the profit and loss account is a statement of the firm's performance over the period and provides a measure of how successful that performance has been.

9.3.2 *Balance sheet*

The balance sheet (see Figure 9.15) is a statement of the financial position of a firm as at a particular instant, usually the end of the accounting period of the firm: its year end. In economic terms, it may be regarded as a statement of the wealth of the firm at a particular time, whereas the profit and loss account is a statement of the change in the wealth of the firm over the stipulated period. Hence the 'snapshot' (financial photograph) concept of the balance sheet.

A balance sheet comprises two distinct sections, one detailing the assets of the firm and the other detailing the liabilities, including owners' investment. Traditionally, the liabilities (capital from owners, reserves, creditors, etc.) comprised the left-hand side of the balance sheet and the assets (cash, investments, buildings, etc.) comprised the right-hand side. The modern layout is shown in Figure 9.15 which is linear and details liabilities beneath assets. Moreover, following detailing the fixed assets as

D.M. O'Lition (Contractors) plc

Balance sheet as at 1 April 2001

	(£000)	(£000)	2001 (£000)	2000 (£000)	Notes
ASSETS EMPLOYED					
Fixed assets					
Intangible assets					
Goodwill			100	100	
Tangible assets					
Land and buildings		12,500		11,700	
Plant and equipment		3,200	15,700	3,510	
			15,800	15,310	
Investments					
Properties			4,000	4,000	
			19,800	19,310	
Current assets					
Stocks					
Materials	1,200			1,150	
Work in progress	1,350	2,550		1,280	
Debtors		1,470		1,100	
Cash at bank and in hand		1,120		900	
		5,140			
Current liabilities					
Creditors: amounts falling due within one year					
Bank loans	700			700	
Trade creditors	1,600			1,550	
Debenture interest	4			4	
Proposed dividend	975			500	
Current taxation	480	3,759		350	
Net Current Assets			1,381		
Total assets less *current liabilities*			21,181	20,636	
FINANCED BY					
Creditors: amount falling due after more than one year					
Bank loan		2,500		2,500	
Debentures: 4 10% debentures at £10,000 each		40	2,540	40	
Capital and reserves					
Called up ordinary share capital			15,000	15,000	
Reserves:					
General		2,121		1,696	
Profit and loss account		1,520	3,641	1,400	
			21,181	20,636	

Fig. 9.15 Balance sheet.

tangible and intangible, the 'assets' section includes a statement of current assets less current liabilities to give the net current assets or the net working capital before arriving at the statement of the firm's total assets less its current liabilities (the net assets employed). The 'liabilities' part of the balance sheet comprises details of the owners' investment in the firm and of the firm's long-term liabilities.

Whatever format of presentation is adopted, *the two parts of the balance sheet must balance* which must be the case if the principles of double entry bookkeeping have been followed.

The assets of a firm comprise tangible fixed assets, current assets, investments and intangible assets. The fixed assets include such items as buildings, land, plant, machinery, equipment, fixtures and fittings, lorries and cars; items which an economist would regard as components of fixed costs. These assets should be considered subject to depreciation and/or periodic revaluation, dependent upon the circumstances: land and buildings tend to be periodically revalued, plant, machinery, cars, etc., should be depreciated over their useful lives.

Current assets are liquid assets often of short duration, such as cash, debtors, stocks, and work in progress. The current assets are those used in the operating cycle of the firm.

Intangible assets, such as goodwill and patents, are assets of no physical substance (except the paper of patents). Their inclusion in the assets of a firm is justified on the basis that their existence enhances the value of the firm. However, the valuation of intangible assets is a difficult exercise and it is probable that the sums included in accounts for such items are in reality only educated guesses.

The liabilities of a firm include both long-term and current (short-term) liabilities. The long-term liabilities comprise debentures, long-term loans and future taxation provisions. The current liabilities, those associated with the firm's day-to-day operations, including creditors, current taxation and proposed dividends to shareholders, are subtracted from the firm's total current assets to determine the net working capital. It is the inability of construction firms to meet their current liabilities which is the most common cause of bankruptcies and liquidations within the industry; thus both solvency and liquidity are vital considerations.

The owners' investment comprises the subscribed share capital, both preference and ordinary shares, and the reserves which act to increase the owners' investment in the firm. (Unpaid dividends are also attributable to the owners but are part of current liabilities.)

9.3.3 Directors' report

The directors' report must include a review of the company's principle activities for the year and any significant changes in them. The report must also include a review of the development of the business for the year and the position of the company at the year end. Important events which occurred after the date of the balance sheet will be included.

The directors must forecast likely developments of the company in the future – this is an area of projection of intentions rather than a report of the past!

The report will provide information on research and development undertaken, and significant changes in fixed assets: notably differences in book values and market values of land and buildings. The names of the directors must be given with details of any holdings of shares in the company.

The recommendation of the directors concerning dividend distribution and intended transfers to reserves must be stated.

Details must be provided if the company has purchased any of its own shares. If the average number of employees was 250 or more during the year, the report must include employment policy information concerning disabled persons and employee involvement in the company's activities.

Details of political and charitable donations must be given if the total of such donations exceeds £200. If a political donation of over £200 has been made, the identity of the recipient must be given.

Quoted companies will provide a chairman's statement and a chief executive's review. Usually these are at the front of the publication containing the annual accounts and, although not audited, are likely to contain much useful information on the company's activities and prospects. However, there may be a tendency to emphasize the good aspects of the company's performance and so, the statements should be considered in the light of the other, especially the audited, information.

9.3.4 Auditors' reports

The Companies' Acts require company accounts to be audited by qualified, independent examiners: registered auditors, usually chartered or certified accountants. The purpose is to determine whether the accounts give a '*true and fair view*' and a statement must be included accordingly in the audit report attached to the accounts.

The auditors are appointed annually by a proposal and vote at the company's annual general meeting.

The auditors' report comprises a heading stating that it is to the shareholders of the named company. It proceeds to identify the accounts on which the report is based and the accounting convention and policies adopted in their preparation. The responsibilities of the directors and the auditors are expressed. The basis of the auditors' opinion is given (usually in the form that the audit comprised checking of appropriate, representative samples of recordings of transactions). Next comes the most important element, the statement of the auditors' opinion – whether the accounts give a true and fair view and have been prepared properly in accordance with the requirements of the Companies' Acts. Finally, the name and address of the auditors is given and the date of the report.

If the auditors have given a qualified opinion, the nature of the qualification should be considered in the context of the consequences for the accounts giving a true and fair view, i.e. the nature of the discrepancy should be investigated to examine any distortions in the accounts.

Half-yearly accounts of quoted companies are reviewed by auditors but are not audited. Provisions to relax audit requirements apply to small companies: very small private companies are exempt from audit requirements; other small private companies must provide an accountant's report with the accounts.

9.3.5 Other financial statements

Although the profit and loss account and balance sheet are the best known financial statements, many more statements are available. An income and expenditure statement takes the place of a profit and loss account for certain types of organization, e.g. professional institutions.

The other financial statements which are prepared periodically within a firm are dependent upon the type and activities of the firm. Most construction firms will prepare manufacturing accounts, trading accounts, profit and loss appropriation accounts and cash flow statements.

A manufacturing account (Figure 9.16) shows the change in value of stock of materials (and goods) to determine the cost of those used in the

D.M. O'Lition (Contractors) plc

Manufacturing account for the year ended 1 April 2001

	(£000)	(£000)
Materials		
Opening stock	1,150	
Purchases	14,800	15,950
Less: closing stock		1,200
Cost of materials consumed		14,750
Wages	4,200	
Small tools	1,100	
Sub contractors	13,800	19,100
Prime cost of production		33,850
Site overhead expenses	120	
Depreciation of plant and equipment	600	720
		34,570
Opening work in progress	1,280	
Less: Closing work in progress	1,350	(70)
		34,500

Fig. 9.16 Manufacturing account.

year; adding the other items of direct (variable) costs determines the prime cost of production. Addition of the indirect costs of production and adjustment of the work in progress between the year start and end give the firm's costs of production.

A trading account (Figure 9.17) is a statement of the sales of the firm for the period (excluding capital items which are treated separately elsewhere) and the costs of those sales. The resultant figure is the gross trading profit (or loss), which is transferred to the profit and loss account. As this account is concerned with the trading operations of the firm, not only direct production costs but also any distribution, marketing, research and development, administrative, etc. costs are also considered in the calculation of the trading profit.

D.M. O'Lition (Contractors) plc

Trading account for the year ended 1 April 2001

	(£000)
Sales	41,750
Less: costs of production	34,500
Gross profit to profit and loss account	7,250

Fig. 9.17 Trading account.

A profit and loss appropriation account (Figure 9.18) is a statement of the deployment of the net profit after tax: the profit available for distribution to the owners. The account provides details of the distribution to the owners of the firm and the amount (if any) retained within the firm as reserves. Normally, the distribution of profits in this account is the distribution *proposed* by the directors and is subject to acceptance by the shareholders at the Annual General Meeting.

Cash flows have been examined in Chapter 8 and are of use for evaluation of individual projects as well as for considerations of the firm's

D.M. O'Lition (Contractors) plc

Profit and loss appropriation account for the year ended 1 April 2001

	(£000)
Net profit for year	1,520
Proposed dividend of 6.5 pence per share	975
	545
Transfer to general reserve	545
Balance carried forward	0

Fig. 9.18 Profit and loss appropriation account.

overall position and prospects. A cash flow statement (Figure 9.19) for a firm for a given period should distinguish internal and external sources of cash flow. Internal sources comprise such items as profit before tax, depreciation, and investment income while external sources comprise cash flow from outside the firm such as loans to the firm. The statement should also distinguish sources and applications of cash flow arising from the sale or acquisition of fixed assets (capital items) from those arising from changes in the firm's working capital (revenue items). Statements of this type commonly appear in sets of accounts as statements of 'sources and applications of funds' and denote the changes in balance sheet items which have occurred during the accounting period.

D.M. O'Lition (Contractors) plc

Sources and applications of funds for the year ended 1 April 2001

	(£000)	(£000)
Sources of funds		
Profit before taxation		2,000
Adjustments for items not involving the movement of funds:		
Depreciation		800
Total generated from operations		2,800
Sale of fixed assets		30
		2,830
Applications of funds		
Dividends paid	500	
Taxation paid	350	
Purchase of fixed assets	1,320	2,170
		660
Increase/decrease in working capital		
Decrease in stocks and work in progress	120	
Increase in debtors	370	
Increase in trade creditors	(50)	
Increase in cash balances	220	660

Fig. 9.19 Sources and applications of funds statement.

9.3.6 *Accounting terminology*

Certain words and phrases have particular meanings in an accounting context. The following is a glossary of some common accountancy terms:

Accruals (usually expenses): items which have been utilized by the firm prior to payment therefor, but the appropriate cost or value is included in the firm's accounts to give an accurate record of the financial consequences of the firm's activities in the period. In construction, income is accrued, as indicated by interim payments, in order that the profit

earned by the project is shown in the appropriate periods in which it is earned.
Asset: an item, owned by an individual or a firm, which has a monetary value.
Capital: the goods which are used in the production and have, themselves, been produced; the investment in the business by the owners.
Capital employed: the capital in use in a business; net assets.
Capital reserve: a reserve created to provide for future capital gains tax liabilities, etc. (Capital gains tax is payable *on the sale* of an asset - a fixed asset - which has appreciated in value, e.g. a building.)
Credit: a bookkeeping entry denoting an increase in a liability account or a decrease in an asset account.
Creditor: someone (or another firm) to whom the firm owes money.
Debit: a bookkeeping entry denoting an increase in an asset account or a decrease in a liability account.
Debtor: someone (or another firm) who owes money to the firm.
Depreciation: the process of reducing the value of an asset (as shown in the firm's accounts) for its wearing out and obsolescence. The annual reduction in value of the asset is also a cost to the firm and appears as such in the profit and loss account. This does not mean that cash is set aside for future replacements but depreciation is an important source of capital as it keeps the capital in the firm which otherwise would be appropriated to the owners as profits. Depreciation applies to fixed assets and is charged on either a 'straight line' or a 'reducing balance' basis (see Chapter 7).
Dividend: the residual profits distributed to the shareholders.
External liabilities: debenture loans, other loans, and creditors of a firm.
General reserve: the residual profits retained within the firm for reinvestment. This represents an increased stake in the firm by the owners.
Goodwill: an intangible asset, the amount by which the value of a business exceeds the value of its total net tangible assets plus any other intangible assets (e.g. patents). This arises through the reputation of the firm and its potential future earnings.
Liquidation: the winding-up of a company; its termination.
Liquidity: the ease with which an asset may be exchanged for (turned into) cash; the ability to meet short-period requirements for cash.
Net assets: net current assets plus fixed assets.
Net worth: the net assets employed by a firm minus the external liabilities.
Over-trading: the situation where current creditors cannot be paid fully out of the receipts from current sales.
Profit: *gross profit* is the total sales revenue minus costs incurred directly in the operations of the firm; *net profit* is the gross profit minus interest on loans and depreciation.
Revenue: income.
Revenue expenditure: expenditure on items other than fixed assets; current expenditure.

Solvency: the ability to raise cash to meet debts; the ability of a firm to meet its current liabilities by realizing its current assets.
Turnover: the total revenue from sales of a business (excludes capital items); net of discounts and VAT.

9.3.7 Accounting concepts and conventions

Accountancy is based upon the application of a common set of rules for the preparation of all accounts. This is essential in order that comparisons may be made and, more importantly, to facilitate understanding and interpretation and to avoid frauds. Many of the concepts and conventions of accountancy in the UK are embodied within statutes, particularly the Companies Acts, but others are applied due to accountants adopting standard procedures in the practice of their profession, notably those laid down by the accounting professional institutions and the Stock Exchange. The accounting standards are embodied in *Statements of Standard Accounting Practice* (SSAPs) and *Financial Reporting Standards* (FRSs); *Generally Accepted Accounting Practices* (GAAPs) are normal accounting practices which are followed by most accountants but are not formal standards. An appreciation of the main concepts and conventions is vital for the understanding and the correct interpretation of accounts:

Accruals: costs and revenues are accrued (taken into account) as they are incurred or earned (as noted under accounting terminology), rather than when they are paid or received.
Business entity: each business is treated as a distinct and separate entity in its own right.
Conservatism (or *prudence*): provision is made in the accounts for liabilities (costs, expenses) as soon as they become apparent whereas profits are not usually included in the accounts until they are realized.
Consistency: identical or very similar items must be treated in the same way in different accounting periods and in different sets of account.
Cost concept: initially assets are recorded in the accounts of the business at their cost and thereafter are depreciated on that basis. Revaluations may occur periodically for fixed assets so that more realistic values are shown.
Current cost accounting: assets and liabilities are included in the accounts at their current market value or current cost.
First in first out (FIFO): the convention of considering that the oldest stocks of materials, etc. are used first in production.
Going concern: a business is regarded as an entity which has a continuing life of infinite length.
Historic cost accounting: assets are included in the accounts on the basis of their original cost. (Fixed assets may be subject to occasional revaluation.) Liabilities are treated in the same way. This convention is used for taxation purposes.

Last in first out (LIFO): the convention of considering that the newest stocks of materials, etc. are used first in production.
Monetary measurement: accounts record monetary effects upon the business; events, the effects of which cannot be measured in monetary terms, are therefore excluded from the accounts. This concept implies a stable monetary unit to facilitate direct comparisons and so leads to problems during inflationary periods.
Prudence: losses are recognized at the earliest opportunity; profits are recognized only when realized as cash (or near cash).

9.3.8 *Recording transactions*

Each transaction in which a firm is involved must be recorded in its monetary effects. The recording will occur in the various books of account of the firm and will be executed by the process of double-entry bookkeeping. The bookkeeping reflects the nature of transactions – usually one party sells a good or service to the other so that the former gains money and the latter gains the good or service through the exchange. The double-entry aspect is fundamental, as every transaction will constitute an increase in one account of the firm and a decrease in another so that the worth of the firm remains constant but the constituents of that worth change (subject to value-revenue-accumulation in sales). Thus the purchase of 1,000 bricks by a contractor for a cash sum of £50 will be recorded in the contractor's books as an increase in the stock of bricks of £50 and a decrease in the firm's cash of £50; thereby the entries in the accounts reflect the change in the nature of the assets held by the firm and that the monetary value of the firm has remained constant.

It is useful to consider the accounting aspects of transaction by the use of 'T-accounts' (so named due to their format). Note that the *left-hand* side of a T-account is *always* the *debit side* and the *right-hand* side *always* the *credit side*. Thus the example given above would be recorded in T-account format as:

Stock – bricks

(Dr)	(Cr)
50	

Cash

(Dr)	(Cr)
	50

As all transactions may be recorded by the use of T-accounts, the following example illustrates the establishment and first year of trading of a new contracting company (taxation is ignored):

1. The owners invest £15,000.
2. The company purchases £4,500 of materials on account; the account is subsequently paid.

3. The company purchases £500 of small tools for cash.
4. The company purchases a yard and office building for £21,000 for which it pays cash of £7,000 and issues a debenture of £14,000 which bears interest at 10 per cent per annum.
5. During the trading period, the company executes work which costs in wages £5,000, salaries £1,000, materials £4,000, and plant hire £1,000 and for which it receives £15,000.
6. At the end of the period, an ongoing contract has cost £400 materials, £100 salaries and £200 wages, is 50 per cent completed and the contract sum is £2,000; no payments have been received for this work.
7. Depreciation is charged on a straight-line basis at 10 per cent per annum, except for buildings, etc. which are depreciated at 2 per cent per annum.
8. Dividends of £1,000 are paid.

Please see Appendix D for the T-accounts, profit and loss account and the balance sheets.

Because a firm undertakes many transactions during any single accounting period, T-accounts are not suitable for recording purposes; instead various 'books' of account are employed. Now, with the widespread use of information technology, it is normal for the books of account to be in electronic form. The book in which transactions are recorded initially is the *journal*, the book of original entry. This book may take various forms but traditionally was a bound volume with the pages of columnar layout, as illustrated in Figure 9.20 which records the credit purchase of reinforcement from Y-Bar Steels Ltd and the subsequent cash payment. Journal entries are subsequently posted (transferred) to the ledger accounts, the ledger folio column being for cross-referencing.

Date	*Account*	*Ledger folio*	*Debit*	*Credit*
2001				
Sept. 3	Materials	12	731 52	
	Y-bar Steels Ltd	71		731 52
Sept. 29	Y-bar Steels Ltd	71	731 52	
	Cash	5		731 52

Fig. 9.20 Traditional journal format.

The ledger is the main book of account. Traditionally, it is in the form of a bound volume in which each page is devoted to recording transactions affecting one individual account. The traditional format gave way to loose-leaf and card systems which have now been superseded by electronic data storage systems. Figure 9.21 shows the ledger accounts appertaining to the transaction recorded in the journal (Figure 9.20).

Both these books of account record all the transactions of the firm but in

Cash — Folio 5

Date	Description	Ref.	Debit		Date	Description	Ref.	Credit	
2001					2001				
Sept. 1	Balance b/d	✓	3,062	05	Sept. 29	Purchase reinf. bars	J23	731	52

Materials — Folio 12

Date	Description	Ref.	Debit		Date	Description	Ref.	Credit	
2001					2001				
Sept. 1	Balance b/d	✓	2,710	17					
3	Pur.reinf. bars	J23	731	52					

Y-bar Steels Ltd — Folio 71

Date	Description	Ref.	Debit		Date	Description	Ref.	Credit	
2001					2001				
Sept. 29	Sept. 3 a/c paid	J23	731	52	Sept. 1	Balance b/d	✓	92	01
					3	Reinf. bars	J23	731	52

Fig. 9.21 Ledger accounts.

rather different ways. Each firm will develop individualities in the recording procedures and processes but the principles of accounting will be followed in every instance.

The posting of the journal records to the ledger accounts will occur at intervals determined by the number and frequency of transactions. Often this will be done on a daily or weekly basis. Prior to the posting of entries, it is sound practice to check the arithmetic accuracy of the entries; this is conveniently done by totalling the debit column and the credit column to ensure that they balance (the totals are equal). This process is called 'proving'. It does not substantiate the theory of the entries as recorded but arithmetic errors are easily detected and may be corrected with a minimum of difficulty (if errors are not found until a much later stage, the correcting process is considerably more problematic). Given the common use of electronic systems, the posting and proving processes can be executed frequently and automatically to provide a continuous check on the accuracy of the entries.

At the end of each accounting period, the ledger accounts are balanced and closed (for that period), the balances being carried down to the next period (thus becoming the opening balances of the next period). These closing ledger balances are also abstracted to a separate list, called the unadjusted trial balance, on the basis of which the statements of account of the firm are prepared. The expenses and revenues of the period just ended will be abstracted to an expense and revenue summary from which the profit and loss account is obtained.

Several ledger accounts are used frequently: the cash account, the materials purchases account, and the sales account. For convenience, these accounts are commonly kept separate from the general ledger, each in its own book: the cash book (usually, with a subsidiary book for petty cash items), the purchases daybook, and the sales daybook. The cash book is somewhat different from the daybooks in that it is the entire cash account, both debits and credits, whereas the daybooks record only the debits of the purchases and sales accounts (the credit entries being recorded in the ledger against the suppliers' and customers' personal accounts) and are totalled and posted to the ledger sales and purchases accounts annually.

9.4 Financial management

Successful financial management is founded upon an appreciation of accounts, capital provision and the ability to apply that expertise in the context of the firm's business, notably its objectives and the environment in which it operates.

It therefore remains to consider the financial effects of the main operating mechanisms and how interpretation of financial statements may be assisted (notably by the use of ratios).

Given a knowledge of capital provision, financial reports and their interpretation and the operations and objectives of the firm, successful management is the ability to make decisions which lead the firm towards the realization of its objectives, commonly evaluated by examination of measures of performance, especially financial measures.

It is perhaps interesting to consider the paradox in which contracting firms operate; each contractor wishes to continue in business, to be profitable, and (usually) to grow, objectives which are achieved by successfully managing each individual contract out of business! However, it should be recalled that contracting is not an area of activity in which profitability is great, despite the risks involved, but is an activity which generates cash flow.

9.4.1 Net working capital

Net working capital is calculated by subtracting current liabilities from current assets; it is thus that portion of current assets which is financed from long-term funds. The flow of working capital in a business is of a cyclical nature, as illustrated in Figure 9.22. Due to the nature of construction, stocks and work in progress form a relatively large proportion of a building firm's working capital.

The cycle, as shown in Figure 9.22, may be modified to allow for the payment provisions under the typical system of contracting. Completed items of work, not just complete buildings, are paid for by the client in the

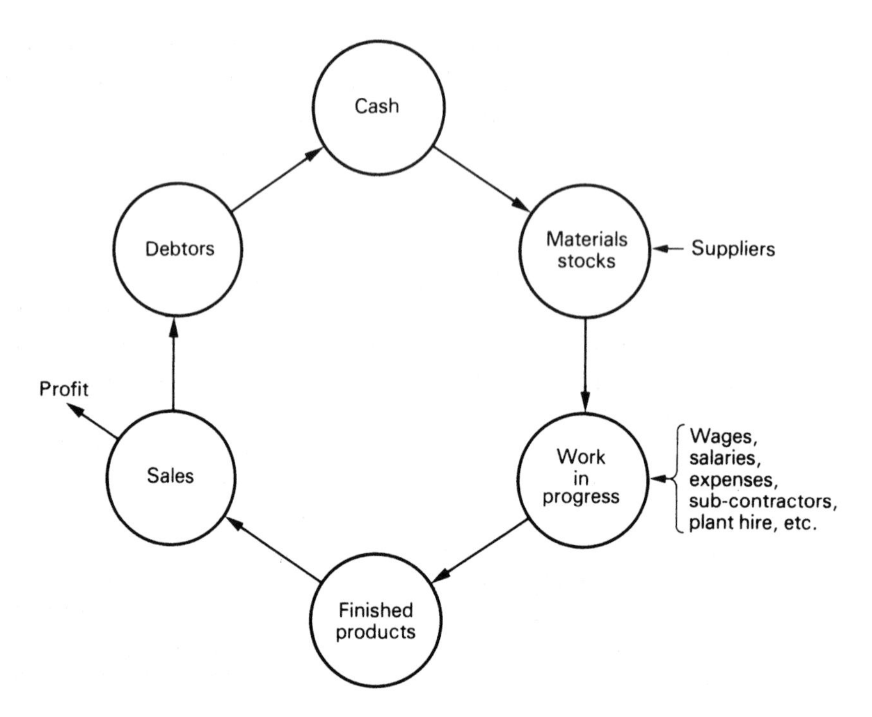

Fig. 9.22 The working capital cycle.

honouring of the interim certificates which, are usually, monthly and based upon progress of the projects' work. Stocks of materials on-site, and sometimes off-site, are included in the payments together with sums for fluctuations and variations. It is apparent that the cycle, as shown, applies for each individual work item but that there is also a direct route from materials to sales when materials on and off-site are included in interim certificates. One further amendment occurs at the sales point where retention is withdrawn from the cycle to constitute a subsequent two-stage cash injection at practical completion (first moiety released) and at final certificate (second moiety released).

However, medium and long-term capital is more expensive than short-term capital and so, profitability may be enhanced by:

- maintaining the proportion of short-term to long-term liabilities at as high a level as possible commensurate with the working capital ratios required
- keeping net working capital to a minimum (net working capital is financed from long-term funds)
- ensuring that the working capital cycle is as short as possible

Good management of working capital is critical to financial success. Hence it is helpful to relate the cycle to the payment processes under the standard contracts employed to ensure that the need for working capital is kept to a minimum and that the cycle is kept as short as possible, i.e. that the working capital circulates quickly, thereby minimizing the use of 'own' funds and maximizing the use of 'others' funds.

9.4.2 Cash

Cash is an element of working capital but is of major importance in its own right. Cash, of course, is more than notes and coins - primarily it is balances in bank accounts (usually current account) and cash flows are effected by transfer mechanisms between project participants' accounts (cheques, etc.). Cash is not wanted for its own sake (as money is merely the medium of exchange) but to 'oil the wheels' of business. No firm wishes to hold more than the minimum amount of cash as this is an asset which earns (virtually) nothing, despite some current accounts with banks earning nominal interest on positive balances. It is important to a firm's profitability, therefore, that it manages its cash efficiently to minimize the cash held. Deposit accounts offer a useful form of investment for short-term cash surpluses and automatic transfers between accounts can be effected, based upon minimum and maximum prescribed balances.

Successful cash management is dependent upon the ability to manage the financial consequences of the elements of the business cycle of the firm. The more rapidly the cash cycle revolves, the lower the firm's requirement for working capital.

9.4.3 Debtors

Clearly, debtors represent a delay to the cash inflow of the firm. Their existence constitutes a lengthening of both the cash and working capital cycles. However, in construction, as in most industries, the offering of a period of credit to customers is a normal and accepted business practice. Most construction contracts (main, subcontracts, supply contracts, etc.) stipulate the period of credit to be applicable, sometimes also specifying discounts to be offered for payment within the period of credit allowed (e.g. NSC/C - the JCT 98 associated standard form of nominated subcontract). For main contractors, the main element of debtors shown on the balance sheet is certified payments awaiting honouring by clients.

It is possible to speed up the debtors' section of the cycle by the process of factoring the debts. In factoring a debt, the creditor sells the debt to a bank or other financial institution (some specialize in this activity) at a discount (dependent upon the period prior to payment becoming due and the credit worthiness of the debtor). The creditor receives a reduced amount of cash but at an earlier time. After the debt has been factored, the risks and collection responsibilities lie with the factoring agent.

Debt factoring is, perhaps rather surprisingly, not very common in construction.

9.4.4 *Stocks*

Stocks of materials and goods may represent a considerable 'tie-up' of the firm's financial resources and hence a delay to the cash and working capital cycles. However, the possibility is mitigated by two major procedures in the construction industry: the purchasing of materials and goods on credit (and, usually, at a discount also) and the inclusion of 'materials on-site' as an element of interim payments. Further, major materials and goods which are produced and stored off-site until required for installation may be included in interim payments.

Thus stocks of materials, etc., shown on the balance sheet are items for which no payment certificate has been issued but which are owned by the contractor (irrespective of their location) and irrespective of whether the contractor has paid for them or not. (The prices of these items for which the contractor has not paid will form a major part of the creditors sum in the balance sheet.)

9.4.5 *Work-in-progress*

Work-in-progress is included in the accounts of most business at a valuation which accords with the principle of 'cost or market value, whichever is the lower' and is included to reflected fairly the value of the goods which are actually within the production process. Clearly this is reasonable for manufacturing and similar industries, but construction represents an exception.

Building contractors produce finished articles (completed, projects) in which there are very many components, and although the finished article (the building) takes a long time to produce, individual components (such as those identified as separate items in BQs) are completed over much shorter time periods. Further, with all but small contracts, the client makes periodic (usually monthly) progress payments to the contractor in respect of the work completed.

As, in effect, the building is sold prior to its construction and the construction period is long, contractors' work-in-progress is included in the accounts at cost plus profit less allowances for contingencies such as patent and latent defects. Such procedure enables the accounts to reflect the true profit earned and the financial position of the contractor at the end of the accounting period. The balance sheet will thus include work-in-progress as the total value of work executed to date on uncompleted contracts, less the total payments certified to date in respect of those contracts.

9.4.6 Depreciation

Depreciation, the process of reducing the value of a fixed asset (writing off) for obsolescence and wear and tear (the proportion of depreciation due directly to the utilization of the asset in the production process), appears as a cost in the profit and loss account. It is included in the balance sheet by its effect of reducing the value of the assets; details of the provisions for depreciation are given in the notes to the accounts. Thus, the values of fixed assets to which depreciation has been applied, as shown in a balance sheet (their written down values), are the proportions of the costs of those assets which remain available to the firm.

It is a fallacy to consider depreciation to be a fund set aside for the replacement of fixed assets which wear out and become obsolete. The appearance of depreciation as a cost in the profit and loss account is a key to its true purpose – to ensure that the costs of use and ownership of capital by the firm are included in the accounts (and thus the profit shown) and that the capital of the firm is not consumed. An example of capital consumption would be where an excavator, which cost £100,000, had a life of ten years and no depreciation had been included in the accounts of the firm over that period. Ignoring tax, if all profits were distributed, the owners would have received £10,000 per year extra dividend (assuming a straight-line basis for depreciation) but at the end of the ten years, when the excavator was scrapped, there would be an imbalance in the accounts of £100,000. The balance sheet would show the firm's assets to be £100,000 less than the liabilities and owners' investment, the remedy to redress the balance being to reduce the owners' investment by £100,000.

9.4.7 Reserves

Reserves, as has already been noted, may be established against a specific future liability (e.g. a capital reserve for the future payment of capital gains tax on the sale of a building, the value of which is appreciating), or may be of a general nature. In either case, reserves represent increases in the owners' investment in the firm.

Firms grow via increased investment and this is obtained most easily by retaining profits. Thus reserves tend to grow quite rapidly; however, their rate of growth is constrained by the liquidity preference – as dividends – of the owners of the firm. As the funds (undistributed net profits after tax) from which the reserves have been created have been spent by the firm in purchasing fixed assets, etc., the reserves do not represent stocks of money that are available for distribution to the owners. Hence public companies periodically have scrip issues of shares; the shareholder's investment (as per the accounts) is increased by the value of the issue and the reserves are reduced by a corresponding amount. Such a process alters the capital structures of the companies and maintains a proportionality between issued shares and reserves.

9.4.8 *Accounting ratios*

Figure 9.23 shows the interrelationships of the main accounting ratios. Accounting ratios are used to aid the analysis of sets of accounts which are themselves often rather complex. It is important that not too much reliance for drawing conclusions is placed upon any individual ratio but that several ratios are considered together to produce a meaningful analysis.

Ratios can be broadly classified under three headings: working capital ratios, operating and profitability ratios, and supplementary ratios.

Working capital ratios

Current ratio

$$\text{Current ratio} = \frac{\text{Current assets}}{\text{Current liabilities}}$$

This ratio, given normal operating, indicates the ability of the firm to meet its short-term liabilities. This ratio should be approximately from 1.5:1 to 2:1.

Quick ratio (or acid test)

$$\text{Quick ratio} = \frac{\text{Very liquid assets (cash plus debtors)}}{\text{Current liabilities}}$$

This ratio indicates the ability of the firm to meet its (short-term) immediate liabilities in the worst circumstances. Thus, this ratio should be 1:1 or possibly slightly greater. If the ratio is much greater than 1:1 funds which could be gainfully employed are idle; if it is much less than 1:1 the firm may find it difficult to pay bills and so may be forced to wind-up its operations.

Working capital ratio

$$\text{Working capital ratio} = \frac{(\text{Current assets}) - (\text{Current liabilities})}{\text{Turnover}}$$

This ratio should be established for each firm (together with permissible variability) and should be maintained. This control will ensure that sufficient working capital is maintained to allow the firm to operate smoothly. The ratio will vary between firms (and between periods) due to the nature of the items which constitute working capital.

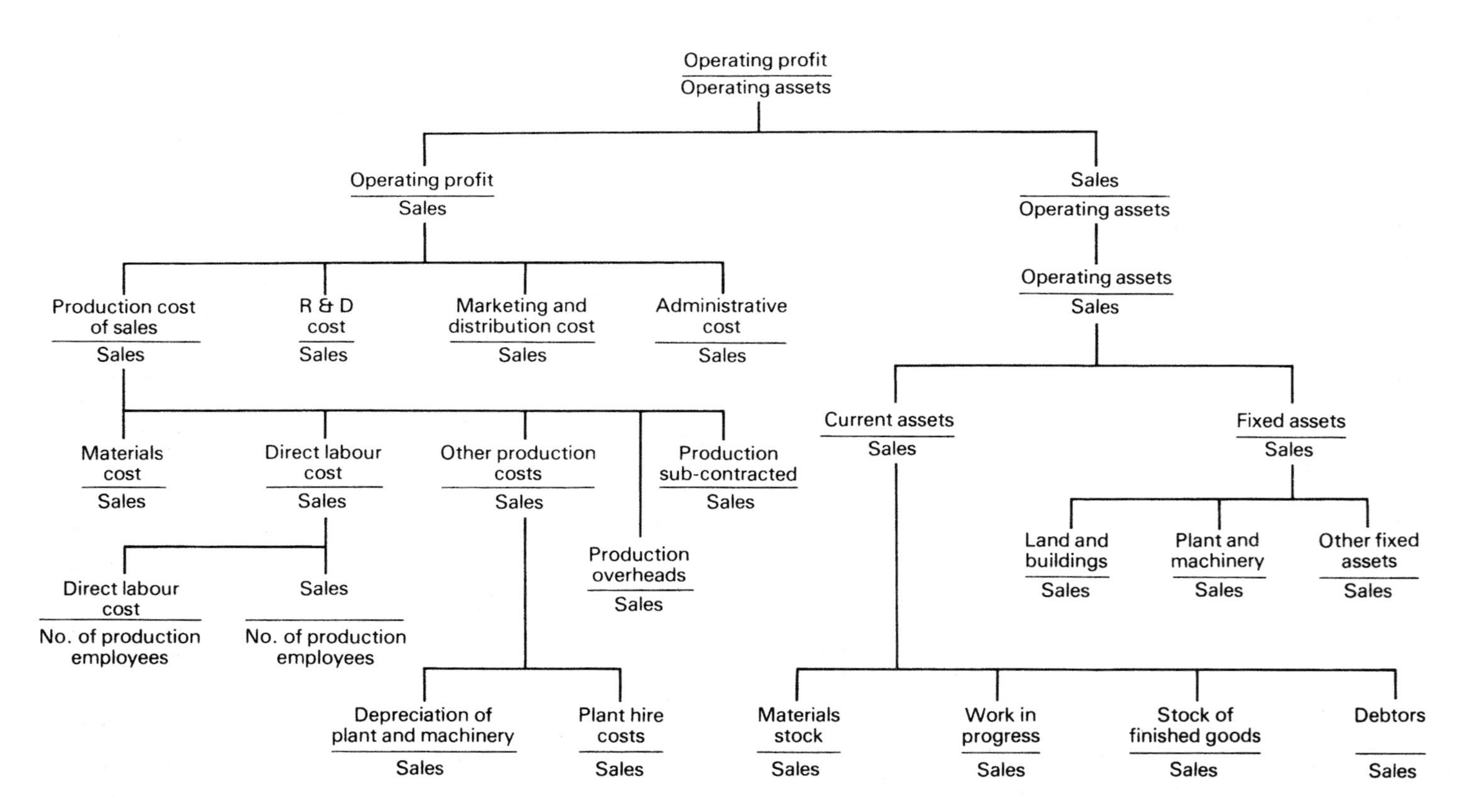

Fig. 9.23 Pyramid of ratios.

Operating ratios (usually expressed as percentages)

$$\frac{\text{Total costs, expenses and taxes}}{\text{Total income}}$$

This is the main operating ratio.

$$\frac{\text{Net profit before tax}}{\text{Turnover}}$$

Investment income should be excluded from this analysis as it is not included in turnover and would produce a false result if included in the net profit before tax.

$$\frac{\text{Net profit before tax}}{\text{Capital employed}}$$

It is useful to compare the ratio obtained when investment income is included in the net profit before tax with that obtained when investment income is excluded. Such an analysis will assist a decision of whether to sell the investment, to maintain the investment or to acquire further similar investments.

$$\frac{\text{Turnover}}{\text{Capital employed}}$$

In this case also, the effect of including any investment income should be examined. A low ratio indicates that the firm is capital-intensive in its operations. A firm which owns its own plant, for instance, will have a lower ratio than a similar-sized firm which hires plant.

There are many more operating ratios which are used to examine the detailed performance of firms. Sizer (1989) provides a full discussion of such ratios.

Supplementary ratios

Periods of credit:

$$\frac{\text{Debtors}}{\text{Turnover}} \times 12$$

This gives the period in months for which credit is offered by the firm. If the ratio is preferred in weeks or in days, them the multiple should be 52 or 365, instead of 12.

$$\frac{\text{Creditors}}{\text{Purchases}} \times 12$$

This gives the period of credit which the firm enjoys in months (or weeks or days).

Both these ratios are only quick and approximate indicators. As the sums involved for debtors and creditors may fluctuate significantly and rapidly, the periods obtained may not be very meaningful. It is preferable to calculate the periods of credit offered and enjoyed on a weighted average basis, as discussed in Chapter 8.

Productivity: Productivity is a measure of the output obtained from given amounts of input. The two factors of production whose productivity is most frequently considered are labour and capital. In the construction industry it is usual to consider primarily labour productivity, as construction is a labour-intensive industry. However, the situation in the industry has become more complex with the increasing incidence of subcontracting to the extent that productivity on a project is largely dependent on the productivity of the subcontractors employed; of course, the management of the project is very important too in selecting, coordinating, and controlling the subcontractors' performance and in ensuring that information and other resources are available as and when required to progress the works and that the workplaces are ready for the succeeding operations.

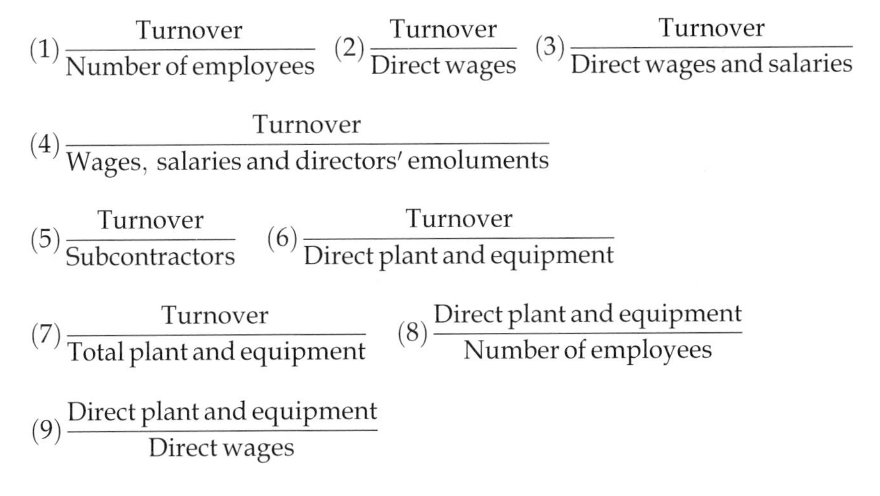

$$(1)\ \frac{\text{Turnover}}{\text{Number of employees}} \quad (2)\ \frac{\text{Turnover}}{\text{Direct wages}} \quad (3)\ \frac{\text{Turnover}}{\text{Direct wages and salaries}}$$

$$(4)\ \frac{\text{Turnover}}{\text{Wages, salaries and directors' emoluments}}$$

$$(5)\ \frac{\text{Turnover}}{\text{Subcontractors}} \quad (6)\ \frac{\text{Turnover}}{\text{Direct plant and equipment}}$$

$$(7)\ \frac{\text{Turnover}}{\text{Total plant and equipment}} \quad (8)\ \frac{\text{Direct plant and equipment}}{\text{Number of employees}}$$

$$(9)\ \frac{\text{Direct plant and equipment}}{\text{Direct wages}}$$

For many of the productivity ratios, it is useful to recast them in terms reflecting the use of subcontractors, thus the denominators comprise operatives on site, payments to subcontractors, etc. as appropriate – and shown specifically in (5).

Many ratios may be constructed to render monitoring information. It is useful to consider ratios which evaluate the performance at the workplace (site) with those which evaluate the overall performance of the firm. Such analyses will indicate, over a period, the sources of productivity changes, i.e. whether they are due to on-site performance or managerial performance, etc.

Returns to investors:

$$(1)\ \frac{\text{Net profit}}{\text{Capital employed}}$$

(Net profit is profit before interest payments and taxation.) This ratio indicates the overall return the firm has earned on its total capital employed and thus the efficiency of the operations.

$$(2)\ \text{Return on shares} = \frac{\text{Net profit after tax}}{\text{Shareholders' investments}}$$

The net profit after tax is the residue attributable to the shareholders. The shareholders' investments comprise the capital subscribed by the shareholders plus the reserves attributable to the shareholders.

$$(3)\ \text{Return on equity} = \frac{\text{Net profit attributable to holders}}{\text{Equity holders' investment}}$$

Ratios (1), (2), and (3) reflect the capital structure of a firm and are subject to considerable influence from the gearing of the organization and the system of taxing firms' earnings.

$$(4)\ \text{Interest cover} = \frac{\text{Earning before interest and taxation}}{\text{Interest payable}}$$

Ratio (4) shows the vulnerability of shareholders' earnings – if the ratio is low, the shareholders may receive no dividend, etc. if the earnings of the firm fall; further, in such circumstances, the firm may have difficulty in meeting its interest charges. Hence firms and industries establish norms for the ratio which reflect their financial structures and risks.

$$(5)\ \text{Earning per share} = \frac{\text{Profit on ordinary activities after taxation and minority interests}}{\text{Number of ordinary shares issued}}$$

Ratio (5) assumes no preference shares are issued. Quoted companies must show the earnings per share on the profit and loss account. Earnings per share may change for two main reasons: changes in the firm's profitability and changes in the number of shares issued.

$$(6)\ \text{Price} - \text{earnings} = \frac{\text{Share price}}{\text{Earnings per share}}$$

The price-earnings (PE) ratio is a very common measure for judging the share price of a quoted company. The ratio indicates how the stock market rates the company and its performance; a high ratio reflects a good rating although a low ratio might reflect nothing more than a temporary fall in the company's profits.

Numerous ratios may be used in the analysis of published accounts but many firms, usually the smaller firms in an industry, are not required to publish such information (see Chapter 7). Also, as much more internal information is available for the analysis of performance, ratio analysis forms an integral part of a firm's internal system of control. Construction firms often use ratio analysis to monitor productivity on each site as well as for the firm as a whole.

9.4.9 Interpretation of financial statements

Interpretation of financial statements usually revolves around the profit and loss account and the balance sheet, supplemented by the other published accounts, the notes to the accounts and the directors' and auditors' reports. Internally firms have more information and details to use for decision-making. It is, of course, useful to study several periods so that 'trends' may be detected rather than to view each set of accounts in isolation. However, it should be recalled that accounts portray the financial consequences for the firm of its transactions. It is the underlying economic variables which govern changes. Hence five years of accounts should be employed to establish trends, and the validity of trends in accounts, even if so established, is likely to be about three years only.

Of particular value in providing amplifying detail of the figures which appear on the face of the accounts are the notes to the accounts. The notes will include details of investment income, depreciation charges, audit fees and directors' remuneration.

It is important for analyses to consider all available information. Trends in sums in the accounts, accounting ratios and trends therein should be reviewed against each other and in the light of the sums for individual items (and groups of items) in the accounts. The sums in the accounts should be examined for realism – especially, in consideration of the sums necessary for the firm to operate reasonably – thus considering the normal procedures (notably concerning credit periods and cash holding requirements) in the industry.

In examining sums, it is important to consider 'real' (constant price) as well as 'cash' amounts. The 'deflator' used to convert cash to real sums should be appropriate and noted; it may be advisable to use different deflators for conversion of different sums, e.g. tender price index, building cost index.

As for any analysis, simple and fundamental aspects – especially liquidity – should be considered initially; if such examinations show that the firm is worth examining further, the analyses should proceed with progressively more detailed examinations.

The profit and loss account will provide information about bank charges, interest on loans and bad debts which the firm has incurred during the period. These items provide an insight into the firm's financial management capabilities and its credit control, including assessments of customers' credit worthiness.

Ratios may be extracted and examined: income from each source to turnover; profit (gross, net, net of tax) to turnover, etc. Ratio analysis is particularly valuable in determining whether the firm is benefiting as it should from an increase in turnover, and if not, why not. Here it is essential that several consecutive sets of accounts are examined in the context of prevailing economic conditions and the policies and objectives of the firm.

As the balance sheet is a picture of the firm's assets and liabilities at a particular instant, the importance of comparisons with previous accounts is enhanced. Ratios are again used for the analyses. The working capital ratios indicate the solvency of the firm, the quick ratio indicating the firm's liquidity. The ratios of debtors:turnover and creditors:turnover indicate the credit control of the firm. The return on capital employed (profit expressed as a percentage of net assets) provides the answer to the primary question, 'Is it worthwhile remaining in business or could the capital earn a better return if invested elsewhere?' In a recession, a contractor's return on capital employed may be very low but over a long period (which will hopefully include some buoyant times) should be significantly higher, sufficient to keep the firm in business.

A firm must consider, monitor and control its performance in both the long and the short-periods as inadequacies in either (or both) instance may cause the firm to be wound up. In the short-period, solvency and liquidity requirements dominate; in the long-period it is profitability, in particular the returns available to the owners for their investment when compared with alternative investments (the opportunity cost of the investment), and growth which dominate. Trends provide vital warnings of impending failure; the earlier unfavourable trends are detected, the easier is their correction. All firms prepare accounts annually and all (except small) companies must produce interim accounts at mid-year. Building contractors monitor performance on their projects at one or three-month intervals: a sound and essential policy in an industry where economic fluctuations occur rapidly and often unpredictably and tend to be of great amplitude.

As with the monitoring techniques discussed in Chapter 8, the more frequent and sophisticated the monitoring system, the more expensive it is (itself adding to overheads and thereby reducing profit). Again a cost-benefit evaluation should be carried out to determine the optimum frequency and detail of the system. Failure of an individual project to earn the required profit is serious but failure of the firm to earn adequate long-period profits or, more particularly in building, the inability of a firm to meet its short-term demands for cash (a liquidity failure) is untenable.

Summary

A prime function of management is decision-making. Many decisions will be made in connection with the financing of the firm and its activities. Such decisions must be based upon sound data and act towards the realization of defined objectives. There is much subjectivity in decision-making but it is possible to introduce a measure of objectivity by the application of decision-making techniques such as decision analysis. The personality of the decision-maker cannot be ignored but the application of scientific decision-making techniques does promote consistency, thoroughness, and objectivity, even in the evaluation of unknowns.

The data upon which financial decisions are made are obtained from financial reports and statements. To interpret such information it is necessary to appreciate the system under which the information is produced. The accounting system in the UK has its own terminology, concepts and conventions which are used to record the transactions entered into by firms and ultimately to produce financial statements, the best known of which are the profit and loss account and the balance sheet.

Thus financial management is the ability to understand and interpret financial statements and reports and to make reasoned and objective decisions in controlling the operations of the firm. This will inevitably involve the successful management of working capital and all its components. Ratio analyses are very useful in interpreting accounts, particularly for comparisons with standards, competitors, and to identify trends. The ability to correctly interpret financial statements is a first, but significant, step towards successful financial management.

Questions

1. (a) 'Building is a risky business.' Discuss.
   (b) Why are builders more likely to be optimists than pessimists?
2. How may a contractor assess the best mark-up to apply in bidding competitively for a project?
3. Describe and discuss a technique by which a developer may select the most appropriate method of letting a new building project.
4. What is depreciation and why is it necessary? How may depreciation be applied to (a) plant and (b) buildings? Discuss the effects of depreciation provisions on the profit and loss account and the balance sheet.

5. In what ways are accounting conventions, when applied to building contractors, at variance with their usual application?
6. What is working capital and why is it considered to be of major importance to building firms?
7. Record the necessary accounting entries and produce a profit and loss account and a balance sheet for the following transactions which occurred in the first year of trading of A Bodger Ltd.
   (a) Jan. Cash sale of 6,000 shares of £1 each (total authorized share capital).
   (b) 3 Jan. Purchase of builder's yard on three months' credit for £5,000; cash deposit of 20 per cent paid.
   (c) 21 Jan. Issue of debenture – 10 years, 15 per cent, £3,000.
   (d) 28 Jan. Purchase of mortar mixer and sundry small items of plant for £1,500, cash deposit 10 per cent, credit for balance of one month.
   (e) 4 Feb. Cash purchase of equipment and small tools, £500.
   (f) 28 Feb. Purchase materials £700 on one month credit.
   (g) 30 April Cash receipt £2,200.
   (h) 21 May Purchase materials £900 on one month credit.
   (i) 1 July Cash receipt £3,200.
   (j) 1 July A Bodger: half-year salary £2,000.
   (k) 8 July Purchase materials £650 for cash.
   (l) 10 Aug. Cash receipt £3,000.
   (m) 1 Sept. Cash receipt £2,100.
   (n) 3 Nov. Purchase materials £1,000 on one month credit.
   (o) 1 Dec. Cash receipt £2,000.

   From 1 Mar. Wages paid at £500 per month.

   At 31 Dec. a contract of value £4,600 is 50 per cent complete, £2,000 having been received as the single-stage payment; stocks of materials is £400. Depreciation charged on plant and equipment at 25 per cent per annum.
8. Draft a report for A Bodger Ltd on its first year of trading and its position at 31 Dec. (information given in Question 7).

References and bibliography

Baumol, W.J. (1959) *Business Behaviour, Value and Growth*, Macmillan.

Bennett, J. (1982) 'Cost Planning and Computers', from Brandon, P.S. (Editor), *Building Cost techniques: New Directions*, E & F N Spon.

Bierman, H. Jr (1963) *Financial and Managerial Accounting – An Introduction*, Macmillan.

Blockley, D.I. (1992) 'Engineering from reflective practice', *Research in Engineering Design*, 4 (pp. 13–22).

Button, K.J. (1985) 'New Approaches to the Regulation of Industry', *The Royal Bank of Scotland Review*, Number 148, December (pp. 18–27).

Fellows, R.F. and Langford, D.A. (1980) 'Decision theory and tendering', *Building Technology and Management*, October.

Fellows, R.F. (1996) 'The Management of Risk', *Construction Papers*, Number 65, Chartered Institute of Building.

Fine, B. (1975) 'Tendering Strategy', from Turin, D.A. (Editor), *Aspects of the Economics of Construction*, George Godwin.

Flanagan, R. and Norman, G. (1993) *Risk Management and Construction*, Blackwell Science.

Friedman, L. (1956) 'A Competitive Bidding Strategy', *Operations Research*, 44 (pp. 104–112).

Gates, M. (1967) 'Bidding Strategies and Probabilities', *Journal of the Construction Division of the American Society of Civil Engineers* (ASCE), 93, CO1, March (pp. 74–107).

Gray, C. (1983) 'Estimating Preliminaries', *Building Technology and Management*, April, 5.

Green, S.D. and Popper, P. (1990) *Value Engineering: The Search for Unnecessary Cost*, Occasional Paper Number 39, Chartered Institute of Building, Ascot.

Hanson, J.L. (1970) *A Textbook of Economics*, Macdonald and Evans.

Harris, F. and McCaffer, R. (1989) *Modern Construction Management* (Third Edition), BSP Professional Books.

Hillebrandt, P.M. (1985) *Economic Theory and the Construction Industry* (Second Edition), Macmillan.

Kaufman, G.M. and Thomas, H. (Editors) (1977) *Modern Decision Analysis*, Penguin.

Kelly, J. and Male, S. (1993) *Value Management in Design and Construction: the economic management of projects*, E & F N Spon.

Lipsey, R.G. (1979) *An Introduction to Positive Economics* (Fifth Edition), Weidenfeld and Nicolson.

Moore, P.G. and Thomas, H. (1976) *The Anatomy of Decisions*, Penguin.

Morris, D.J. (Editor) (1979) 'The behaviour of firms', *The Economic System in the UK* (Second Edition), Oxford University Press.

Morrison, N. (1984) 'The accuracy of quantity surveyors' cost estimating', *Construction Management and Economics*, 2 (pp. 57–75).

Norton, B.R. (1994) 'Value Management Techniques in Construction', *CIOB Directory 1994/5*, Chartered Institute of Building, Ascot.

Perry, J.G. and Hayes, R.W. (1985) 'Construction Projects – know the risks' *Chartered Mechanical Engineer*, February (pp. 42–45).

Perry, J.D. and Hoare, D.J. (1992) 'Contracts of the Future: risks and rewards', *Proceedings Construction Law 2000, Fifth Annual Conference*, Centre of Construction Law and Management, London, September.

Schofield, C.D.A. (1975) 'What is the cost of capital?', *The Quantity Surveyor*, April.

Sizer, J. (1989) *An Insight into Management Accounting* (Third Edition), Penguin.

Tah, J.H.M., Thorpe, A. and McCaffer, R. (1994) 'A survey of indirect cost estimating in practice', *Journal of Construction Management and Economics*, volume 12, Number 1 (pp. 31–36).

Uher, T.E. (1990) 'The Variability of Subcontractors' Bids', *Proceedings, CIB 90: Building Economics and Construction Management*, 6, University of Technology, Sydney (pp. 576–586).

Venkataramanan, S.S. (1991) *V.E. Basics in a Nutshell* (Third Edition), Venconvave PVT (Ltd.), New Delhi.

Willenbrock, J.H. (1973) 'Utility Function Determination for Bidding Models', *Journal of the Construction Division of the American Society of Civil Engineers* (ASCE), CO1, July (pp. 133–153).

10 Quantitative Decision-making

Many management problems are described in numerical terms and call for numerical answers. In assessing the cost or time for a construction project it is not sufficient to use qualitative terms such as 'cheap' and 'expensive' or 'fast' and 'slow'. We have to express our forecasts and measure our progress in numbers and there are examples throughout the earlier chapters of this book. In everyday life, too, we use numerical measures to judge such diverse activities as the punctuality of transport services, the academic achievements of schools and universities, and the prowess of athletes and sportsmen. At the highest level, the economic health of a nation is measured by statistics such as exchange rates, bank rates and unemployment figures.

The concept of *scientific management* was put forward at the beginning of the twentieth century in the context of factory production (Taylor 1911) and has since been developed to analyse every aspect of management. From the early techniques of time and motion studies in workshops, the subject has developed so that mathematical techniques are now brought to bear on problems that were previously tackled intuitively or by trial and error. The use of numerical techniques is now so widespread that some authors (Anderson *et al.* 1994) use the term *quantitative decision-making* as being synonymous with scientific management.

The general approach is to set up a 'model' – that is some representation, usually on paper, that possesses certain properties of the project, system or organization in which we are interested. The model might be a graph, a network, a table of values, a mathematical formula, or a computer program. In the field of construction it could be a time chart or a bill of quantities. By investigating the behaviour of the model we can predict to some extent what will happen in practice, and this can help us in making decisions. Computer spreadsheets, such as Microsoft Excel and Lotus 1-2-3, with their facility for tackling 'what if?' questions, are widely used as models in assessing financial examples.

An analysis of computer technology in the 1990s has shown (Bedford 1999) that clock speed, memory and hard disk capacity have grown logarithmically during the decade. If this pattern continues, the computer of the year 2010 is likely to have a clock speed of 100 GHz compared to the typical 500 MHZ in the year 1999 and a hard disk capacity of 3200 Mb instead of the current 10 Mb. In contrast, it is conjectured that the price of desktop and portable machines will not increase. The construction manager is not likely to be handicapped by the limitations of his or her computer!

Nevertheless, however much computer power is available, no model can reproduce every feature of the original system and the results must be interpreted in the light of the data on which the models are built and the approximations inherent in constructing them.

Furthermore the manager is generally dealing with future activities and a measure of uncertainly is inevitable. For this reason it is often necessary to apply probability theory and statistical techniques, and to regard the results of the analysis as 'average' or 'expected' values.

10.1 Cost models

Several examples of cost modelling are to be found in earlier chapters. In Chapter 5, a model in the form of a graph is given (Figure 5.4) for the costs of accident prevention and Chapter 8 describes the S-curve (Figure 8.5) that represents the accumulation of expenditure throughout the life of a construction project. The same chapter introduces the discounted cash flow model for investment appraisal. The following models are used in the next problems:

- fixed and constant marginal costs straight-line model
- two cost components that are inversely proportional
- reducing capital sum mortgage model
- life costs of equipment allowing for maintenance and depreciation

Problem 10.1

A contractor is considering the purchase of a van that is expected to cover 20,000 miles per annum. He examines the running costs for two vans in his existing fleet that perform similar duties but are of different types. He finds that in the last year van A covered 15,000 miles at a total cost of £6,120, the corresponding figures for van B being 25,000 and £7,700. Can a decision be made on the basis of this information as to which of these types is likely to prove the cheaper?

In a more detailed analysis the contractor divides the costs into two groups fixed and variable. The 'fixed' annual charges such as tax, insurance depreciation and the replacement of components that wear due to age rather than mileage are estimated to be £3,120 for A and £4,200 for B. The remaining costs arise from those items – such as routine servicing and tyre replacements – that are related to the distance covered. Which type of van is likely to prove the cheaper for the expected annual mileage of 20,000?

Solution

It is not possible to forecast which van will be cheaper for an annual mileage of 20,000 from the data given in the first paragraph. The gross annual cost for A (£6,120) is less than for B (£7,700) but the average cost per mile is smaller for B (£7,700/25,000 = 30.8p as against 40.8p for A). We are not justified in applying these average values to other mileages because part of the total cost is attributed to tax, insurance and other items that are independent of the distances covered.

The second paragraph of the question suggests that the total cost may be divided into two groups; the 'fixed' or 'standing' charges that are independent of how much the vehicles are used and those costs that are related to the distances covered. Assuming that the mileage-related costs are directly proportional to distance, a simple and familiar model is obtained. If a is the annual standing charge and b is the additional cost per mile, or *marginal cost*, the total cost for a year in which x miles are covered is given by:

$$C = a + bx \qquad [10.1]$$

Working in pounds and using the figures for van A, we have $a = 3{,}120$ and $b = (6{,}120 - 3{,}120)/15{,}000 = 0.2$. For this van, therefore, the cost model is:

$$C = 3{,}120 + 0.2x$$

The corresponding result for van B is:

$$C = 4{,}200 + 0.14x$$

Figure 10.1 presents these results in graphical form. The equations give straight lines which intersect at a point corresponding to 18,000 miles. For smaller annual mileages van A has the lower total cost but above this figure van B is the cheaper. With the assumptions made in this analysis, and the data given in the question, van B has the lower total cost if the annual mileage is 20,000. The limitations of this analysis must be borne in mind, however. The cost information for the vehicles relates to a single mileage in each case and the assumptions made in such a model should be assessed carefully. We shall return to these matters in Section 10.4.

It is instructive to consider the variation in average cost per mile. Dividing the equation through by x, the average cost is seen to be:

$$\frac{C}{x} = \frac{a}{x} + b \qquad [10.2]$$

Figure 10.2 shows the variation for the two types of van. Each graph is a curve and approaches a limiting value or *asymptote* as the mileage x increases. These two graphs also intersect at the point where $x = 18{,}000$.

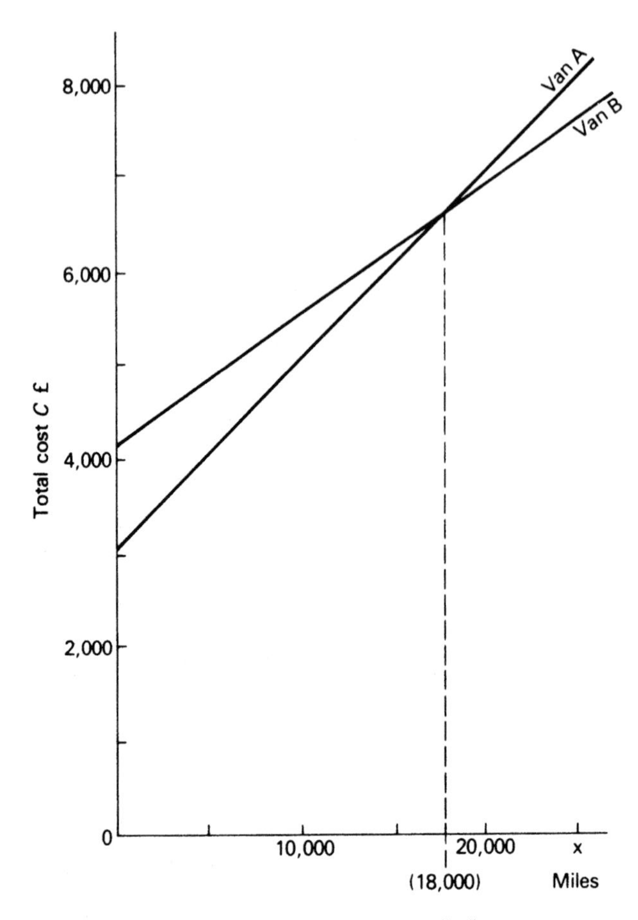

Fig. 10.1 Linear solution (mileage cost and distance travelled).

The cost model represented by equations [10.1] and [10.2] has wide applications. In the case of industrial production, the quantity a represents the overheads or fixed costs and b is the marginal cost for each unit produced. The total cost of producing x units is given by equation [10.1] and the average cost per unit by [10.2]. As before, the average unit cost decreases as x increases, an effect known as the *economy of scale*. However, there is a limit set by the asymptotic value b, the marginal cost, and no increase in the number of units can bring the average cost below this figure. There have been instances of large, long-established organizations finding themselves in severe financial difficulties by ignoring this basic truth.

The same cost model given by Equation [10.1] applies to many electricity, gas and telephone charges. These are normally based on a standing charge per month or quarter to which is added a sum for each unit used.

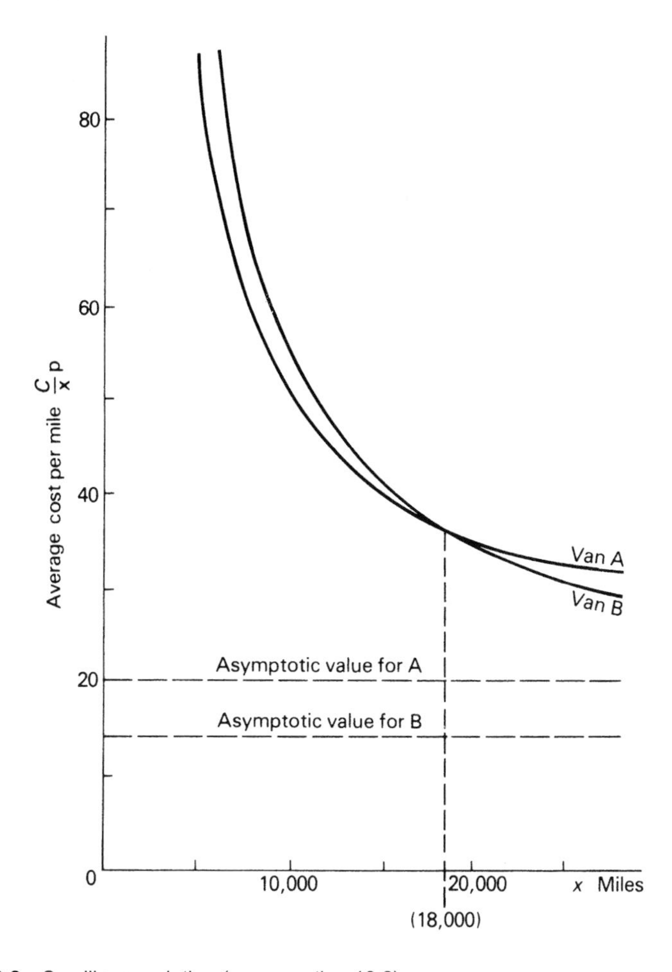

Fig. 10.2 Curvilinear solution (see equation 10.2).

Problem 10.2

In reviewing the operation of plant on a construction site it was found that in one quarter £2,000 was spent on preventive maintenance, and the total cost of breakdowns, including repairs and consequential losses, was £12,000. Advise on the scale of the maintenance programme, basing your decision on the assumption that the costs under these two headings are inversely proportional to one another.

Solution

There are many activities in which an increase in expenditure under one heading can lead to a cost reduction elsewhere. One example – accident prevention – is analysed in Chapter 5 and another – inventory of stock control – is examined later in this chapter, Section 10.5.

The exact relationship between the costs under the two headings is likely to vary from one case to another and the example given in the question is only one of many possibilities. It leads to a very simple model in which an increase in preventive cost by a given factor is accompanied by a reduction in breakdown cost by the same factor. Taking factors of 1–6 and tabulating costs we have:

Table 10.1

Factor	*1*	*2*	*3*	*4*	*5*	*6*
Preventive cost (£)	2,000	4,000	6,000	8,000	10,000	12,000
Breakdown cost (£)	12,000	6,000	4,000	3,000	2,400	2,000
Total cost (£)	14,000	10,000	10,000	11,000	12,400	14,000

If the results are plotted on a suitable base such as the value of the factor, it is found that the preventive costs result in a straight line and the breakdown costs produce a curve not unlike those of Figure 10.2. The total cost is represented by a curve that has a minimum just below 10,000 between factors 2 and 3.

The minimum total cost occurs at a point where the separate preventive and breakdown costs are equal – a result which can be shown by calculus to be universally true for this particular cost model. Furthermore, the product of the two separate costs is the same for each of the factors (24×10^6) and, at this minimum point, each is equal to the square root of this product. Thus, for minimum total cost, the preventive cost is:

$$\sqrt{24 \times 10^6} = £4{,}900$$

On the basis of the assumption suggested in the question, the maintenance programme should be expanded from its present level of £2,000 per quarter to £4,900. Several points should be noted. The minimum total cost is £9,800 but, for preventive costs in the range £4,000 to £6,000, the variation is only 2 per cent. Other factors can therefore be considered in making a decision. For example, the expenditure on maintenance could be increased towards the £6,000 figure for reasons of morale or prestige without substantially raising the total cost. It must also be remembered that, in seeking the minimum total cost, the incentive to find a balanced judgement will be increased if one manager is responsible for expenditure under both headings.

Problem 10.3

What is the annual mortgage repayment on a building society loan of £78,000 at 6 per cent over 20 years under the reducing sum agreement if interest is added annually? How long will it take to discharge the debt if the interest rate is increased to (a) 7 per cent and (b) 9 per cent, the annual repayments being unaltered?

Solution

This is an example of *present value* (or *worth*) introduced in Chapter 8. From Table A.4 in Appendix A the present value factor for unit sum per annum over 20 years at 6 per cent ($i = 0.06$) is found to be 11.470. Hence for the loan of £78,000, we have:

$$\text{Annual repayment} = \frac{78{,}000}{11{,}470} = £6{,}800$$

The total repayments during the life of the mortgage are $20 \times 6800 =$ £136,000, a considerable increase on the sum initially borrowed. However, the rise in house property values in UK has generally outstripped such increases in recent decades to the benefit of the borrower. It should be noted that many building societies and other lenders add interest half-yearly or quarterly and their figures will differ from those calculated with the factors given in Table A.4. The results also depend on whether the interest is added at the beginning or the end of each period.

If the interest rate changes but the repayments are unaltered we look for the same factor, 11.470, in a different column. For (a), 7 per cent, the table gives almost exactly the same figure, 11.469, at 24 years. The increase in the interest rate would therefore extend the term of the mortgage by four years and increase the total sum repaid to £163,200. Finding an almost exact factor in such cases is fortuitous and it is usually necessary to interpolate the column of figures.

In (b), an inspection of the table shows that it is not possible to achieve a factor as high as 11.470 and the mortgage is never discharged. The interest added in the first year, 9 per cent of £78,000 = £7020, is greater than the annual repayment and the debt grows, rather than diminishes, year by year.

Problem 10.4

An item of earth-moving equipment has a capital cost of £50,000 and Table 10.2 shows the estimated maintenance costs and depreciation for the first eight years of its life:

Table 10.2

Year	*1*	*2*	*3*	*4*	*5*	*6*	*7*	*8*
Maintenance cost (£)	1,500	2,000	3,000	4,000	5,000	6,000	11,000	15,000
Depreciation (£)	8,000	8,000	7,000	6,000	6,000	5,000	4,000	3,000

If there is a permanent need for this item and the cost of capital (the interest rate for borrowing) is 15 per cent per annum, when should it be replaced?

Solution

This is another example in which the total cost is made up of two components that vary in opposite directions. When new, the maintenance costs of vehicles and other items are likely to be small but repair bills can grow as it wears out. On the other hand, depreciation will be high in the early years and will diminish when the residual value is small.

Present value techniques are used to effect a comparison between the costs for various life periods. It is convenient to set out the calculations in a table, such as Tables 10.3 and 10.4, where *PV* = present value and all cash values are in £. The task is very well suited to a computer spreadsheet and many commercial versions have the *PV* factors built in.

Table 10.3

① Year	② PV factor for single payment $i = 0.15$	③ Cash payments	④ PV of cash payments ② × ③	⑤ Cumulative total of PV of payments	⑥ Resale value	⑦ PV of resale value ② × ⑥	⑧ Net PV of life costs ⑤ − ⑦
0	1.0000	50,000	50,000	50,000	50,000	50,000	—
1	0.8696	1,500	1,304	51,304	42,000	36,523	14,781
2	0.7561	2,000	1,512	52,816	34,000	25,707	27,109
3	0.6575	3,000	1,973	54,789	27,000	17,753	37,036
4	0.5718	4,000	2,287	57,076	21,000	12,008	45,068
5	0.4972	5,000	2,486	59,562	15,000	7,458	52,104
6	0.4323	8,000	3,458	63,020	10,000	4,323	58,697
7	0.3759	11,000	4,135	67,155	6,000	2,255	64,900
8	0.3269	15,000	4,904	72,059	3,000	981	71,078

The present value factors for single payments, shown in column ①, are taken from Appendix A, Table A.2. Column ③ shows the capital expenditure at year 0 and the maintenance costs for each of the eight years. The present value of these items, obtained as the product of columns ② and ③, is shown in column ④. Column ⑤ gives the cumulative total of the present value of the outgoings obtained by summing the values in column ④ up to the required year in each case.

The resale values shown in column ⑥ are obtained by deducting the depreciation amounts from the original capital cost, and their present value amounts are shown in column ⑦ as the product of the corresponding figures in columns ② and ⑥. Finally in Table 10.3 the present

Table 10.4

⑨ Year	⑩ PV factor for equal annual payments	⑪ Net PV of life costs	⑫ Annual sum required in perpetuity ⑪ ÷ ⑩
0	—	—	—
1	0.870	14,781	16,990
2	1.626	27,109	16,672
3	2.283	37,036	16,223
4	2.855	45,068	15,786
5	3.352	52,104	15,544
6	3.784	58,697	15,511
7	4.160	64,900	15,601
8	4.487	71,078	15,841

value of the net life costs are shown in column ⑧, being the differences between the values in columns ⑤ and ⑦.

The figures in column ⑧ are not directly comparable because they apply to different life periods. It is not obvious, for example, whether the expenditure of £37,036 every three years is preferable to, say, the expenditure of £64,900 every seven. However, if we calculate the equal annual payments that would produce these amounts, we have the annual sums that would cover the costs in perpetuity. The calculations are shown in Table 10.4.

The values in column ⑩ are taken from Appendix A, Table A.4 and those in column ⑪ are reproduced in column ⑧. Finally, in column ⑫ we have the annual sums to cover the costs of this item for the various life periods.

The minimum value corresponds to replacement after six years, but the figure for five years is only slightly higher. With the given data, the item should be replaced every five or six years.

10.2 Probability

Many problems faced by managers are fraught with uncertainties. Success cannot be guaranteed and may well depend on matters outside our control such as the weather, the state of the economy and the decisions made by our competitors. In these circumstances we have to speak of probabilities rather than certainties.

Probability theory is used to access the chances of success and failure and the consequent financial implications of delays in construction projects. A method known as Risk Analysis and Management of Projects (RAMP) has been developed jointly by the Institute of Actuaries and the Institution of Civil Engineers (1998) to assess all kinds of risks and uncertainties so that they can be identified, evaluated, reduced and controlled.

Mathematicians denote probability by p and measure it on a scale of 0 to 1, with higher values representing greater probabilities. Thus $p = 1$ represents absolute certainty and $p = 0$ corresponds to impossibility. When two or more different outcomes are possible the probability of each may be regarded as the proportion of occasions on which each occurs *in the long run*. This concept of a proportion is familiar in everyday life. We talk about 'nine times out of ten', 'a fifty-fifty chance' or 'a chance in a million' to describe the likelihood of various outcomes but we can be more precise in some cases. Thus the probability of a tossed coin coming down heads is $p = 0.5$ and the probability of drawing, say, the ace of diamonds from a pack of cards is $p = 1/52 = 0.0192$.

We often need to combine the probability values of two or more outcomes and we can do this using two rules. If A and B are alternative outcomes in a given case, the probability of one or the other occurring is the sum of their separate probabilities. This is known as the *addition rule* and may be stated thus:

If A and B are mutually exclusive,

$$p(A \text{ or } B) = p(A) + p(B)$$

It follows from this rule that the sum of the probabilities of all the possible outcomes in a given case is 1. One particular example is the chances of success and failure. If a given project has a 0.7 probability of success then its probability of failure is $p = 0.3$. The rule is useful in calculations in which failure can occur in several different ways but success in only one.

The other rule concerns different cases or events. The probability of outcome A, in one case, and outcome B, in another, both occurring is the product of their separate probabilities. This is known as the *multiplication rule* and may be stated thus:

If A and B are independent,

$$p(A \text{ and } B) = p(A) \times p(B)$$

Returning to the example of the pack of cards, the probability of drawing *either* the ace of diamonds *or* the queen of hearts is obtained by the addition rule since each excludes the other. The result is:

$$p = 1/52 + 1/52 = 1/26 = 0.0385$$

Suppose, however, we have two packs of cards. The probability of drawing the ace of diamonds from the first and the queen of hearts from the second is given by the multiplication rule since the two events are independent. The result is:

$p = 1/52 \times 1/52 = 0.000370$

The result would be the same if each of the specified cards were drawn from the other packs and, by the addition rule, the probability of drawing the two cards, one from each pack, is therefore 0.000370 + 0.000370 = 0.000740.

Some examples involving the concept of probability are to be found in Chapter 9; other applications are illustrated by the following problems.

Problem 10.5

Observations over 100 days on the use of dumper trucks on a construction site gave the following pattern of demand:

Number of dumper trucks required	0	1	2	3
Number of days	23	36	27	14

It is expected that this pattern of demand will continue. What is the average daily cost if the trucks are hired by the day when required at an all-in cost of £25 per day?

The trucks are also available on long-term leasing at a basic charge of £5 per day with an additional cost of £10 for each day they are used. Will it pay to use long-term leasing to meet part of the demand and, if so, how many trucks should be leased?

Solution

From the figures given in the question the probability of requiring none, one, two, or three trucks on a particular day are $p = 0.23, 0.36, 0.27$ and 0.14 respectively, and these four values add up to 1. If the required trucks are all hired by the day the corresponding daily bills are £0, £25, £50 and £75. Combining the probabilities with these values gives the result:

$$\textit{average daily cost} = 0.23 \times £0 + 0.36 \times £25 + 0.27 \times £50 + 0.14 \times £75$$
$$= £33.00$$

If one truck is leased long-term, the daily costs when none, one, two and three trucks are required are, respectively, £5, £15, £40 and £65. The calculation becomes:

$$\textit{average daily cost} = 0.23 \times £5 + 0.36 \times £15 + 0.27 \times £40 + 0.14 \times £65$$
$$= £26.45$$

It will be found that with two trucks leased the result is £25.30 and with three it becomes £28.20. Leasing therefore pays and two trucks should be leased.

Problem 10.6

A major company is planning its business strategy for the next five years. It estimates that there is a 0.3 probability of high growth in the construction market during this period, a 0.5 probability of low growth, and a 0.2 probability of zero growth. Three courses of action are considered:

(a) Expansion now. By recruiting staff immediately and developing regional offices the company would be in a strong position if growth were high but this policy would reduce profitability in the event of zero growth.
(b) No expansion. This would put the company at a disadvantage should the market grow but would enable it to operate economically if the size of the market were static.
(c) Reappraisal after two years. Under conditions of high growth the policy would lead to lower profits than (a) but it would enable the company to keep its options open and to choose an appropriate level of expansion for the remaining years.

The estimated outcomes, in terms of percentage profitability, are shown in Table 10.5. Analyse the problem and recommend a course of action.

Table 10.5

	High growth	*Low growth*	*Zero growth*
Expansion now	20	13	0
No expansion	10	9	8
Reappraisal after two years	16	14	8

Solution

A convenient model for this problem is the 'decision tree' introduced in Chapter 9, Figure 9.11. Figure 10.3 gives the details, with decision points indicated by [A], [B], [C] and [D], and chance points shown by circles. At each of the decision points [B], [C] and [D] there is a choice of no expansion or an expansion appropriate to the state of the market.

In the case of zero growth [D], the reappraisal merely confirms the policy of zero expansion, but at [B] and [C] the expansion choices are preferred. With these choices the decision tree is reduced to the form of Figure 10.4, the three choices at A being labelled (a), (b) and (c).

The 'average' profitability corresponding to initial decision (a) is:

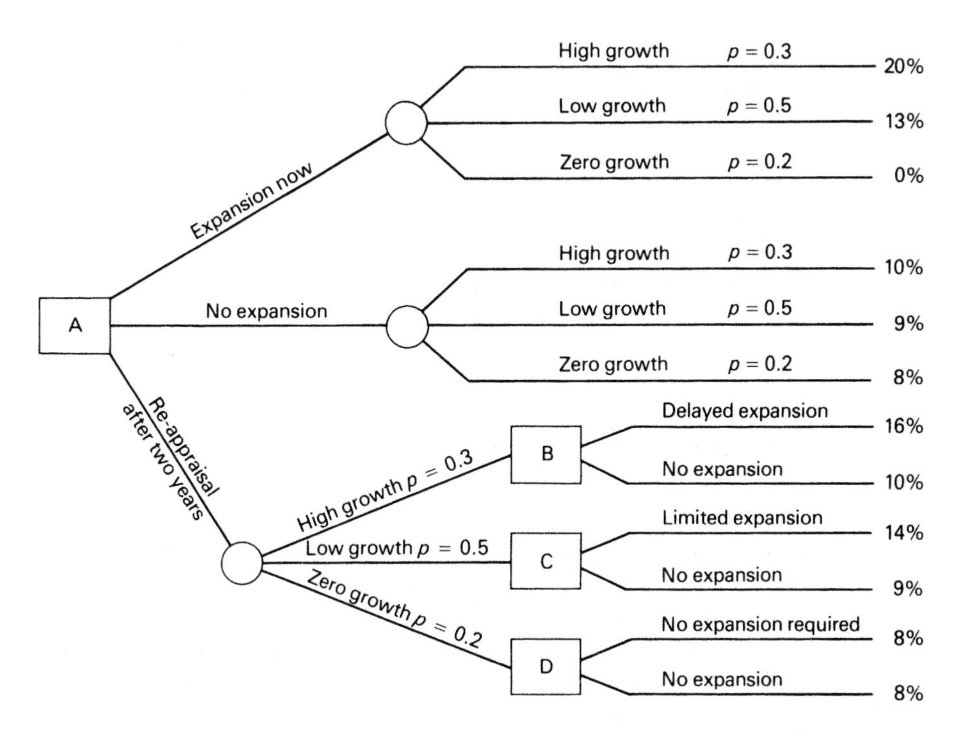

Fig. 10.3 Decision tree.

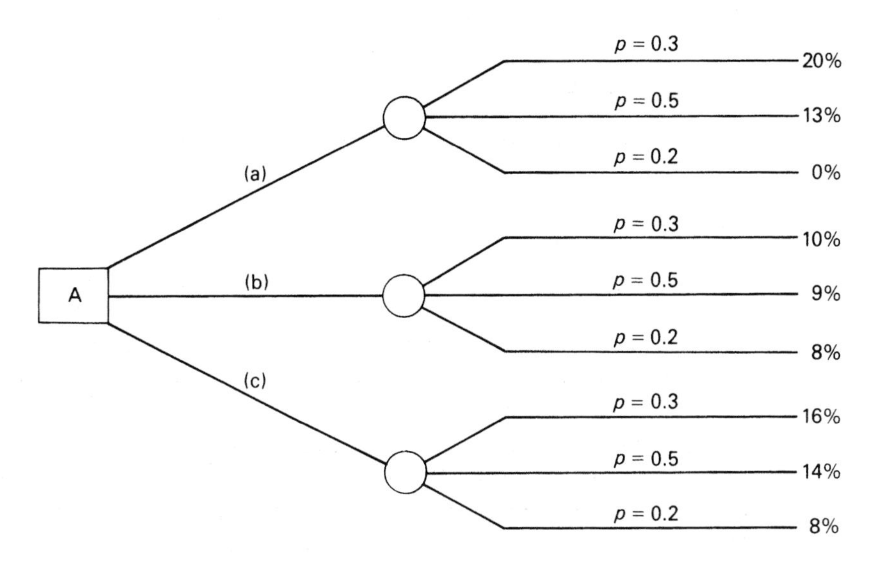

Fig. 10.4 Reduced form decision tree.

$$0.3 \times 20 + 0.5 \times 13 + 0.2 \times 0 = 12.5\%$$

This result, found by combining the separate profitability figures in the proportions of their probabilities is often called the 'expected' value but, in one sense, it is a misnomer. With the assumptions and data given in the question, the outcome will be 20%, 13% or 0%. The argument is that we will face many such problems over a period of time and we can expect the average outcome to be 12.5%.

The corresponding results for initial choices (b) and (c) are 9.1 and 13.4, respectively. The highest of the three values is given by (c) and the initial decision should be to reappraise the policy after two years.

Problem 10.7

A safety protection system comprises three units: detector, microprocessor and shutdown valve, linked in series as shown in Figure 10.5. The system will only operate correctly if all three units are working satisfactorily. In a given period the probabilities of failure of these units are 0.08, 0.06 and 0.10 respectively. What is the probability that the system will fail during this period?

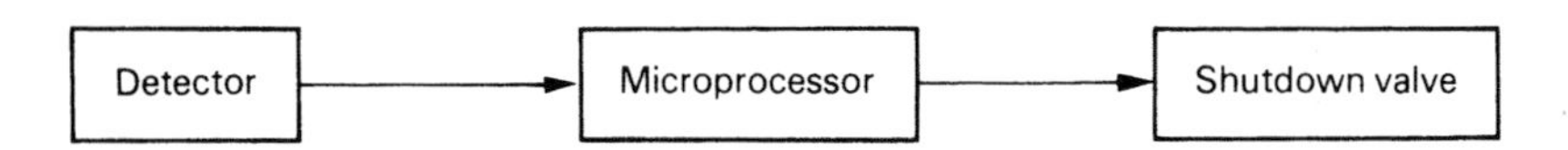

Fig. 10.5 Series system.

The reliability is to be improved by duplicating one or more of the devices by equally reliable units as shown in Figure 10.7. At (a) a second detector is added, at (b) two identical systems are used and, at (c), cross-links are provided so that signals can flow by alternative routes. What will the probability of system failure be in each case?

Solution

If success and failure are denoted by s and f respectively then for each of the units or the complete system:

$$p(s) + p(f) = 1$$

Using this result, the success probabilities of the individual units are $p(s) =$ 0.92, 0.94 and 0.90. A tree diagram for the system, Figure 10.6, shows that there are eight possible routes and outcomes. Seven of these contain one or more failures and only one, shown at the bottom, constitutes success of the system. The multiplication rule can be applied to this path and thus for the system:

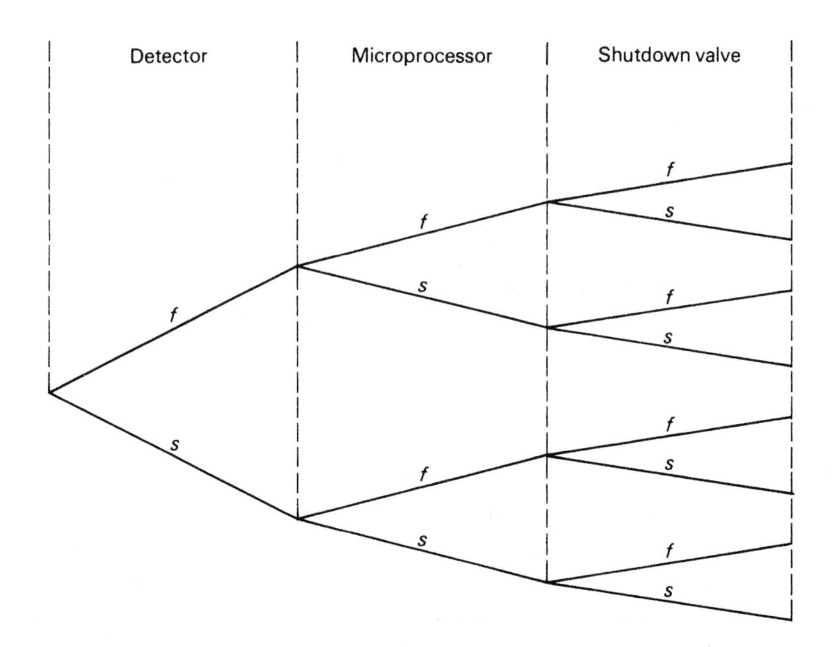

Fig. 10.6 System tree diagram.

$p(s) = 0.92 \times 0.94 \times 0.90 = 0.778$

Hence

$p(f) = 1 - p(s) = 1 - 0.778 = 0.222$

This result can also be obtained (but more tediously) by calculating the probabilities of the seven routes through tree diagrams that contain at least one failure, and adding the results. It is also worth noting that when the failure probabilities are small a good approximation to the result can be found by adding the separate values (0.08, 0.06 and 0.10). These total 0.24 compared with the 'exact' value of 0.222.

Figure 10.7(a) shows the modified system with the detector duplicated. Failure at the detector stage will only occur if both units fail. Hence, by using the multiplication rule, the probability of failure of both detector units is:

$p(f) = 0.08 \times 0.08 = 0.0064$

and the probability of success of *at least* one detector unit is:

$p(s) = 1 - 0.0064 = 0.9936$

The probability of success of the whole system is now:

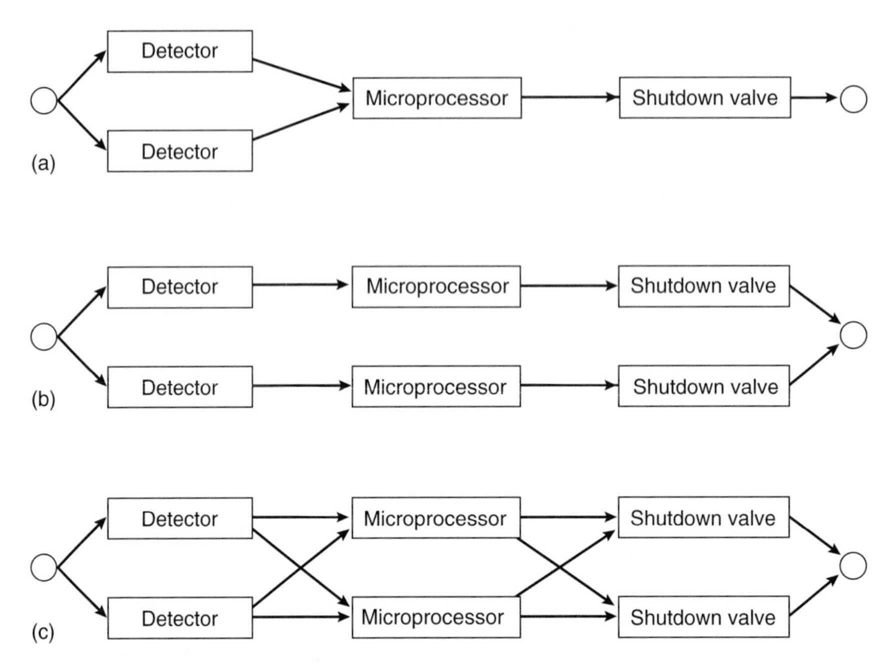

Fig. 10.7 Modified system with detector duplicated.

$$p(s) = 0.9936 \times 0.94 \times 0.90 = 0.8406$$

and the probability of overall system failure is:

$$p(f) = 1 - 0.8406 = 0.1594$$

If the original system is duplicated (b), the probability of both branches failing is:

$$p(f) = 0.222 \times 0.222 = 0.0493$$

In (c), the probability of at least one detector unit succeeding is 0.9936. The probability of both microprocessors failing is:

$$p(f) = 0.06 \times 0.06 = 0.0036$$

and the probability of at least one succeeding is:

$$p(s) = 1 - 0.0036 = 0.9964$$

By similar reasoning the probability of at least one shutdown valve succeeding is:

$$p(s) = 1 - (0.10 \times 0.10) = 0.9900$$

Overall success depends on success at each of the three stages, and combining the success probabilities for each:

$$p(s) = 0.9926 \times 0.9964 \times 0.9900 = 0.9791$$

and the probability of failure for the system as a whole is:

$$p(f) = 1 - p(s) = 1 - 0.9791 = 0.0209$$

Although the configurations (b) and (c) contain the same units, the cross links in (c) result in a marked improvement in reliability.

10.3 Some statistical ideas

Problem 10.5 gave the results of 100 observations on the number of dumper trucks required on a construction site. Figure 10.8 presents a convenient way of displaying such information and a diagram of this kind is called a *histogram*. The number of days on which a given number of trucks was required is called the *frequency* and is represented by the height of the corresponding column. If the widths of the columns are equal, these frequencies are proportional by their areas.

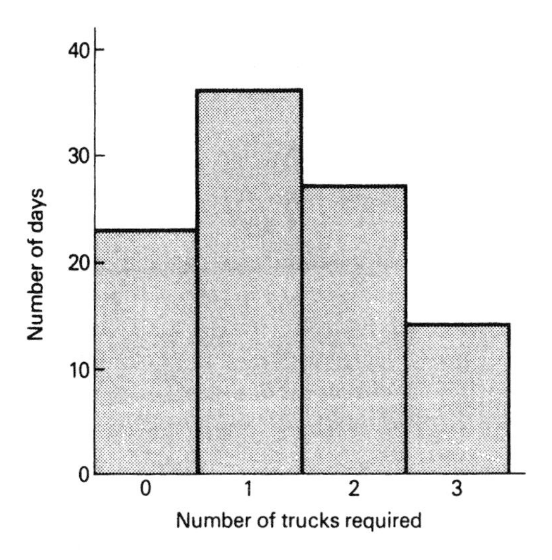

Fig. 10.8 Histogram.

In this example there are only four possibilities for the number of trucks required on a given day and there are distinct 'steps' in the histogram. If there were many it might approximate to a continuous curve of the type shown in Figure 10.9. It is called a *frequency distribution* curve. The ordi-

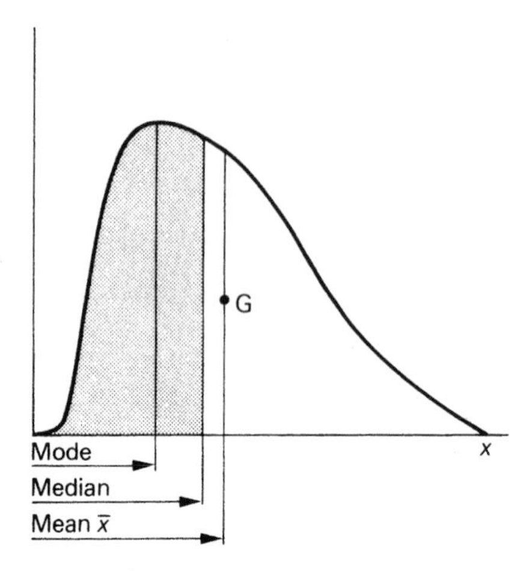

Fig. 10.9 Frequency distribution curve (with positive skew).

nate (height) of this curve at a given point represents the relative frequency at which the corresponding value of x (plotted horizontally) occurs.

For many purposes it is convenient to have some central value of x as a point of reference somewhere between the extremes but the value we obtain depends, in general, upon the definition used. Three possibilities are illustrated in Figure 10.9.

The *mode* or *modal value* is the value of x that occurs most frequently and it corresponds to the highest point on the curve. The *median* is the value that would come exactly half-way if all the observations were listed in order of size. Thus, half of the observations are of values greater than the median. Since the area under the graph represents the frequency of the observations, the median occurs at the value of x whose ordinate divides the area in half.

The *mean*, denoted by $\bar{x}$, is the average of all the observed values of x, that is, the sum of all the observed values divided by the number of observations n. In mathematical notation $\bar{x} = \Sigma x/n$. Strictly speaking it should be called the *arithmetic* mean to distinguish it from the *geometric* and *harmonic* means but the last two need not concern us here. Its ordinate passes through the point G, which corresponds to the centre of gravity of a uniform sheet cut to the shape of the curve.

The curve of Figure 10.9 is asymmetrical and is described as a *skew* distribution. If the distribution is symmetrical, the mode, median and mean coincide. Two symmetrical distributions are shown in Figure 10.10. They have similar shapes but differ in position and 'proportions', the one on the left, A, being more slender. Comparisons between such curves are made more easily if they are shifted horizontally so that they are located on the same axis of symmetry. This is achieved by plotting values of $x - \bar{x}$ instead of x, as shown in Figure 10.11. The vertical axis through the origin

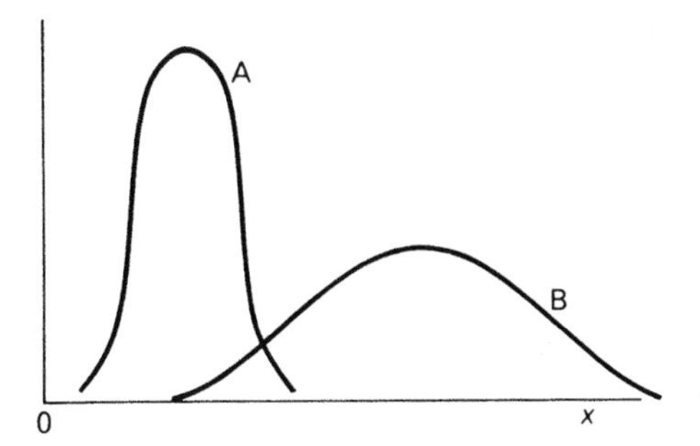

Fig. 10.10 Symmetrical distributions.

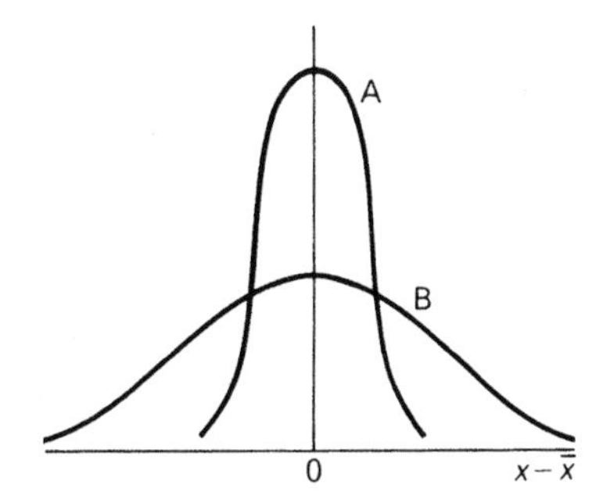

Fig. 10.11 Symmetrical distribution (as Fig. 10.10) plotted on a common axis.

is now the axis of symmetry for both distributions. The shapes indicate that A has more observations than B that are close to the mean value, and it is useful to have a numerical measure of this spread or dispersion. Taking the average of all the values of $x - \bar{x}$ leads to a difficulty because this difference is negative for x values below the mean and will cancel out corresponding positive values. We can overcome this difficulty by averaging the values of $(x - \bar{x})^2$, all of which are positive. This average is called the *variance* and it is a measure of the dispersion of the values about the mean. The square root of the variance is the *root-mean-square deviation* and is generally known as the *standard deviation*. It is denoted by s or σ (sigma) and, in symbols:

$$s = \sqrt{\frac{\Sigma\,(x - \bar{x})^2}{n}}$$

where n is the number of observations.

For many purposes statisticians calculate on the basis of *sampling* and it can be shown that the above formula underestimates the standard deviation. A better estimate is obtained by replacing the denominator n by $n-1$. For our purposes, the difference in the results will be negligible. Pocket calculators with statistical functions are able to compute the mean x and standard variation s of a set of values directly and the details of the calculations will be omitted here.

It would be very time-consuming to treat each distribution in isolation and we saw above how the two curves of Figure 10.10 can be brought to the same axis of symmetry (Figure 10.11) by changing the variable x to $x-\bar{x}$. To relate distributions having different amounts of dispersion or scatter, we can express the difference between an observed value x and the mean $\bar{x}$ as a proportion of the standard deviation s. We speak of a value as being, say, two standard deviations above or below the mean. This idea leads to a new variable z, defined as:

$$z = \frac{x - \bar{x}}{s}$$

The distribution curves obtained by applying this result to different sets of observations will not necessarily coincide, but there is a particular curve that approximates to many results found in nature and human affairs. Examples include the heights and weights of people, the variation in sizes of manufactured components, and human intelligence. The curve is known as the *normal* or *Gaussian distribution* and it is defined by a mathematical formula that involves the exponential function. The ordinates to the curve represent the relative frequency at which values of z occur. Its general shape is indicated by Figure 10.12 and it should be noted that the curve continues indefinitely in both directions.

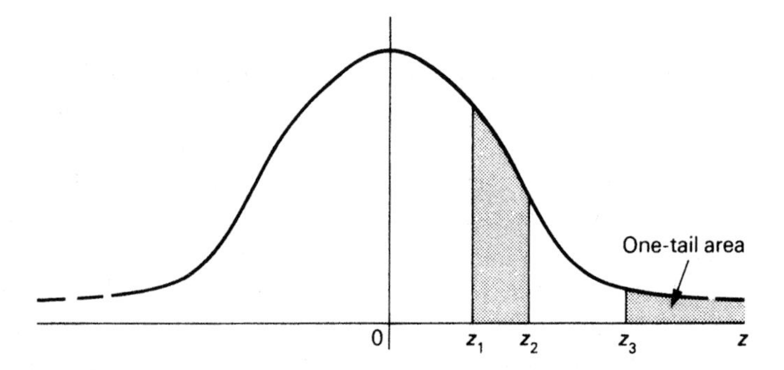

Fig. 10.12 Normal (Gaussian) distribution.

The equation to this curve is such that the total area under it, from $-\infty$ to $+\infty$ is equal to 1. It is equivalent to a histogram for a quantity that is varying continuously instead of by steps. As a consequence, the area under the curve between two values z_1 and z_2 represents the frequency with which values of z fall within this range. Since the total area is 1, the area between z_1 and z_2 is the proportion of all the observations that fall within this range. It therefore gives the *probability* of an observation being between these two values.

A quantity of considerable importance is the *one-tail area*. This is the area under the curve between a value such as z_3 and infinity. It equals the probability that an observed value is greater than z_3. Table B.1 of Appendix B shows one-tail areas for values of z in the range 0.0 to 3.0. The one-tail areas for $z = 1$ and $z = 2$ are 0.1587 and 0.02275. This means that, for a normally distributed set of values, just under 16 per cent are more than one standard deviation away from the mean. For two standard deviations it is just over 2 per cent.

We often need to use the reverse process – finding the value of z for a given one-tail area, especially when we are dealing with small areas or probabilities. Appendix B includes a second table of z values for selected one-tail areas.

Problem 10.8

The lighting system in a company's new office block requires 500 fluorescent tubes. These have an average life of 3,000 hours with a standard deviation of 220 hours. The tubes are switched on for 40 hours per week. Assuming that tube life is normally distributed, how soon should the company expect to start replacing tubes? Within what period are the first replacements likely? How many tubes are likely to require replacement within 80 weeks?

Solution

Since the normal curve extends to infinity in both directions it would appear that any individual tube life is possible, including negative values! However, a life of zero hours would correspond to about 14 standard deviations below the mean, and an examination of the table in Appendix B will show that this theoretical objection to the use of the normal distribution is of no consequence in the present case.

The first failure in a set of 500 tubes corresponds to a probability of p = 1/500 = 0.002. From the table of one-tail areas (or from a calculator with statistical functions) this corresponds to a z-value of 2.878. Thus:

$$z = \frac{x - \bar{x}}{s} = 2.878$$

and

$$x - \bar{x} = z.s = 2.878 \times 220 = 633\,\text{hours}$$

Since the normal curve is symmetrical, this result applies above and below the mean. Hence the tube life at which the first failure is expected to occur is 633 hours below the mean, i.e. 2,367 hours and this corresponds to 2,367/40 = 59 weeks approximately.

For 50 failures the corresponding one-tail area is 50/500 = 0.1, and the table shows that this gives a z-value of 1.282. We can therefore expect 50 failures by 1.282 × 220 = 282 hours below the mean life. This corresponds to a life of 3,000 – 282 = 2,718 hours or 68 weeks.

A period of 80 weeks or 3,200 hours is above the mean tube life of 3,000 hours and:

$$z = \frac{x - \bar{x}}{s} = \frac{3{,}200 - 3{,}000}{220} = 0.909$$

From the table in Appendix B, the corresponding one-tail area is 0.1817, and this is the probability of a tube having a life above the 3,200 value. The number of tubes that are expected to exceed this life is $0.1817 \times 500 = 91$ and the number expected to fail within the 80-week period is $500 - 91 = 409$.

10.4 Linear regression

The importance of the straight-line graph as a model of costs was emphasized in Section 10.2 and equation [10.1] showed the algebraic relationship between cost C and the number of units x produced or sold, or the number of miles covered in the transport examples. The constants are a and b in this equation; a being the intercept on the vertical axis and b the slope or gradient of the line. Establishing the values of a and b is therefore the key step in setting up a model of this kind.

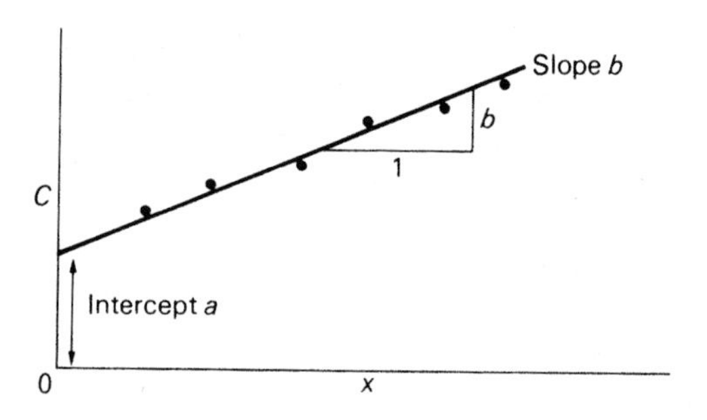

Fig. 10.13 Linear regression ('straight' line of best fit).

Suppose a set of values for C and x are plotted as in Figure 10.13. A straight line can be drawn 'by eye' to fit the data and the values of the constants determined by measurement of the graph. There is a statistical technique known as *linear regression* for finding the straight line that best fits the set of points. It determines the line that minimizes the scatter of the points and pocket calculators with statistical functions are programmed to determine the intercept and slope values directly from the data. Linear regression presumes that the relationship between the C and x values can be approximated by a straight line and it is always worth checking that this is the case. How close the data fits a straight line is measured by a *correlation coefficient* (r) in the range $r = +1$ to $r = -1$, the value +1 indicating that the points fall exactly on a straight line with positive slope.

Problem 10.9

A transport manager keeps records over a one-year period for a fleet of six vans of a particular type. He finds that the total cost and the distance covered for the vans are as follows:

Table 10.6

Van	*1*	*2*	*3*	*4*	*5*	*6*
Total cost (£) *C*	6,380	7,120	6,890	8,370	8,050	7,850
Distance covered (miles) *x*	13,400	15,000	10,900	19,500	21,100	16,400

Determine the annual 'fixed' or 'standing' charge and the marginal cost per mile to be expected for this type of van and estimate the total cost for an annual mileage of 25,000.

Solution

The reader is advised, as a first step, to plot the values (C vertically and x horizontally), draw a straight-line graph chosen 'by eye', and find the intercept and slope by measurement. Using the linear regression key on a pocket calculator with statistical functions, the intercept and slope are found to be 4,684 and 0.172 respectively. The expected annual fixed or standing charge is £4,684 and the marginal cost per mile is £0.172 or 17.2 pence. The cost equation becomes:

$$C = 4,684 + 0.172x$$

and this can be used to estimate the costs for other mileages. In particular, the expected total annual cost for 25,000 miles will be:

$$\begin{aligned} C &= 4{,}684 + (0.172 \times 25{,}000) \\ &= £8{,}984 \end{aligned}$$

The correlation coefficient in this example is found to be 0.853. This accords with the trend found in the table that the higher mileages lead to higher costs but there are individual anomalies. Van 5, for example, covered more miles than van 4 but cost less. The analysis should be compared with that used in the solution to Problem 10.1. Note that the last part of the present question requires the application of the cost equation to a value of x outside the range of the set of observed values. This is an example of *extrapolation*, a process that should be regarded with caution. *Interpolation*, the estimation of results within the original range of values, may be carried out with more confidence.

10.5 Stock control

Every organization needs to maintain a stock or *inventory* of materials if it is to operate without interruption and the principles of stock control apply equally to components needed on a construction site, the stationery supplies used in an office, and the food stored in a larder. Stocks may be classified under three headings, as follows:

active: the stock required to meet the expected demand for a given item
buffer: the additional stock needed to cope with fluctuations in demand
strategic: those stocks that are held in anticipation of shortages or price rises

The time interval between taking a decision to place an order and having the goods available for use is known as *lead time*. It includes the time taken to prepare the order, delays in despatch, the time for delivery, and the checking of items on arrival. If the lead time is underestimated, or the buffer stocks prove inadequate, an item may become out of stock, a condition usually referred to as *stockout*. Stock (or inventory) control is generally organized in one of two ways. Under the *fixed order quantity* system the size of the order is always the same but the interval between successive orders is varied to meet demand. The alternative is the *fixed interval* system under which orders are placed at regular intervals but their size varies according to demand.

An important parameter in stock control analysis is the *batch size* or *order quantity*. Its size affects the overall stock cost in two ways: the holding cost and the ordering cost. If the batch size is very large, there will be heavy holding costs associated with storage, insurance and the interest on capital. On the other hand, small batches mean frequent orders and this can lead to high costs in the buying department. The batch size that minimizes the total cost is known as the *economic batch size* or *economic order quantity* (EOQ).

A simple theory can be developed from the following assumptions:

- the demand is uniform
- the supplier is completely reliable
- the unit cost of the item is constant (i.e. no discounts for specified quantities)
- there is sufficient storage capacity
- there is sufficient ordering capacity

The first assumption leads to a variation in stock level of the kind shown in Figure 10.14. The level falls steadily until stocks are exhausted. A new batch is then received from the supplier and the process is repeated. On this basis the fixed ordering and fixed interval systems lead to the same outcome. It is convenient to derive a general formula for the economic batch size:

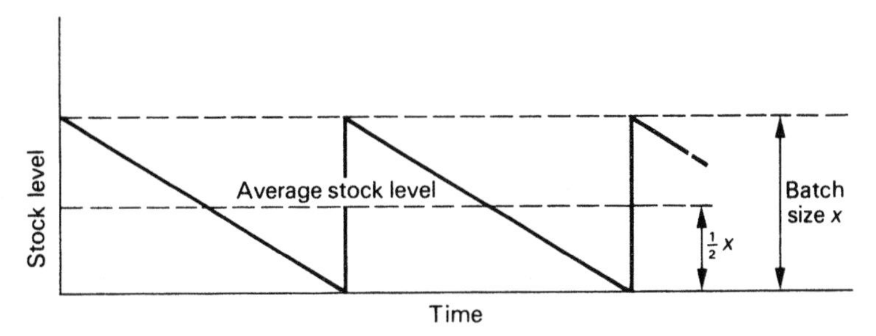

Fig. 10.14 Stock level time series plot.

Let: x = batch size
Q = quantity used per annum (or month or week)
p = unit price of the item
i = the cost per annum (month or week) of holding the stock including warehousing, insurance, and interest on capital as a proportion of the average value of the stocks held
S = cost of placing each order

From the graph it can be seen that the average stock level is $\frac{1}{2}x$ and the cost of holding stock is therefore $\frac{1}{2}xpi$ per annum (month or week). The number of orders placed during the same period is Q/x and the corresponding cost is QS/x. The total cost C (holding plus ordering) is therefore given by:

$$C = \tfrac{1}{2}xpi + \frac{QS}{x} \qquad [10.3]$$

Differentiating this expression, and equating to zero, to find a minimum (or maximum), we have:

$$\frac{\mathrm{d}C}{\mathrm{d}x} = \tfrac{1}{2}pi - \frac{QS}{x} = 0$$

from which:

$$x = \sqrt{\frac{2QS}{pi}} \qquad [10.4]$$

The positive root corresponds to a minimum and the value of x given by [10.4] is the economic batch size. The two terms in [10.3] are respectively proportional to, and inversely proportional to, x. The cost model is therefore similar to that of Problem 10.2 and the minimum occurs when the two terms are equal. This leads to the same result as [10.4]. The variation of the separate holding and ordering costs is shown in Figure 10.15, together with the total cost. The diagram shows that the two components are equal

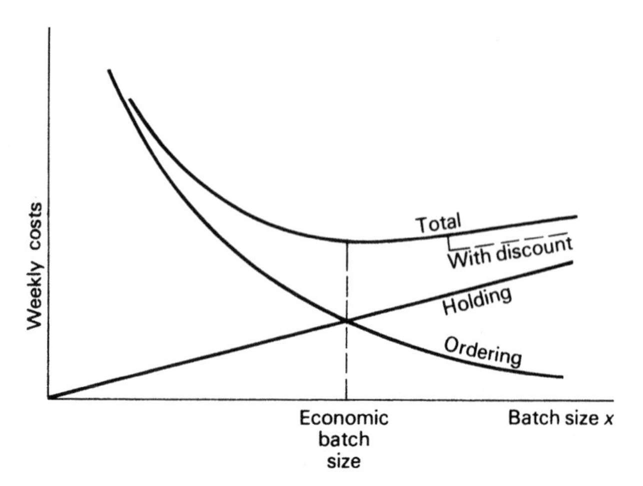

Fig. 10.15 Batch size costs.

and the total cost is a minimum at the economic batch size. The following problems illustrate the principles introduced in this section.

Problem 10.10

A stationery store supplies an office in which there is a steady demand for paper at the rate of 20 reams per week. It is estimated that the cost of processing each order is £4.80 and the cost of holding the paper in stock is 3 pence per ream per week. Tabulate the weekly holding and ordering costs for this item over a range of ordering intervals and batch sizes. What is the economic batch size (EOQ) for orders and the minimum weekly cost for stocking this item? What would the results be if four such offices were served by the same store?

Solution

Table 10.7 shows the weekly costs for ordering intervals between 1 and 6 weeks. Since the demand is steady the average stock level is half the batch size in each case. Since the ordering cost is spread over the period between one order and the next, its weekly average decreases as the interval increases.

It is seen from the values in the table that the minimum total cost is achieved with a four-week order interval. The economic batch size (EOQ) is 80 and the minimum weekly cost is then £2.40. The same result can be obtained using equations [10.3] and [10.4], by noting that the product *pi* is equivalent to a figure of £0.03 per week.

The second part of the problem can be solved by a revised tabulation. With four times the demand and the same order intervals the batch sizes and average stock levels are four times as great. The weekly holding costs are also four times as great but the ordering costs per week are the same as

Table 10.7

Order interval	Batch size (reams)	Average stock level (reams)	*Average weekly costs (£)*		
			holding	*ordering*	*total*
1	20	10	0.30	4.80	5.10
2	40	20	0.60	2.40	3.00
3	60	30	0.90	1.60	2.50
4	80	40	1.20	1.20	2.40
5	100	50	1.50	0.96	2.46
6	120	60	1.80	0.80	2.60

before. The minimum total in the final column becomes £4.80 and occurs with an order interval of two weeks. The result can also be obtained by noting, from the second equation, that the economic size is proportional to the square root of the quantity used, Q. If Q is increased four times, therefore, the economic size is increased by the factor $\sqrt{4}$, i.e. 2, and the optimum order becomes 160 reams every two weeks at a weekly cost of £4.80. In general it pays to combine smaller stores into bigger ones.

Problem 10.11

Suppose, in the first part of the previous problem, the supplier offers discounts of 1 per cent on orders of 200 reams and 5 per cent on orders of 400 reams on the basic price of £4.00 per ream. What should then be the ordering policy?

Solution

As the batch size increases beyond the economic quantity there is a rise in the total stockholding cost as indicated by the upper curve of Figure 10.15. If a discount is offered at some larger batch size, this should be deducted from the total cost as shown by the broken curve of the diagram.

In the present case, the weekly cost of the paper used (20 reams) is 20 × £4 = £80.00. Discounts of 1 per cent and 5 per cent therefore represent weekly savings of £0.80 and £4.00 respectively. By extending Table 10.7 we obtain the results shown in Table 10.8.

The table shows that there is no advantage in changing the batch size to 200 but a small saving results with an order quantity of 400.

Table 10.8

Order interval (weeks)	Batch size (reams)	Average stock level (reams)	*Average weekly costs (£)*			
			holding	*ordering*	*discount*	*total*
4	80	40	1.20	1.20	—	2.40
10	200	100	3.00	0.48	0.80	2.68
20	400	200	6.00	0.24	4.00	2.24

Problem 10.12

On a construction site, the number of 50 kg bags of cement used per week was recorded over a period of 50 weeks and the following results were obtained:

Weekly demand	53	54	55	56	57	58	59	60	61	62	63	64	65
Number of occurrences	1	2	3	5	6	8	7	6	5	3	2	1	1

Determine the mean demand and, assuming weekly deliveries and a normal distribution of demand, the necessary buffer stocks if the stockout risk is not to exceed 1 in 100.

Solution

Using the results of Section 10.3, the mean demand is:

$$x = \frac{\Sigma x}{n} = \frac{2{,}931}{50} = 58.6$$

and the standard deviation is:

$$s = \sqrt{\frac{\Sigma\,(x - x)^2}{n - 1}} = 2.67$$

The stockout risk of 1 in 100 represents a probability of $p = 1/100 = 0.01$ of being out of stock and, from Table B.1 in Appendix B, the corresponding value of z is 2.326. Hence the buffer stock $(x - \bar{x})$ is equal to $zs = 2.326 \times 2.67 = 6.21$ (say 7).

Adding the buffer stock to the mean demand figure suggests that the stock level should be set at 58.6 + 6.2 = 65 bags at the time of each delivery.

Problem 10.13

In a joinery shop a particular type of window frame is made in batches. The annual demand for this item is 600 at a unit cost of £30 and the cost of changeover is £135 per batch. The cost of holding stock is 24 per cent per annum of the average value of the stocks held.

Calculate the economic manufacturing batch size and the corresponding total annual cost of changeover and stockholding if the production rate is (a) very large compared to the rate of demand, and (b) three times the rate of demand.

Solution

(a) If each batch is produced very rapidly the stock level variation is similar to that shown in Figure 10.14. As a consequence, the batch size that minimizes cost can be found by the methods used in Problem 10.11. The ordering cost S is replaced by the changeover cost associated with the interruption and rearrangement of jigs that arises each time a new batch is made. In the notation of equation [10.4], and working in pounds, $S = 135$, $P = 30$, $Q = 600$, and $i = 0.24$. The economic batch size is therefore:

$$x = \sqrt{\frac{2QS}{pi}} = \sqrt{\frac{2 \times 600 \times 135}{30 \times 0.24}} = 150$$

This leads to four batches a year and an average stock level of 75. Hence:

$$\begin{aligned}\text{Annual cost} &= 4 \times £135 + 75 \times £30 \times 0.24 \\ &= £1{,}080\end{aligned}$$

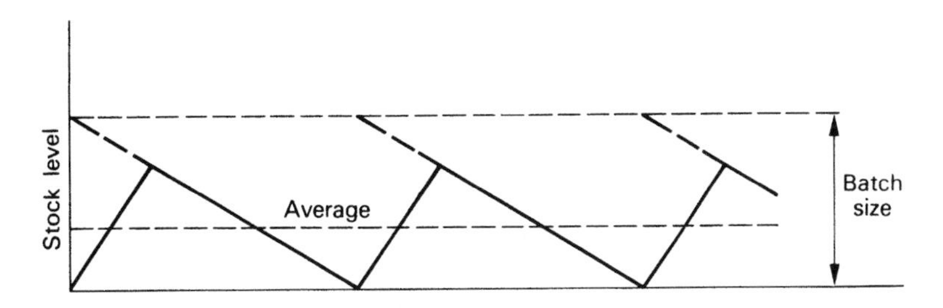

Fig. 10.16 Stock level variation.

In case (b) the stock level variation is modified as shown in Figure 10.16. The peak level of stock is now lower than the batch size because part of the batch is sold or delivered during the production period. In the present example this level, and the associated holding costs, are two-thirds of the values corresponding to very rapid production. The required result can therefore be obtained by reducing the denominator of the fraction in the equation [10.4] in the same proportion. The economic batch size becomes:

$$x = \sqrt{\frac{2 \times 600 \times 135}{30 \times 0.24 \times (\frac{2}{3})}} = 184$$

This result is equivalent to 600/184 = 3.26 batches per year. Using these 'exact' results, the peak stock level is $(\frac{2}{3}) \times 184 = 122$ and the average stock level is 61.

The corresponding annual cost $= 3.26 \times £135 + 61 \times £30 \times 0.24 = £880$

It would simplify the production schedule to manufacture three batches a year, each of 200 units. The peak stock level is then $(\frac{2}{3}) \times 200 = 134$ and the average level becomes 67. With these values:

annual cost $= 3 \times £135 + 67 \times £30 \times 0.24 = £887$

This approximation is close to the so-called exact value because the variation is small near the optimum batch size as Figure 10.15 shows.

10.6 Just-in-time

The cost of maintaining stocks, particularly in periods of high interest rates, has prompted managers to analyse their inventory systems. In the manufacturing industry, much attention had been given to the elimination of all sources of waste and a concept known as *just-in-time* (R W Hall 1989) has been adopted to avoid the accumulation of stocks in advance of demand.

Its success depends on careful scheduling because the cost of stockout in critical areas can far outweigh the benefits gained by reducing stock levels under steady conditions. 'Penalty' clauses in contracts can reduce profitability and premium prices may be charged on last-minute orders. Probability values can be attached to these factors and the corresponding expected costs assessed but there is no formula that covers all cases and each project must be analysed individually.

Summary

This chapter has surveyed a number of models and numerical techniques that are used in the analysis of management problems, notably those associated with costs. It must be remembered that each model is an idealization and the results it produces are, at best, approximations. However detailed the analysis and however powerful the computing techniques, it is impossible to compensate for imperfections in the data. Furthermore, the results in some cases are but average or 'expected' values.

Three cost models are of particular importance. The first of these applies to systems or operations in which total cost may be divided into two parts, one of which is fixed and the other is proportional to the number of units produced or received. Under these conditions the average cost per unit decreases as the number of items increases but it cannot fall below the marginal cost.

In the second type, part of the cost is proportional to the number of units and the rest is inversely proportional to that number. For this model the minimum total cost occurs when the two parts of the cost are equal. Stock control and batch production provide examples of this particular model.

The third model is concerned with the present equivalent of future expenditure or income, allowing for interest payments. The technique of present value analysis or discounted cash flow is used in making decisions on the choice of competing strategies and the replacement of plant.

Probability theory can be applied in assessing the outcome of complex sequences that involve several steps for which the individual probabilities are known. Statistical models such as the Gaussian or normal distribution are used to determine expected results when a range of values is possible.

Questions

1. A transport manager estimates that the cost of running a particular truck owned by the company amounts to £3,000 per annum plus 30p per mile. He is offered a similar vehicle on contract hire at £8,000 per annum, which includes maintenance, repairs, tax and insurance, the only additional cost being fuel at 9p per mile.

   What are the total annual costs for annual mileages of (a) 20,000, (b) 30,000 in the two cases? For what circumstances is the hired vehicle cheaper?
2. A manufacturer of architectural ironmongery finds that the sales per week s of a certain item varies with the selling price £p as shown in the following table:

p	4	6	8	10	12	14	16	18
s	115	95	79	63	50	37	25	15

   Illustrate the weekly revenue by a suitable graph, using the sales value as the base.

   (a) Estimate the maximum weekly revenue and the level at which the price should be set to achieve it.

   (b) If the production cost is £3 per item, plus a fixed overhead cost of £300 per week, for what range of sales is the operation profitable?

   (c) What is the maximum profit achievable under these conditions and what are the corresponding sales per week?
3. Use Table A.1 in Appendix A to find the present value of £80,000 in 10 years' time if compound interest is added (a) annually at 16 per cent, (b) half-yearly at 8 per cent and (c) quarterly at 4 per cent
4. (a) What is the annual repayment on a 20-year building society reducing sum mortgage of £60,000 if interest, added annually, is at 8 per cent? (b) Assuming no change in the interest rate, what is the outstanding debt

after (i) 10 years, (ii) 15 years? (c) If, at the outset, the interest is increased to 9 per cent, how long will it take to discharge the debt?

5. Two types of boiler are under consideration for a factory heating installation. The system is likely to be replaced in eight years' time. Type A has a capital cost of £25,000 and an estimated fuel cost of £1,000 per annum. For B the figures are £22,000 and £1,100 respectively.

   Depreciation for both types is 25 per cent per annum of the value at the beginning of the year. The maintenance costs for A and B in the first year are estimated to be £600 and £800 respectively, both figures rising at the rate of 10 per cent per annum thereafter. (Assume that the annual maintenance charge is payable at the end of each year.)

   Using this information and the present value criterion, which type should be selected? Take the interest on borrowed capital at 15 per cent per annum.

6. A contractor runs a fleet of lorries costing £40,000 each when new. Records on similar vehicles show that the annual depreciation amounts to 25 per cent of the value at the beginning of the year. Repairs and maintenance amount to £1,250 in the first year of the life of such lorries, increasing annually by 40 per cent thereafter. The cost of capital is 15 per cent. At what intervals should the lorries be replaced if a permanent need is envisaged for vehicles of this type. Consider periods up to 10 years.

7. A civil engineering contractor is reviewing the next five years of its activities and needs to make a decision about earthmoving equipment. It considers three possible courses of action: (a) purchasing a large fleet at a capital cost of £800,000; (b) purchasing a small fleet at a cost of £250,000 and (c) relying on hired equipment.

   It is estimated that there is a 0.6 probability of obtaining enough work to fully use the large fleet. This level of activity would show an annual return of £400,000 if the company possessed the large fleet but, because of rental charges, the net return would be reduced to £200,000 if the company possessed the small fleet, and £80,000 if it relied entirely on hired equipment.

   On the other hand, there is a 0.4 probability that the level of work will be sufficient only for a small fleet. This would produce an annual return of £120,000 if the company possessed the equipment but only £50,000 if it had to hire. Excess equipment could be sold after two years at 60 per cent of its initial cost.

   The scrap value of equipment after five years is 20 per cent of the initial cost. Take the cost of capital as 15 per cent per annum. On the basis of this information what should the initial decision be?

8. (a) A builders' merchant finds that there is a steady demand at the rate of 5,000 per annum for an item that costs £0.80. The cost of placing each order is £4 and the stockholding cost is 20 per cent of the value of the

stocks held. Calculate the economic ordering quantity and the corresponding annual cost of stocking this item.

(b) For another item, the demand is variable with a mean demand of 250 per week. It is found, on average, that the weekly demand is between 220 and 280 for four weeks out of five. Assuming that the demand is normally distributed and the stock is replenished weekly, find the buffer stock required to give a stockout risk of 0.02, i.e. once in 50 deliveries.

9. On a housing development the demand for bricks (in thousands) was recorded over a period of 100 weeks and the following results obtained:

Weekly demand	40–50	50–60	60–70	70–80	80–90	90–100	100–110	110–120
Number of weeks	2	8	15	23	27	14	6	5

Determine the mean weekly demand and the necessary buffer stocks if the stockout risk is not to exceed 1 in 100. Assume weekly deliveries, zero lead time and a normally distributed demand.

References and bibliography

Anderson, D.R., Sweeney, D.J. and Williams, T.A. (1994) 'An Introduction to Management Science' (*Quantitative Approaches to Decision-Making*), Eighth Edition, West Publishing Company.

Bedford, M. (1999) 'Life Begins at 50', *Computer Shopper*, February (pp. 674–7).

Hall, R.W. (1989) *Zero Inventories*, Dow Jones-Irwin.

Institute of Actuaries and Institution of Civil Engineers (1998) *Risk Analysis and Management of Project*, Thomas Telford.

Paulos, J.A. (1990) *Innumeracy*, Penguin.

Taylor, F.W. (1911, republished 1947) *The Principles of Scientific Management*, Harper.

11 Operational Research

Operational research has its origins in World War II when the methods of science and mathematics were applied to problems of organization and management associated with military operations. The subject has since developed rapidly and is now concerned with a wide range of problems facing managers in industry, business, and government. In particular it deals with resources - manpower, machines, materials, and money - and how they can be used to best effect. As stated in the definition adopted by the Operational Research Society (UK):

> The distinctive approach is to develop a scientific model of the system incorporating measurements of factors, such as chance and risk, with which to predict and compare the outcomes of alternative strategies or controls.

There are a number of characteristics common to many operational research problems:

- they are described and analysed in numerical terms
- there are constraints such as limitations of resources
- the objectives are expressed as optimizations
- they involve uncertainties

A solution that satisfies the constraints is termed *feasible* and there may be several or many feasible solutions to a particular problem. The aim is to determine the one that is 'best' by whatever criterion is adopted: it can be maximum production, or earliest completion, or least cost, or the optimum value of some other measure.

From its inception, operational research has made use of existing mathematical techniques, particularly from the field of statistics, and the topics covered in the previous chapter might be classified as parts of the subject. On the other hand, some of the problems encountered over the years by operational researchers could not be solved by existing techniques and new methods of analysis, such as *linear programming*, have been developed as a result. This process has gained such momentum that the theory of the subject has grown quickly and become highly mathematical. However, the present chapter is an introduction to operational research and concentrates on some of its applications using elementary mathematics only.

11.1 Network analysis

In most construction contracts time is important and delays can be costly. It is essential, therefore, to achieve the earliest possible completion and various models are used to plan and control the progress of complex projects. The best-known of these is the *bar chart* (Figure 11.1) whose base is a timescale of days or weeks; alternatively, it can be labelled in calender form. Each job that contributes to the total project is shown as a horizontal bar whose length is proportional to its estimated duration and whose position indicates the period within which it is due to take place. What is less obvious from this diagram is the sequence in which jobs follow one another, but this shortcoming can be overcome by the use of a *network* or *arrow diagram* (Figure 11.2), in which each job, or *activity* as it is called, is represented by an arrow whose direction shows how it leads to subsequent activities. A junction or node where activities begin or end is called an *event* and is usually represented by a circle. In the form shown here, the length and slope of each line are unimportant, and the one rule for constructing diagrams of this kind is that all activities approaching an event must be completed before any activity leaving it can commence. There can be only one starting event and one finishing event, and no other loose ends.

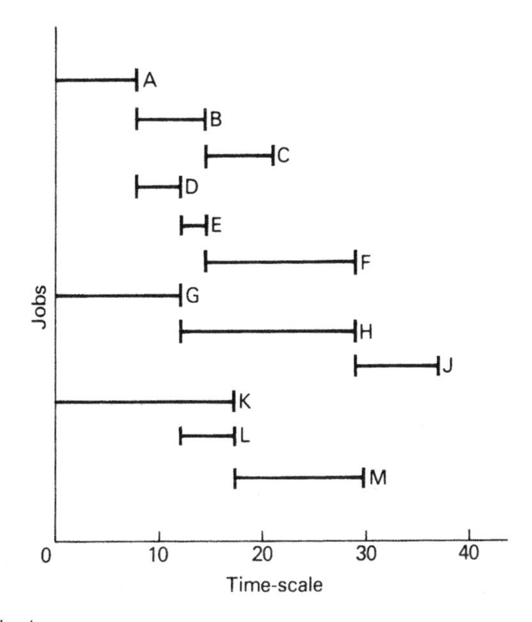

Fig. 11.1 Bar chart.

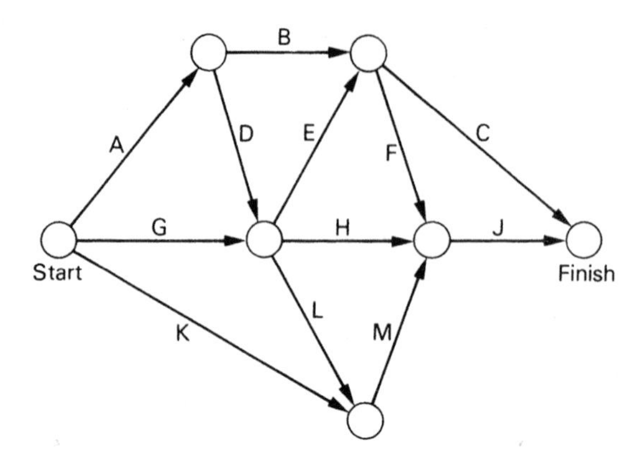

Fig. 11.2 Network: activity on arrow diagram.

Three possible sequences of four activities are shown in Figure 11.3. At (a) there are two separate relationships; activity A must be completed before B can commence, and C must be completed before D can commence. The logic illustrated by (b) is that neither B nor D can commence until both A and C are completed. There are other possibilities: suppose B depends on the completion of A only but D requires the completion of A and C. This is achieved in (c) by the introduction of an extra activity called a *dummy*. This arrangement satisfies the requirements because B can start when A is finished, whereas D depends upon the completion of C and the dummy, and the dummy in turn depends upon the completion of A. The dummy is assigned zero duration.

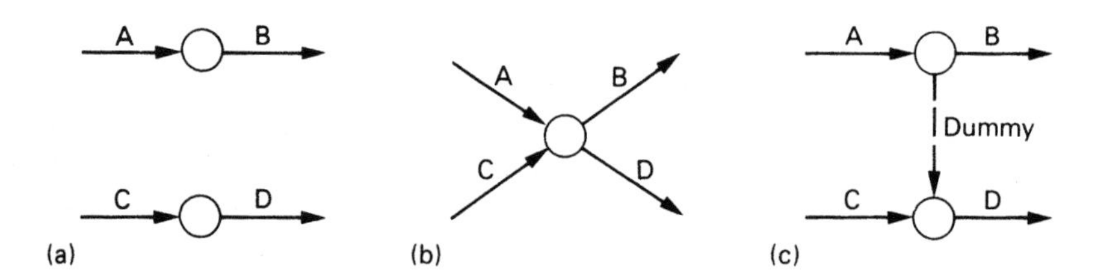

Fig. 11.3 Activity sequences (with different logic relationships).

Once the network is constructed and the estimated duration of each activity established, the minimum time for completing the project is easily calculated. As can be seen in Figure 11.2 there will be a number of routes through the network from the starting to the finishing of events. The total time for each route is the sum of the durations of the activities along it and the greatest of these totals is the minimum time for completion. The route having the longest duration is called the *critical path* and the activities along it are called *critical activities*. Any delay in one of these will affect the time of completion. Other activities have time to spare, and this spare time is called *float*. There is more than one type of float and the distinctions between them are given later.

The details of constructing and analysing such networks are explained in the numerical problems that follow. A thorough treatment of the subject with much useful advice on its application to construction projects is to be found in the paper by Nuttall and Jeanes (1963). Among other matters it indicates how a timescale can be applied to a network, thus retaining one of the advantages of the bar chart. A number of computer programs are available for the analysis of these networks, one of the best known being PERT (Programme Evaluation and Review Technique). It uses a duration for each activity based on the weighted mean of the optimistic, most likely, and pessimistic time estimates, the weightings being $\frac{1}{6}$, $\frac{4}{6}$, and $\frac{1}{6}$ respectively. The basis of this approach is the same as that referred to in Chapter 9 for obtaining a mean probability when determining a bid strategy. The application of probability theory to network analysis is described in the paper by Mitchell and Willis (1973).

Problem 11.1

Suppose the durations (in weeks) of the activities in the network of Figure 11.2 are: A,7; B,1; C,10; D,3; E,2; F,3; G,12; H,13; J,8; K,17; L,4; and M,12. Find (a) the minimum project time, (b) the earliest and latest times for each event, (c) the critical path, and (d) the total float, free float, interfering float and independent float on each non-critical activity.

Solution

(a) The minimum project time can be found, albeit laboriously, by considering every possible route from start to finish that follows the directions of the arrows. There are eleven altogether, and they have the following sequences of activities (see Figure 11.2): ABC, ABFJ, ADEC, ADEFJ, ADHJ, ADLMJ, GEC, GEFJ, GHJ, GLMJ, and KMJ. Their durations are, respectively, 18, 19, 22, 23, 31, 34, 24, 25, 33, 36, and 37. Since every activity is essential to the project, it is the greatest of these, 37 weeks, that corresponds to the minimum project time.

(b) The more usual method of analysing networks is shown in Figure 11.4. Each event circle is divided by a horizontal diameter and the upper half carries a reference number for that event. The nodes at the tail and head of an activity arrow are called the *preceding* and *succeeding* event, respectively; in computer programs they are usually denoted by i and j. It is convenient to refer to activities by their i and j numbers. In the present network, for example, activity D is (2, 5) and E is (5, 3).

The lower half of each event circle is divided into two quadrants and the left-hand one shows the earliest time at which all the activities approaching the event can be completed. Starting at time 0, event ② can be

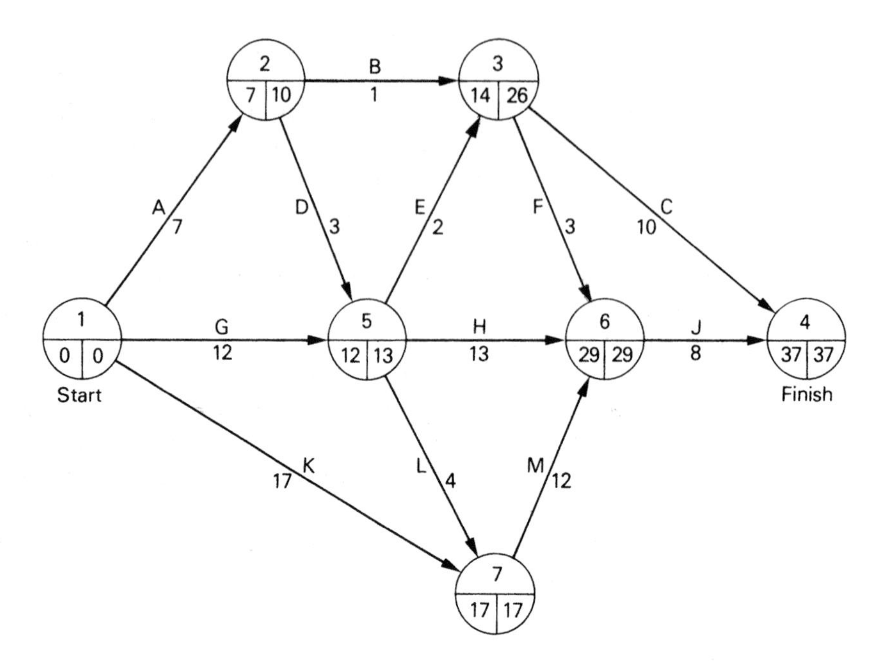

Fig. 11.4 Activity on arrow network (solution to problem 11.1).

reached after 7 weeks, the duration of activity A (1, 2). Event ⑤ is approached by two activities D and G and these have arrival times of 7 + 3 = 10 and 0 + 12 = 12 weeks, respectively. The greater value, 12, is recorded and the procedure repeated at events ③ and ⑦ which have earliest times of 14 and 17 weeks, respectively. Note that event ⑥ is approached by three activities, F, H, and M, giving earliest arrival times of 17, 25, and 29 weeks, respectively. Again the greatest value, 29, is recorded and the process is repeated for event ④ giving an earliest completion time of 37 weeks – a result that tallies with the value found in (a).

The right-hand quadrant shows, in each case, the latest time by which all activities approaching the event must be completed. If, as is usually the case, this project is to be completed in the minimum time, the figure 37 is recorded in the right-hand quadrant at event ④. Working backwards, the corresponding value at event ⑥ is 37 − 8 = 29 and for event ⑦ it is 29 − 12 = 17. At event ③, two routes must be considered: along activity C the result is 37 − 10 = 27 and along F it is 29 − 3 = 26. The lower value is recorded and the process is repeated for events ⑤, ②, and ①. (The reader is invited to check all the results shown in Figure 11.4.)

(c) A critical activity is one for which the two i-values coincide, the two j-values coincide and the difference between the i- and j-values equals the duration. In the present case the critical activities are K, M, and J (bearing out the result found in (a)) and, in terms of event numbers, the critical path is ①–⑦–⑥–④.

(d) *Total float* is defined as the amount by which the maximum time available for an activity exceeds the time needed to perform it. For example, activity B could start after seven weeks (if A were finished at the earliest time) and finish as late as 26 weeks. Thus 19 weeks are available and, since only one week is required, the total float is 18 weeks.

Free float is the spare time available assuming that the activity can start at the earliest time but must finish by the earliest time of the succeeding event. In the case of activity B, therefore, the free float is (14 – 7) – 1 = 6 weeks.

Interfering float is the difference between the total float and the free float. For activity B it is 18 – 6 = 12 weeks. This result can also be obtained as the difference between the earliest and latest times of the succeeding event.

Independent float may be regarded as the spare time available for a given activity without affecting the float of activities that come earlier or later. On this basis, activity B could start after ten weeks and finish at 14 weeks, giving an independent float of (14 – 10) – 1 = 3 weeks.

Table 11.1 shows the collected results for the whole network, with times measured in weeks.

Table 11.1

Activity	*i*	*j*	*Duration*	*Total float*	*Free float*	*Interfering float*	*Independent float*
A	1	2	7	3	0	3	0
B	2	3	1	18	6	12	2
C	3	4	10	13	13	0	1
D	2	5	3	3	2	1	−1
E	5	3	2	12	0	12	−1
F	3	6	3	12	12	0	0
G	1	5	12	1	0	1	0
H	5	6	13	4	4	0	3
J	6	4	8	critical			
K	1	7	17	critical			
L	5	7	4	1	1	0	0
M	7	6	12	critical			

Problem 11.2

School laboratories for biology, physics, and chemistry are to be built in that order using an industrialized building system. Three teams are to be employed in the construction, working in the following order on each building: erectors, plumbers, electricians. No team can commence work on a given building until the previous team has finished. Table 11.2 gives the estimated durations of the various activities in weeks.

Determine the minimum project time and the critical path. State the weeks during which the teams can take holidays without delaying the completion.

Table 11.2

Building	*A (biology)*	*B (physics)*	*C (chemistry)*
Erection	2	4	3
Plumbing	3	2	3
Electrical installation	2	5	2

Solution

In the previous problem the network was presented in the question but in the practical application of critical path analysis it is usually necessary, as in the present case, to construct the network diagram. To do this we must first establish the order in which the various activities follow one another and a convenient way of doing this is shown in Table 11.3.

Table 11.3

Activity reference	*Activity name*	*Preceding activities*	*Duration (weeks)*
1	Erection A	—	2
2	Erection B	1	4
3	Erection C	2	3
4	Plumbing A	1	3
5	Plumbing B	2, 4	2
6	Plumbing C	3, 5	3
7	Electrical A	4	2
8	Electrical B	5, 7	5
9	Electrical C	6, 8	2

It will be seen that each activity (other than the first) requires the completion of one or two earlier activities. For instance, the plumbing of the second building cannot begin until its erection is completed and the team of plumbers has finished its work in the first building. This table contains all the information needed for completing the analysis and there are 'network generator' computer programs that can handle the data in this form. For a manual analysis, however, a network diagram is needed, and Figure 11.5 shows one of several possible configurations.

Note that two dummies are needed to relate the activity 'plumbing B′ to its predecessors and achieve the correct sequences at events ③ and ④. The earliest and latest event times are shown on the network.

The minimum project time is 15 weeks and, in terms of events, the critical path is: ①–②–③–⑤–⑥–⑧–⑨–⑩. The corresponding activities, in order, are erection A, erection B, plumbing B, electrical B, and electrical C, together with two dummies.

The reader is advised to gather all the results in a table similar to that given in the previous solution and to draw a bar chart to scale. From the

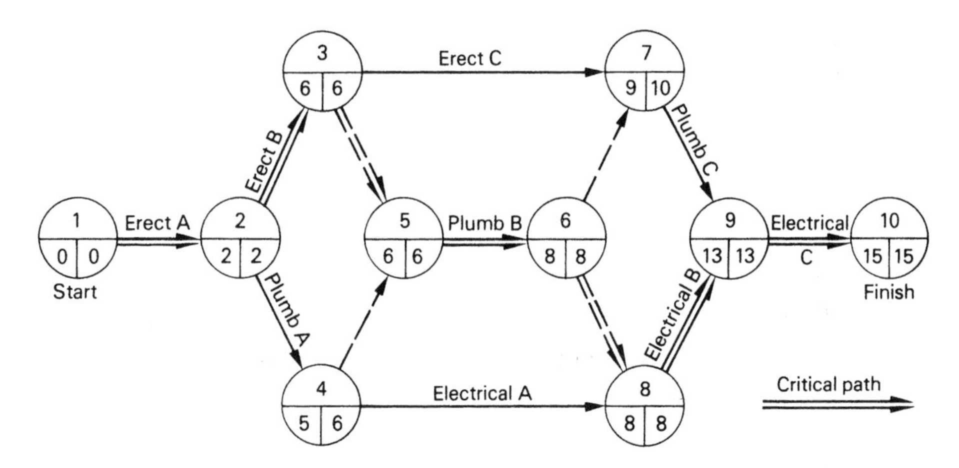

Fig. 11.5 Critical path (activity on arrow network; solution to problem 11.2).

latter it will be seen that if each activity were to begin at the earliest possible time the erectors could be on holiday from the beginning of week 10. They could take one week's holiday (free float) in the period weeks 7–9 without affecting other teams and two weeks' holiday (total float) in period weeks 7–10 but the latter would restrict the possibilities for other teams.

Again, if all activities were completed at the earliest possible times, the plumbers could be on holiday in weeks 1, 2, 6, 9, and 13–15 and the electrical installers in weeks 1–5 and 8.

11.2 Linear programming

The manager is frequently faced with the task of allocating resources to various activities and many problems of this type are examples of linear programming. The word 'programming' is now closely associated with the use of computers and, although computers are frequently used to solve linear programming problems, it has a different meaning in the present context. 'Planning' might be a better word for describing the allocation of resouces. 'Linear' refers to the form of the equations and graphs that arise in the analysis.

Linear programming problems fall into two categories: *transport* and *mixture*. Transport problems are concerned with the scheduling of goods and vehicles from various starting points (or *sources* as they are called) to a number of destinations. The method of solution can also be used to allocate staff to various roles so as to make the best use of their abilities.

Problems of the mixture kind arise when resources have to be shared among two or more products or activities. No special formulae are required at this stage and the methods are best illustrated by numerical examples.

Problem 11.3

A contractor is organizing the supply of ready-mixed concrete to four sites. He estimates that the total daily requirement amounts to twenty-four lorryloads and he finds three suppliers who are able to meet this demand between them. The separate amounts available from the suppliers are (in lorryloads):

A, 4; B, 8; C, 12

and the quantities needed at the four sites are:

K, 5; L, 2; M, 10; N, 7

Show, on a suitable matrix, an allocation schedule that matches these amounts.

In the price negotiation it is agreed that the transport costs will be charged to the contractor in proportion to the mileage incurred. The distances involved are:

	K	L	M	N
A	6	12	2	5
B	18	21	13	12
C	11	16	5	6

Calculate the total one-way daily distance for the allocations made and obtain the schedule that gives the minimum total distance.

Solution

It is convenient to show the allocations on a matrix such as Figure 11.6 (a) in which the sources are listed on the left and the destinations at the top. Each element or cell then represents one of the routes, there being twelve in the present problem. The number of lorryloads available at each source is shown to the right of the corresponding row and the number required at each destination is given at the foot of the appropriate column.

A feasible solution is one in which the numbers in the rows and columns add up to these values. Also, by the nature of this problem, fractional and negative numbers are excluded. Some routes are unused and the corresponding cells in the matrix are left empty. It is easy to find a feasible solution by trial and error and two of the many possibilities are

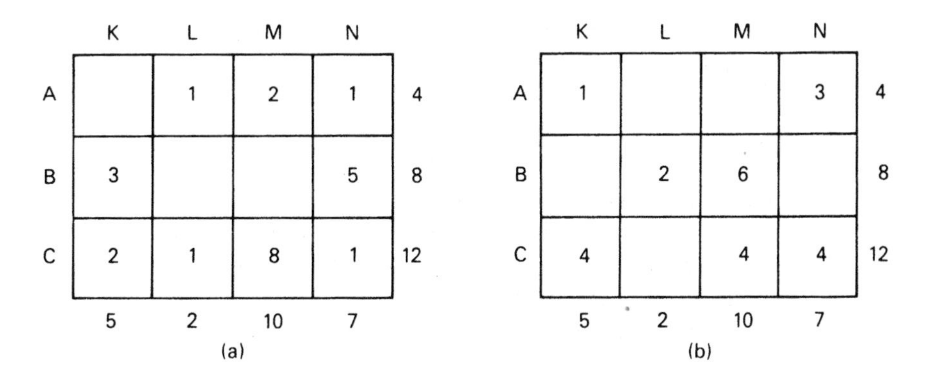

Fig. 11.6 Allocation matrices (two possibilities for problem 11.3).

shown by Figure 11.6 (a) and (b). Indeed, with the data of the present problem, there are nearly a thousand feasible solutions.

If the numbers in the matrix of Figure 11.6 (a) are combined with the appropriate mileages, the total distance is:

$$1 \times 12 + 2 \times 2 + 1 \times 5 + 3 \times 18 + 5 \times 12 + 2 \times 11$$
$$+1 \times 16 + 8 \times 5 + 1 \times 6 = 219 \text{ miles}$$

and the corresponding total for (b) is 229 miles. Although these solutions are both feasible, (a) is to be preferred in that it involves the lower total mileage.

It might be possible to obtain a more economical solution with a 'commonsense' approach of avoiding long routes and making as much use as possible of short ones. However, with such a large number of feasible solutions, this process could take a long time and there would be no way of knowing when the optimum was reached.

A more systematic way of tackling the problem is to modify the initial solution step by step in order to reduce the total mileage while retaining feasibility – a process known as *iteration*. When it is found that no further improvement is possible, the optimum arrangement has been achieved. To facilitate this process the matrix is displayed with the route distances in the corners of the cells, as shown in Figure 11.7.

In seeking an initial solution we first note that if there are m rows and n columns a feasible solution can be found with not more than $m + n - 1$ routes in use. These are called *occupied* elements, the others being *free* elements. The value $m + n - 1$ is called the *critical number* and in the present case, with $m = 3$ and $n = 4$, a feasible solution can be found with not more than six occupied elements. A solution satisfying this rule, and known as the 'north-west corner' solution, is found as follows.

The largest possible number is entered in the top left-hand cell (Figure 11.7). It must be 4, the row total, and the remaining cells in this row

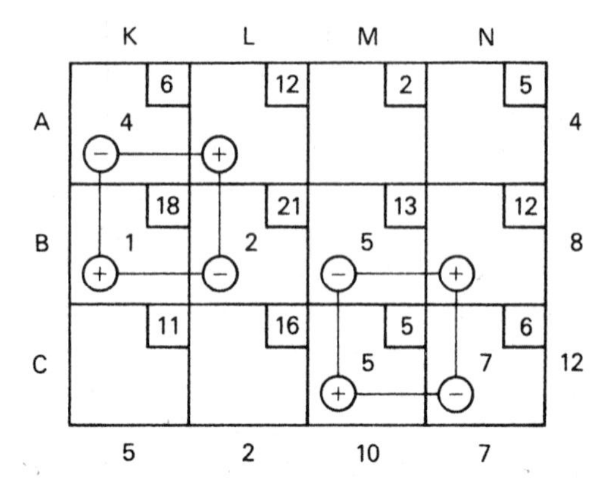

Fig. 11.7 Allocation matrix showing optimum arrangement (from iteration).

remain empty (free elements). In the first cell of the second row the largest possible number is 1 since this completes the requirements of the first column. In the second cell of this row the maximum number is 2 (the column total) and in the third it is 5, thus completing the row total. In the bottom row, 5 lorryloads are needed in the third cell to complete the requirements at destination M and the remaining cell or element has the value 7. This procedure satisfies the critical number rule and is an efficient way of finding an initial solution. At this stage the total distance is

$$4 \times 6 + 1 \times 18 + 2 \times 21 + 5 \times 13 + 5 \times 5 + 7 \times 6 = 216 \text{ miles}$$

The unused routes in this solution are now investigated one by one to see if improvements can be obtained by incorporating them. Consider, for instance, the effect of allocating one lorryload to route AL: this is indicated by placing a $\oplus$ sign in the AL cell as shown in Figure 11.7. This move would upset the totals of the first row and second column, and to preserve the feasibility of the solution corresponding reductions of one lorryload each are made in the cells AK and BL, indicated by $\ominus$ signs. Finally a lorryload would have to be added to the BK route, as shown by the $\oplus$ sign in its cell. Together these four changes would lead to a different, but still feasible, solution. The effect on mileage would be

$$+12 - 6 + 18 - 21 = +3$$

The use of route AL at this stage would therefore increase the mileage. On the other hand, the allocation of one lorryload to route BN with corresponding adjustments to other routes, as shown in the bottom right-hand corner of Figure 11.7, causes a change in total mileage of

$$+12 - 13 + 5 - 6 = -2$$

and this represents an improvement. The change in the allocation to each of the routes involved need not be limited to a single lorryload, but for each one added to BN there is a reduction of one on each of the routes BM and CN. At present BM has 5 and CN has 7, so that the proposed rearrangement can be made for a maximum of 5 lorryloads. The reduction in miles is therefore 10, the new total being 206 and the modified allocations are shown in Figure 11.8. The process, usually called the 'steppingstone' method, is now repeated. An empty cell is selected and linked with three occupied cells by horizontal and vertical steps – 'rook' moves, in chess parlance. If the allocation of one lorryload to the empty cell, and the corresponding adjustments to the others, lead to a reduction in total mileage the change is made for as great a quantity as possible, the limit being reached when one of the occupied cells becomes empty.

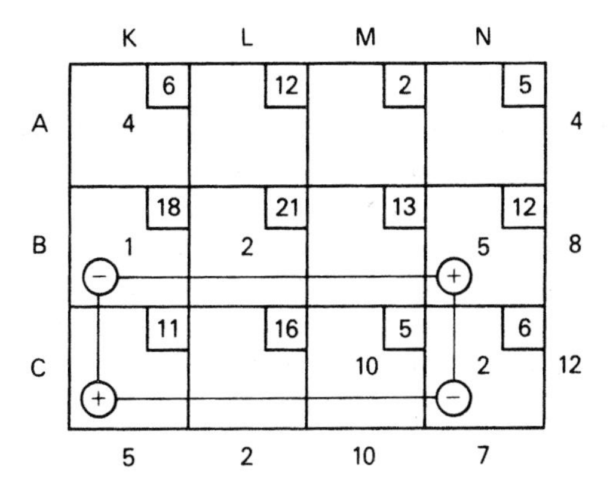

Fig. 11.8 Modified allocation matrix.

The four cells involved do not necessarily form a square. The addition of one lorryload to CK for example, cannot be offset in the bottom row by subtracting one from CL because this route is empty already. Instead, CK is linked with CN, BN, and BK as shown in Figure 11.8. The change in mileage is:

$$+11 - 6 + 12 - 18 = -1$$

but the improvement can only be made for one lorryload because BK is then empty. The total mileage is reduced to 205 and the new allocations are shown in Figure 11.9.

If all the empty routes are tested at this stage it will be found that no further improvement is possible. AL presents a special problem because it cannot be linked with three occupied routes in a square or rectangle. The links form a more complicated path, as shown in Figure 11.9, but the solution remains feasible because each row and column involved contains one $\oplus$ and one $\ominus$ sign. The change in mileage is:

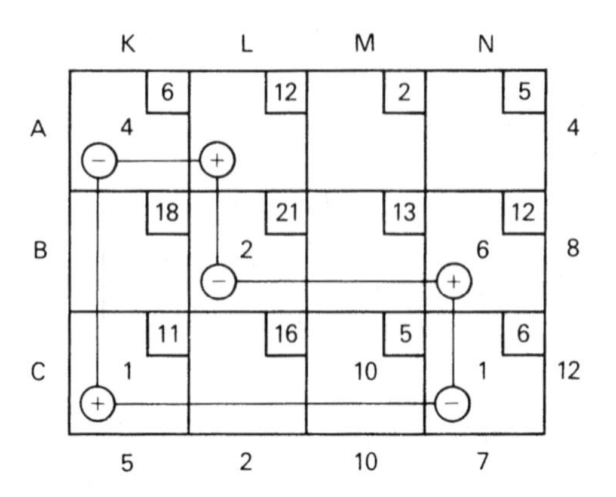

Fig. 11.9 Further modified allocation matrix.

$$+12 - 6 + 11 - 6 + 12 - 21 = +2$$

The minimum total mileage is 205 and the allocations are:

A to K, 4; B to L, 2; B to N, 6;
C to K, 1; C to M, 10; and C to N, 1

Of all the feasible solutions this is the optimum: the one with the minimum total mileage. It is noteworthy in that it does not involve the shortest route, AM, at all and it makes the greatest possible use of the longest route, BL. Site L requires only two lorryloads and they are both coming from supplier B. Furthermore, site N receives all but one of its loads on the longest of the three routes approaching it, BN. These allocations may seem surprising and it might have taken a considerable time to find this optimum solution using a 'commonsense' approach.

Problem 11.4

Suppose, in the previous problem, the quantities available from the sources, in lorryloads, are increased to the following amounts:

A, 6; B, 12; C, 14

the quantities required at the destinations being unaltered.

Determine the new minimum total distance for a feasible solution and list the allocations that achieve it.

Solution

The total of lorryloads is now 32, eight more than are required at the destinations. This imbalance is allowed for by introducing an extra des-

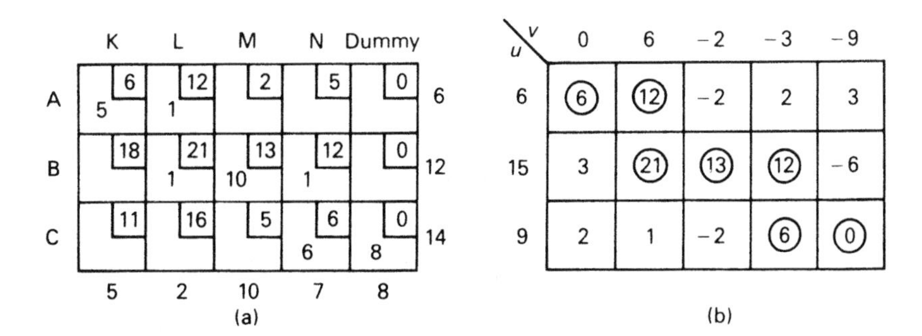

Fig. 11.10 Allocation matrices (problem 11.4).

tination called a *dummy* to receive the excess loads. It is represented by an additional column in the matrix, Figure 11.10 (a), and the dummy is taken to be at zero distance from each of the sources. The optimum arrangement can be found by the methods of the previous solution and loads scheduled to the dummy destination remain, in practice, at the sources.

The north-west corner solution for this extended matrix is shown in Figure 11.10 (a). It gives a total mileage of 241 and contains the critical number of occupied elements, seven. The eight free elements are now tested to see if an improvement can be made by incorporating one of them in the allocation arrangement. Instead of testing each free element separately (as in the previous solution) they can all be tested simultaneously by a technique known as the method of *shadow costs*.

For this a new matrix is drawn up, Figure 11.10 (b). First the lengths of the occupied routes are noted, the circled numbers in the diagram. Next, numbers known as u- and v-values are assigned to the rows and columns such that $u - v$ for an occupied element equals the corresponding distance. The first of these numbers is chosen arbitrarily, the others being derived from it. In the present case, for example, the u-value for the first row was chosen as 6 and the v-values for the first and second columns become 0 and 6, respectively, to satisfy the values of the AK and AL routes. The u-value for the second row can now be deduced from the distance of the BL route. The process is continued until all the u- and v-values are determined. Note that these may be negative. For each of the free elements the appropriate value of $u + v$ is subtracted from the route distance. The results, shown in Figure 11.10 (b), are called the shadow costs and each represents the effect on total mileage of allocating one load or vehicle to the route and making the necessary adjustments to the related occupied routes. Thus the effect of putting unit load on route BK is to increase the total mileage by 3, whereas the allocation of unit load to B/dummy would reduce the mileage by 6.

At this stage the greatest reduction can be achieved by the use of CM, with corresponding adjustments to CN, BN, and BM. The saving is two miles for each load and a maximum of six can be reallocated, thus reducing the total mileage to 229. The revised allocations are shown in Figure 11.11 (a) and the corresponding shadow costs appear in Figure 11.11 (b). The results were obtained by, again, choosing 6 as the *u*-value for the first row but other initial choices lead to the same shadow costs. The use of route B/dummy with adjustments to BM, CM, and C/dummy saves eight miles per load and with a maximum of four loads, the total mileage is reduced to 197.

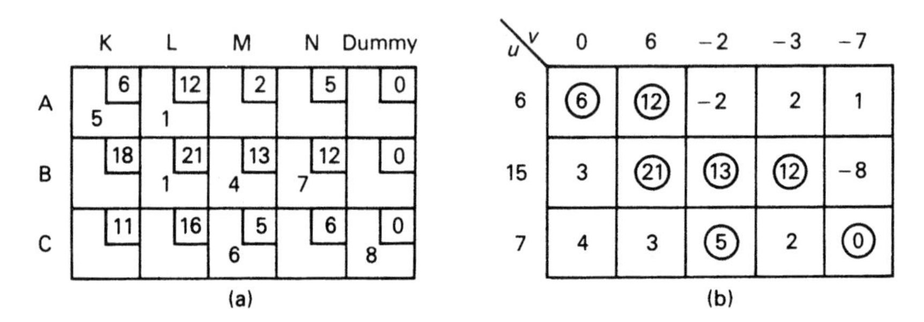

Fig. 11.11 Revised allocations. (b) With shadow costs.

The next stage is shown in Figure 11.12, the details being left to the reader. The shadow cost for CN is now −6, and by allocating four loads to this route the total mileage is reduced to 173. The new solution and corresponding shadow costs are given in Figure 11.13 (a) and (b). There are no negative shadow costs at this stage and no further reduction in total mileage is possible. However, the shadow cost for AM is zero and this means that one or more other solutions can be found having the same total mileage. The relevant changes are indicated by the ⊕ and ⊖ signs of Figure 11.13 (a) and it can be seen that six elements are involved. In the present example these changes can only be made in respect of one load because this will cause AL to become an empty route. The two optimal solutions, therefore, are:

A to K, 5; A to L, 1; B to L, 1; B to N, 3; C to M, 10; C to N, 4

as shown in Figure 11.12 (a) and

A to K, 5; A to M, 1; B to L, 2; B to N, 2; C to M, 9; C to N, 5

obtained by an alteration of one lorryload in each of the cells containing ⊕ or ⊖.

The total mileage is 173 in both cases, the quantities available from sources A and C are fully utilized, but eight of the 12 loads available from supplier B are scheduled to go to the dummy destination and therefore remain at B.

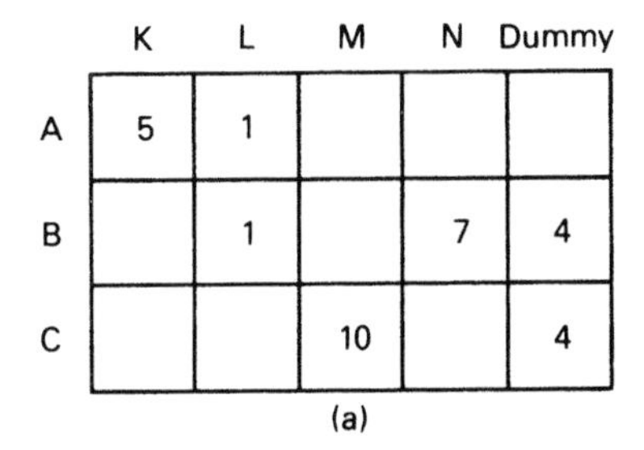

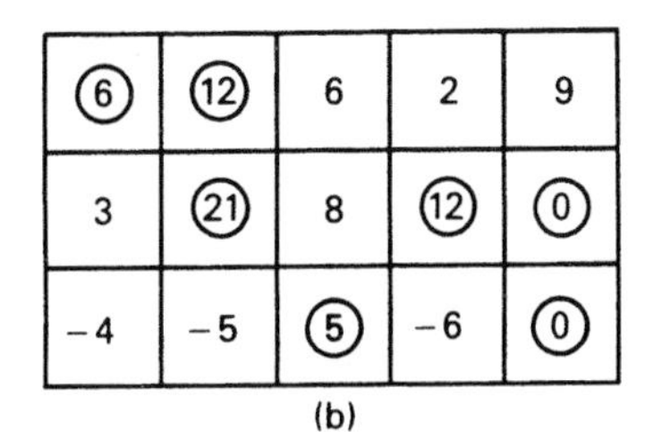

Fig. 11.12 Revised shadow cost matrices.

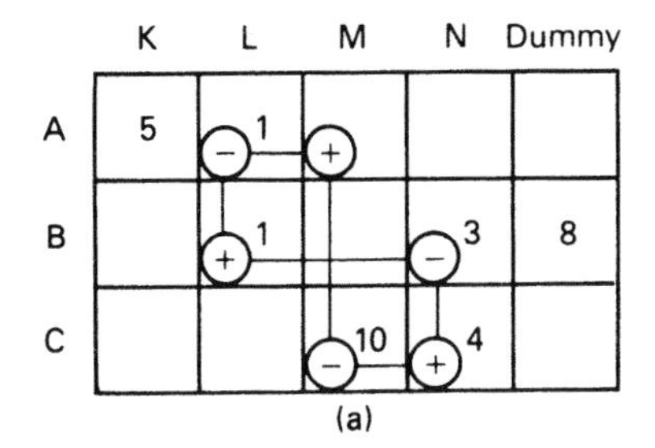

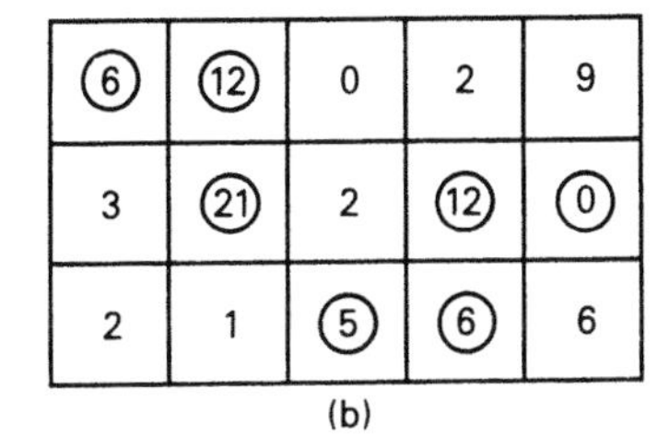

Fig. 11.13 Solution matrices (problem 11.4).

Problem 11.5

Five managers (A, B, C, D, and E), who differ in ability and experience, are to be placed in charge of five projects which are different in type and value. The suitability of each manager for each project is assessed on a numerical scale with a maximum of 20 points, the results being shown in Figure 11.14. To which project should each manager be assigned in order to obtain the highest total points score?

Solution

Questions of this type are related to linear programming problems of the kind dealt with in the last two examples. The matrix of points values (Figure 11.14) is equivalent to the cost or distance tables of those examples except that we are here seeking the maximum total value. The problem is converted to one of minimization by subtracting each element from the maximum value, 20. The results are shown in Figure 11.15.

It is a property of the cost matrix that the optimum solution is unchanged if all the elements in any row or column are increased or decreased by the same amount. This result follows from a consideration of the stepping stone method in which each row and column involved in a rearrangement contains one ⊕ and one ⊖ sign so that the shadow costs are unaffected by such increases or decreases. However, the total cost or distance must be determined from the original values.

		Projects				
		1	2	3	4	5
Managers	A	18	16	11	19	5
	B	14	10	15	8	6
	C	9	13	8	8	6
	D	15	14	10	12	10
	E	11	11	14	10	8

Fig. 11.14 Point values matrix (problem 11.5).

2	4	9	1	15
6	10	5	12	14
11	7	12	12	14
5	6	10	8	10
9	9	6	10	12

Fig. 11.15 Minimisation matrix.

The linking of each manager to a project corresponds to a 'route' in the transport problem. In the present example there will be five routes in use, a smaller figure than the critical number which is $5 + 5 - 1 = 9$. A problem of this kind in which the numbers of sources and destinations are equal, and only one unit is available at each source or required at each destination, is called an *assignment* problem. It can be solved by an algorithm known as the *Hungarian method* and consisting of the following steps:

1. *The smallest element in each row is subtracted from every element in that row.* In Figure 11.15 the smallest row elements are 1, 5, 7, 5, and 6 so that this step leads to the matrix of Figure 11.16.
2. *The smallest element in each column is subtracted from every element in that column.* In the matrix of Figure 11.16, four of the columns already contain zeros and it is only the fifth that requires modification. The result is shown in Figure 11.17.
3. *The zeros in the matrix are covered by the minimum number of horizontal and vertical lines.* In Figure 11.17, at least four such lines are necessary (more than one configuration being possible). If this number equals the number of columns (or rows) an assignment can now be made as explained in step (5) below. If not, as in the present case, the matrix is modified in the following way.
4. *The smallest element not covered is selected. It is subtracted from the elements which are not covered and added to those which are covered twice. The elements which are covered once are unaltered.* In the case of Figure 11.17 the smallest uncovered element is 1 and the modifications lead to the matrix shown in Figure 11.18. If the assignment test of step (3) is now applied it will be found that at least five lines are needed to cover all the zeros and an assignment can therefore be made. (If this were not the case, step (4) would have to be repeated.)
5. *An element of zero value is selected for each assignment.* In some lines there is only one zero and this one must be used. In others a choice presents itself. For example, the zeros in the present case (Figure 11.19) are so

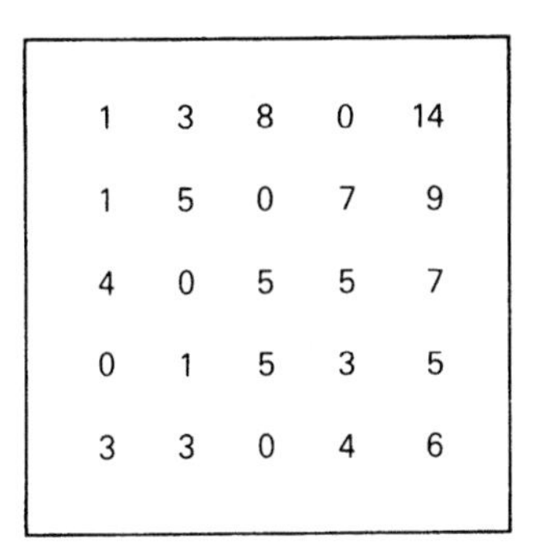

Fig. 11.16 Hungarian method, step (1) matrix.

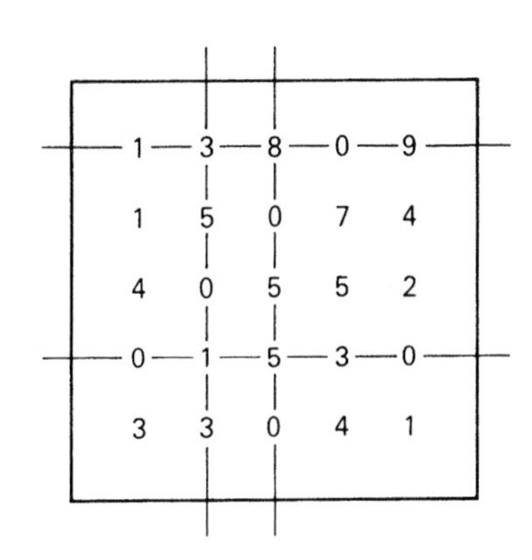

Fig. 11.17 Hungarian method, step (2) matrix.

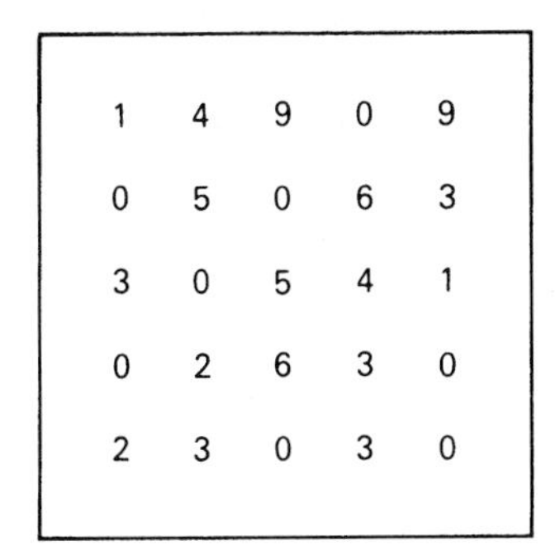

Fig. 11.18 Hungarian method, step (4) matrix.

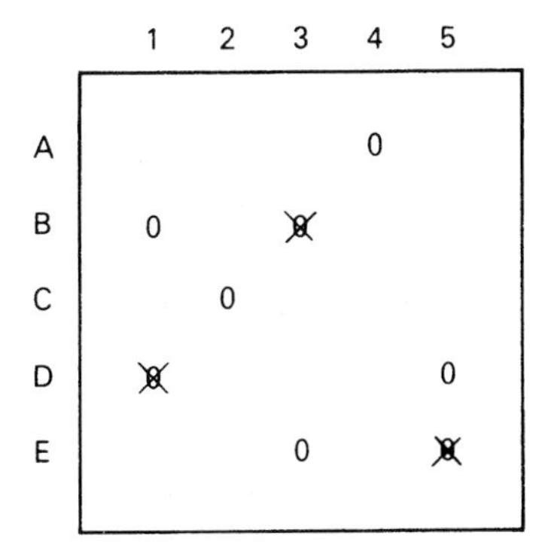

Fig. 11.19 Hungarian method, step (5) matrix.

distributed that C must be assigned to project 2 and A to project 4. Each remaining line contains two zeros and the next choice is arbitrary. Suppose B1 is selected. This eliminates D1 and D must be assigned to project 5. It follows that E5 is crossed out and E is assigned to project 3.

The corresponding points total must be calculated from the original values of Figure 11.14. Thus:

Manager	*Project*	*Points*
A	4	19
B	1	14
C	2	13
D	5	10
E	3	14
	Total	70

Problem 11.6

An estate developer is planning the development of a 25 hectare site with a mixture of bungalows and two-storey houses. The capital available to him is £11,250,000, and the estimated construction cost is £90,000 per house and £75,000 per bungalow.

Planning permission allows for a maximum of six houses or four bungalows per hectare, but there are two additional stipulations: at least 20 bungalows must be included in the scheme, and at least one-third of the units must be houses.

Determine the number of houses and bungalows to be built to achieve each of the following: (a) the maximum number of houses, (b) the maximum number of bungalows, and (c) the maximum total profit (assuming that all units can be sold) if the profit per house is 10 per cent higher than per bungalow.

Solution

This problem, like the three preceding ones, is concerned with the allocation of resources and it is a further example of linear programming. However, a graphical analysis (Figure 11.20) is now more effective than the matrix techniques used in the earlier solutions. If the axes represent houses and bungalows, each combination of units can be shown by a point whose coordinates are the numbers of each type involved. By the nature of the problem, negative and fractional values are excluded.

The capital available, £11,250,000, is sufficient to build 125 houses at £90,000 each or 150 bungalows at £75,000 each. There are other possibilities involving mixtures such as 25 houses and 120 bungalows, or 100 houses and 30 bungalows and the points representing the possible combinations lie on the straight line shown as the *financial constraint*. Points on or below this line represent combinations of houses and bungalows for which there is sufficient capital; points above it lead to schedules for which there is insufficient money.

By similar reasoning the planning densities permit a maximum of 150 houses (and no bungalows), or 100 bungalows (and no houses) or proportions of each that total 1. The straight line marked *planning density constraint* includes the points representing all such combinations. At this stage the diagram can be used to test proposed combinations of houses and bungalows in terms of the capital available and the permitted density of building. A point that is above both lines represents a combination of houses and bungalows that requires more capital than is available and exceeds the planning density. A point below both lines corresponds to a schedule that satisfies both constraints. Points within the triangular regions between these two constraint lines give schedules that satisfy one of the constraints but not the other.

Two other constraints are specified in the question. It is stipulated that there must be at least 20 bungalows and this requirement is represented by a horizontal line through the '20' mark on the vertical axis. Since this is a minimum requirement, acceptable combinations of houses and bungalows are represented by points on or above this line.

The limitation imposed by the specified proportion of houses, one-third of the total number of units, is represented by a sloping straight line through the origin. With 40 bungalows there must be at least 20 houses, with 80 bungalows a minimum of 40, and so on. Hence the required line has a slope of 2, and acceptable solutions are represented by points on or to the right of it.

The four constraints delineate a *feasible region* on the diagram and all feasible solutions lie within it or on its perimeter. The specific solutions required in the question are as follows:

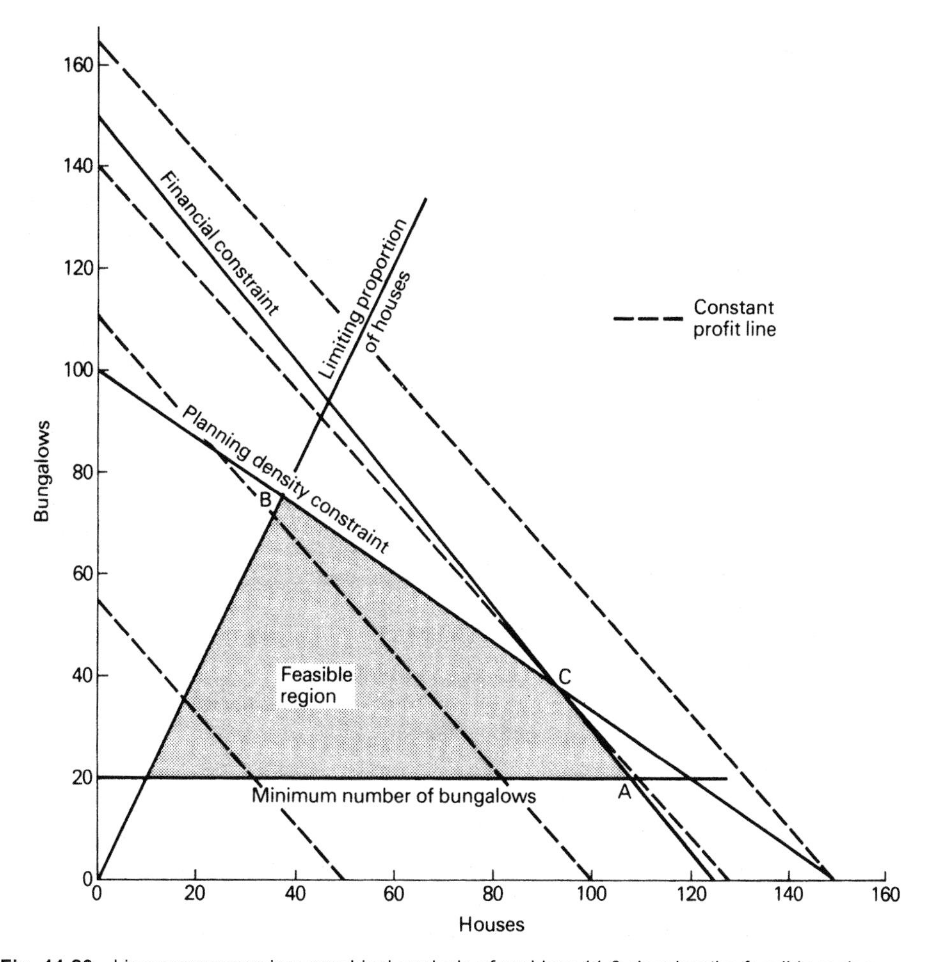

Fig. 11.20 Linear programming; graphical analysis of problem 11.6 showing the feasible region.

(a) The maximum number of houses corresponds to point A and by measurement of the diagram (Figure 11.20), or by calculation, the schedule is 108 houses and 20 bungalows. This combination uses all the capital available (within the limits set by practical considerations) but requires only 23 hectares at the stipulated densities.
(b) The maximum number of bungalows is given by point B, the 'exact' result being 37.5 houses and 75 bungalows. Of the nearest whole-number answers, the solution 37 houses and 75 bungalows does not satisfy the requirement for the minimum proportion of houses. On the other hand 38 houses and 74 bungalows meets this constraint and is just within the stipulated planning density.
(c) If each house produces 10 per cent more profit than a bungalow, 50 houses will produce the same profit as 55 bungalows, 100 houses the same as 110 bungalows, and so on. Lines of constant profit will link these pairs of points on the coordinate axes and it can be seen from the diagram that they are parallel. The highest one of the set to touch the feasible region does so at C, and the required result is therefore 94 houses and 37 bungalows.

The graphical method used in this solution is only suitable when there are two 'products' – houses and bungalows in the present case. However, there is an algorithm known as the *Simplex method* which can deal with any number of products. See, for example, the books by Krekó (1968), Dantzig and Thapa (1997) and Ignizio (1994)

11.3 Queueing

Queueing is a familiar feature of everyday life. We queue at the bus stop, in the post office, and in self-service restaurants. These examples all involve people shuffling forward to a service point but the 'customers' can be lorries waiting to load or unload, letters in an in-tray waiting to be answered, or telephone calls jamming a switchboard.

Queues may be classified by the number of channels, the number of service points, and the queue 'discipline' – the rules by which it operates. For many queues the rule is 'first come, first served' but there are other possibilities and the in-tray may work on the basis of 'last in, first answered'.

A *simple queue* is defined by the following properties:

- single channel, single service point
- first come, first served
- customers are individuals
- no simultaneous arrivals
- no limit to the number of potential customers

- variable service times and variable times between successive arrivals
- the variability of the number of arrivals and of the number of customers being served in given time intervals conform to the statistical distribution known as the Poisson distribution

In the examples that follow it will be assumed that these conditions are satisfied. The key parameter in queueing theory is *traffic density* or *intensity*. It is denoted by ρ (rho) and is defined as

$$\text{Traffic density } \rho = \frac{\text{Mean rate of arrival}}{\text{Mean rate of service}} \qquad [11.1]$$

For the statistical distribution referred to above it can be shown that the mean rate of service is the reciprocal of the mean service time, and that the mean rate of arrival is the reciprocal of the mean time between successive arrivals. From equation [11.1], therefore, we have:

$$\rho = \frac{\text{Mean service time}}{\text{Mean time between arrivals}} \qquad [11.2]$$

If, for example, customers arrived at the rate of four per minute and could be served at the rate of five per minute, the traffic density, by the first definition, would be $\rho = \frac{4}{5} = 0.8$. The corresponding mean service time would be 12 s and the mean time between arrivals 15 s. Thus using [11.2]:

$$\rho = \frac{12}{15} = 0.8$$

Although there will be instances of the service time being greater than the time between successive arrivals, the mean values must be such that ρ is less than 1. Otherwise the queue will grow indefinitely.

In analysing the simple queue it is necessary to distinguish between the queue and the *system*, the latter including the customer being served (see Figure 11.21).

Suppose $p_0, p_1, p_2, p_3 \ldots$ are the probabilities of there being 0, 1, 2, 3 ... customers, respectively, in the system. These probabilities can also be regarded as the proportions of the total time for which the corresponding

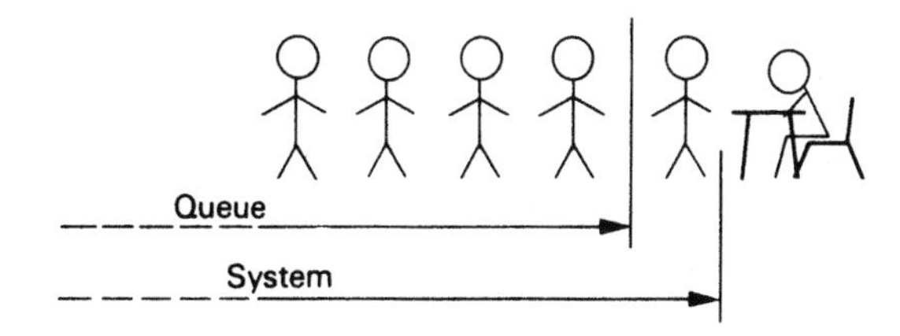

Fig. 11.21 Simple queueing diagram distinguishing the *queue* and the *system*.

Table 11.4

Number of customers in system	*Number of customers in queue*	*Probability*
0	0	p_0
1	0	p_1
2	1	p_2
3	2	p_3

numbers of customers are to be found in the system. Table 11.4 shows how the numbers of customers in the queue and the system are related.

Since the sum of the probabilities of all the possibilities is 1, we have:

$$p_0 + p_1 + p_2 + p_3 + \cdots = 1 \quad [11.3]$$

and the series is infinite since we assume no limit to the number of potential customers. It might be expected that each term in the series will be smaller than its predecessor because, on average, customers can be served more rapidly than they arrive. A full analysis (see, for example, Wilson 1970) shows that each of the probability values is ρ times its predecessor.

$$\text{Thus} \quad p_1 = \rho p_0$$
$$p_2 = \rho p_1 = \rho^2 p_0, \text{etc.} \quad [11.4]$$

and, substituting in equation [11.3], we have

$$p_0 + \rho p_0 + \rho^2 p_0 + \rho^3 p_0 + \cdots = 1$$

This is a *geometric series* or *progression* having a first term $a = p_0$ and common ratio $r = \rho$. If, as in the present case, the common ratio is less than 1, the series converges and the sum to infinity is given by $a/(1 - r)$. Therefore:

$$\frac{p_0}{1-\rho} = 1$$

and

$$p_0 = 1 - \rho \quad [11.5]$$

Thus, the probability of there being no one in the system, and hence the probability of a customer not having to wait, is $1 - \rho$. It follows that the probability of having to wait is ρ since the two values must total 1.

It is useful to have an expression for the average number of customers in the system and this can be determined by noting that there is no one in the system for p_0 of the time, one person in the system for p_1 of the time, and so on. Thus:

$$\text{Average number of customers in system} = \frac{0 \times p_0 + 1 \times p_1 + 2 \times p_2 + \cdots}{p_0 + p_1 + p_2 + \cdots}$$

The denominator is identical to the series in [11.3] and equals 1. The first term of the numerator is zero and the remaining terms form a series which, with the relationship of [11.4], can be written as

$$S = \rho p_0 + 2\rho^2 p_0 + 3\rho^3 p_0 + 4\rho p_0 + \cdots$$

This is not a geometric series, but multiplying through by ρ we have

$$\rho S = \rho^2 p_0 + 2\rho^3 p_0 + 3\rho^4 p_0 + \cdots$$

and, by subtraction,

$$S - \rho S = \rho P_0 + \rho^2 p_0 + \rho^3 p_0 + \rho^4 p_0 + \cdots$$

This is another geometric progression and, summing to infinity,

$$S(1 - \rho) = p_0 \left(\frac{\rho}{1 - \rho} \right)$$

Finally, substituting for p_0 from [11.5], we have:

$$\text{Average number of customers in system} = \frac{\rho}{1 - \rho} \qquad [11.6]$$

By similar methods, expressions can be found for the average number of customers in the queue and for the average times spent by customers in the queue and in the system. The main results are collected below and it will be seen that they all involve traffic density ρ.

Probability of a customer having to wait $= \rho$

Average number of customers in the system $= \dfrac{\rho}{1 - \rho}$

Average number of customers in the queue when there is a queue $= \dfrac{1}{1 - \rho}$

Average number of customers in the queue including times when there is no queue $= \dfrac{\rho^2}{1 - \rho}$

$$\text{Average time a customer is in the queue} = \frac{\rho}{1-\rho} \times \begin{pmatrix}\text{Mean service}\\ \text{time}\end{pmatrix}$$

$$\text{Average time a customer is in the system} = \frac{1}{1-\rho} \times \begin{pmatrix}\text{Mean service}\\ \text{time}\end{pmatrix}$$

Problem 11.7

In a study of the operation of an unloading bay it was observed that 117 lorries and vans made deliveries during a 40-hour week. With two men employed on unloading, the average time to unload each vehicle was found to be 15 minutes. Treating the operation as simple queueing, determine: (a) the probability of a vehicle not having to wait; (b) the average number of vehicles in the system; (c) the average time a vehicle is waiting to be unloaded; and (d) the average time a vehicle is in the system.

For a trial period the number of unloaders was increased. It was found that with 3, 4, and 5 persons involved, the average unloading times were 12, 10, and 9 minutes, respectively. If the all-in rate for each unloader is £18 per hour and the cost of waiting for a van and driver is £60 per hour, how many unloaders should be employed to minimize the total cost?

Solution

With 117 arrivals in 40 hours, we have:

$$\text{Mean rate of arrival} = \frac{117}{40} = 2.925 \text{ per hour}$$

Also,

$$\text{Mean rate of service} = \frac{60}{15} = 4 \text{ per hour}$$

Hence,

$$\text{Traffic density } \rho = \frac{2.925}{4} = 0.731$$

(a) The probability of a vehicle having to wait is ρ. Hence, probability of vehicle not having to wait is:

$$\begin{aligned} 1 - \rho &= 1 - 0.731 \\ &= 0.269 \end{aligned}$$

(b) The average number of vehicles in the system is:

$$\frac{\rho}{1-\rho} = \frac{0.731}{1-0.731}$$
$$= 2.72$$

(c) The average time a vehicle is in the queue is:

$$\frac{\rho}{1-\rho} \times (\text{Average service time}) = 2.72 \times 15$$
$$= 40.8 \text{ minutes}$$

(d) The average time a vehicle is in the system is:

$$(\text{Average time in queue}) + (\text{Average service time}) = 40.8 + 15$$
$$= 55.8 \text{ minutes}$$

Alternatively, by use of the appropriate formula:

$$\text{Average time in system} = \frac{1}{1-\rho} \times (\text{Average service time})$$
$$= \frac{1}{1-0.731} \times 15 \text{ minutes}$$
$$= 55.8 \text{ minutes}$$

Using this result:

$$\text{Total waiting time per week} = \frac{55.8}{60} \times 117$$
$$= 108.8 \text{ hours}$$

Hence,

$$\text{Weekly waiting cost} = 108.8 \times £60.00$$
$$= £6{,}528$$

Also,

$$\text{Weekly unloading cost} = 2 \times 40 \times £18.00$$
$$= £1{,}440$$

When more unloaders are employed the first of these costs is reduced but the second is increased. If the two costs are borne by the same organization, there is an incentive to minimize their sum and, by repeating the calculations for the given numbers, the results shown in Table 11.5 are obtained. The figures in the final column show that the minimum total cost is obtained when four unloaders are employed.

Table 11.5

① No. of unloaders	② Service time (min)	③ ρ	④ $\frac{1}{1-\rho}$	⑤ = ② × ④ Average time in system (min)	⑥ Weekly costs (£)		
					waiting	unloading	total
2	15	0.731	3.72	55.8	6,528	1,440	7,968
3	12	0.585	2.41	28.9	3,381	2,160	5,541
4	10	0.488	1.95	19.5	2,283	2,880	5,163
5	9	0.439	1.78	16.0	1,872	3,600	5,472

Summary

Operational research is largely concerned with the allocation of resources and therefore has many applications in construction management.

Critical path (or network) analysis is used to plan and control the progress of complex projects by considering the sequence in which the contributions of different people take place. It identifies those activities which must be finished on time if the completion of the project is not to be delayed and pinpoints those activities which have time to spare.

Linear programming problems of the transport kind are concerned with the allocation of resources when these are available from several sources and are required at various destinations. The methods of solution can be adapted to assignment problems in which staff or plant are allocated to a number of projects. In mixture problems, resources have to be divided between different products or services subject to overall constraints.

In the construction industry there are many examples of queueing associated with the delivery of materials and the availability of plant. Estimates of queue length and waiting time are based upon probability concepts and the results can be used to predict the most economical way of providing the particular service required.

Operational research does not reduce the manager's responsibility but it aims to present him with the information on which decisions can be based, particularly the returns that can be expected from alternative strategies. It must be borne in mind, however, that operational research models are always approximations and the results they give cannot be more accurate or reliable than the data on which they are based.

In some cases the analysis will confirm the 'commonsense' view, but in others, such as Problem 11.3, the results may be unexpected. Experience is a valuable asset when dealing with familiar processes but the role of operational research is likely to grow in importance as projects increase in size and complexity.

Questions

1. The figures on the lines of the network given in Figure 11.22 represent the durations of the various activities in convenient units. How many paths are there through the network and what is the total duration of each?

   Determine the minimum time for completing the project and locate the critical path. Calculate the total float for each of the non-critical activities.

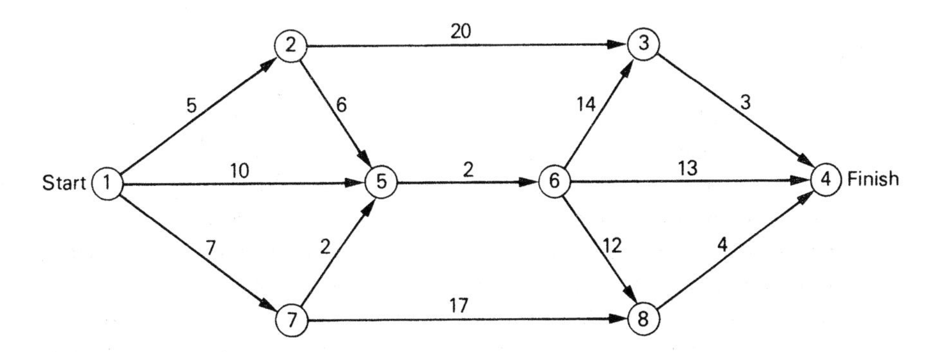

Fig. 11.22 Activity on arrow network.

2. The following table gives the preceding and succeeding event numbers (i and j) and durations (t) for a project with 14 activities. Draw a network, determine the minimum project time, locate the critical path, and calculate the total and free floats for each non-critical activity.

i	j	t
0	1	7
0	2	5
0	4	12
1	3	11
1	2	2
2	3	4
2	4	11

i	j	t
2	5	3
3	5	0
4	5	2
4	6	2
3	7	22
5	7	12
6	7	10

3. Figure 11.23 relates to a factory redevelopment scheme, the present and proposed use of the four areas being as follows:

   (a) A is wasteland ready for building on
   (b) B and C contain scattered workshops that are to be replaced by new buildings on A at twice the present density

(c) D is occupied by an office block that is to be replaced by a new one on B
(d) C and D will be used for a car park and landscaped area

The building programme for A is to be divided into two parts (for the relocation of the units on B and C, respectively), each half requiring a ten months' construction period. The new office block will take 18 months to build.

Building can only take place on one area at a time; the demolition on areas B, C, and D will require two months each, only one area being dealt with at a time. The contract for the car park and landscaping will require seven months.

Draw a network diagram for these activities, allowing one month for each of the three moves to the new buildings. Determine the minimum time for the completion of the project, list the critical activities in sequence, and state the total float on each of the others.

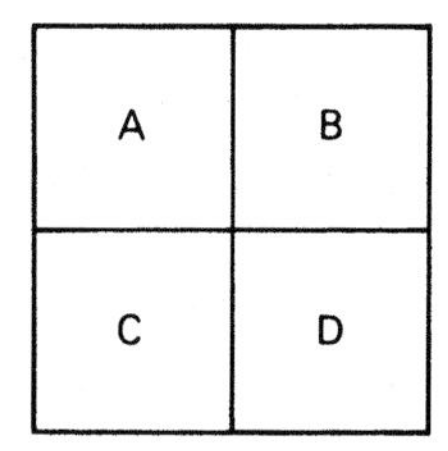

Fig. 11.23 Areas of factory redevelopment scheme.

4. A demolition contractor is working on five sites that are expected to yield the following quantities of hardcore, measured in lorryloads:

   A, 1; B, 6; C, 4; D, 3; E, 5

   He undertakes to supply the same material to three building sites in the following numbers of lorryloads:

   K, 2; L, 7; M, 10

   Show, on a suitable matrix, a schedule that satisfies these numbers.

		To		
		K	L	M
From	A	3	10	19
	B	6	7	22
	C	21	5	4
	D	9	11	13
	E	20	23	26

The table above shows the distances between the sites in miles. Calculate the total of lorrymiles for the schedule you have drawn up, find the feasible solution having the smallest total and state the lorrymiles involved.

5. In a joinery shop five teams are to be assigned to five jobs, one to each. An analysis of the skills, experience and resources of each team, and the complexity of each job leads to the following estimates, in hours, of the times required by the teams to complete the jobs:

		Job 1	2	3	4	5
Team	A	50	52	48	53	52
	B	46	27	50	63	60
	C	39	48	40	62	50
	D	31	42	58	48	51
	E	41	49	46	68	58

Determine the allocations of teams to jobs as to minimize the total team-hours. What is the total time involved in this arrangement?

6. Find an alternative optimum solution to Problem 11.5 Section 11.2 and show that it leads to the same total points score.
7. In the planning of a new town, a site of 400 hectares has been selected for a mixture of housing and industry. The development is subject to the following constraints:

    1. the supply of water and drainage capacity is sufficient for a maximum of 320 hectares of industry alone or 480 hectares of housing alone, or combinations in proportions that total 1
    2. at least 120 hectares must be used for housing
    3. in order to provide enough employment locally, the land allocated to industry must be least one-quarter as much as for housing
    4. at the other extreme the land for industry must not exceed that for housing

    Determine the respective allocations of land to housing and industry that will produce:

    1. the greatest amount of housing
    2. the greatest amount of industry
    3. the greatest rateable value if the value per hectare is 40 per cent higher for industry than for housing

8. A contractor distributes materials to one of his sites by his own lorries where they are unloaded by two men and a mobile crane. A lorry is despatched every 20 minutes and, on average, it requires 15 minutes to

unload each one. (a) Treating the system as a simple queue, calculate the average time a lorry has to wait before it is unloaded.

The labour costs for unloading amount to £15 per hour for each man employed and the cost of a lorry (with driver) is £45 per hour for waiting and unloading time. (b) Assuming that the speed at which a lorry can be unloaded is proportional to the number of men employed, should the number be changed from two and, if so, to what figure?

References and bibliography

Battersby, A. (1967) *Network Analysis*, Macmillan.

Krekó, Béla (1968) *Linear Programming*, Pitman.

Dantzig, G.B. and Thapa, M.N. (1997) *Linear Programming*, Springer.

Ignizio, J.P. (1994) *Linear Programming*, Prentice Hall.

Makower, M.S. and Williamson, E. (1967) *Operational Research*, English Universities Press.

Mitchell, G. and Willis, M.J. (1973) 'The determination of realistic probability levels for project completion dates', *Aeronautical Journal* (December).

Nuttall, J.F. and Jeanes, R.E. (1963) 'The critical path method', *The Builder* (14 and 21 June).

Pilcher, R. (1992) *Principles of Construction Management*, McGraw-Hill.

Singh, Jagjit (1971) *Operations Research*, Pelican.

Wilson, C. (1970) *Operational Research for Students of Management*, Intertext.

Appendix A
Discounted Cash Flow Tables

Table A.1 Amount of unit sum invested for *n* periods – compound interest $A = (1 + i)^n$

Compound interest rate i from 0.03 to 0.10 (3% to 10%)

n	*Interest rate i*								*n*
	0.03 (3%)	*0.04 (4%)*	*0.05 (5%)*	*0.06 (6%)*	*0.07 (7%)*	*0.08 (8%)*	*0.09 (9%)*	*0.10 (10%)*	
1	1.0300	1.0400	1.0500	1.0600	1.0700	1.0800	1.0900	1.1000	1
2	1.0609	1.0816	1.1025	1.1236	1.1449	1.1664	1.1881	1.2100	2
3	1.0927	1.1249	1.1576	1.1910	1.2250	1.2597	1.2950	1.3310	3
4	1.1255	1.1699	1.2155	1.2625	1.3108	1.3605	1.4116	1.4641	4
5	1.1593	1.2167	1.2763	1.3382	1.4026	1.4693	1.5386	1.6105	5
6	1.1941	1.2653	1.3404	1.4185	1.5007	1.5869	1.6771	1.7716	6
7	1.2299	1.3159	1.4071	1.5036	1.6058	1.7138	1.8280	1.9487	7
8	1.2668	1.3686	1.4775	1.5938	1.7182	1.8509	1.9926	2.1436	8
9	1.3048	1.4233	1.5513	1.6895	1.8385	1.9990	2.1719	2.3579	9
10	1.3439	1.4802	1.6289	1.7908	1.9672	2.1589	2.3674	2.5937	10
11	1.3842	1.5395	1.7103	1.8983	2.1049	2.3316	2.5804	2.8531	11
12	1.4258	1.6010	1.7959	2.0122	2.2522	2.5182	2.8127	3.1384	12
13	1.4685	1.6651	1.8856	2.1329	2.4098	2.7196	3.0658	3.4523	13
14	1.5126	1.7317	1.9799	2.2609	2.5785	2.9372	3.3417	3.7975	14
15	1.5580	1.8009	2.0789	2.3966	2.7590	3.1722	3.6425	4.1772	15
16	1.6047	1.8730	2.1829	2.5404	2.9522	3.4259	3.9703	4.5950	16
17	1.6528	1.9479	2.2920	2.6928	3.1588	3.7000	4.3276	5.0545	17
18	1.7024	2.0258	2.4066	2.8543	3.3799	3.9960	4.7171	5.5599	18
19	1.7535	2.1068	2.5270	3.0256	3.6165	4.3157	5.1417	6.1159	19
20	1.8061	2.1911	2.6533	3.2071	3.8697	4.6610	5.6044	6.7275	20
22	1.9161	2.3699	2.9253	3.6035	4.4304	5.4365	6.6586	8.1403	22
24	2.0328	2.5633	3.2251	4.0489	5.0724	6.3412	7.9111	9.8497	24
26	2.1566	2.7725	3.5557	4.5494	5.8074	7.3964	9.3992	11.9182	26
28	2.2879	2.9987	3.9201	5.1117	6.6488	8.6271	11.1671	14.4210	28
30	2.4273	3.3436	4.3219	5.7435	7.6123	10.0627	13.2677	17.4494	30
35	2.8139	3.9461	5.5160	7.6861	10.6766	14.7853	20.4140	28.1024	35
40	3.2620	4.8010	7.0400	10.2857	14.9745	21.7245	31.4094	45.2593	40
45	3.7816	5.8412	8.9850	13.7646	21.0025	31.9204	48.3273	72.8905	45
50	4.3839	7.1062	11.4674	18.4202	29.4570	46.9016	74.3575	117.391	50
55	5.0821	8.6464	14.6356	24.6503	41.3150	68.9139	114.408	189.059	55
60	5.8916	10.5196	18.6792	32.9877	57.9464	101.257	176.031	304.482	60
65	6.8300	12.7982	23.8399	44.1450	81.2729	148.780	270.846	490.371	65
70	7.9178	15.5716	30.4264	59.0759	113.989	218.606	416.730	789.747	70
75	9.1789	18.9453	38.8327	75.0569	159.876	321.205	641.191	1,271.90	75
80	10.6409	23.0408	49.5614	105.796	224.234	471.955	986.552	2,048.40	80
85	12.3357	28.0436	63.2544	141.579	314.500	693.457	1,517.93	3,298.97	85
90	14.3005	34.1193	80.7304	189.465	441.103	1,018.92	2,335.53	5,313.02	90
95	16.5782	41.5114	103.035	253.546	618.670	1,497.12	3,593.50	8,556.68	95
100	19.2187	50.5040	131.501	339.302	867.716	2,199.76	5,529.04	13,780.6	100

Table A.1 (Contd)

Compound interest rate i from 0.11 to 0.20 (11% to 20%)

n	*Interest rate i*								n
	0.11 (11%)	*0.12 (12%)*	*0.13 (13%)*	*0.14 (14%)*	*0.15 (15%)*	*0.16 (16%)*	*0.18 (18%)*	*0.20 (20%)*	
1	1.1100	1.1200	1.1300	1.1400	1.1500	1.1600	1.1800	1.2000	1
2	1.2321	1.2544	1.2769	1.2996	1.3225	1.3456	1.3924	1.4400	2
3	1.3676	1.4049	1.4429	1.4815	1.5209	1.5609	1.6430	1.7280	3
4	1.5181	1.5735	1.6305	1.6890	1.7490	1.8106	1.9388	2.0736	4
5	1.6851	1.7623	1.8424	1.9254	2.0114	2.1003	2.2878	2.4883	5
6	1.8704	1.9738	2.0820	2.1950	2.3131	2.4364	2.6996	2.9860	6
7	2.0762	2.2107	2.3526	2.5023	2.6600	2.8262	3.1855	3.5832	7
8	2.3045	2.4760	2.6584	2.8526	3.0590	3.2784	3.7589	4.2998	8
9	2.5580	2.7731	3.0040	3.2519	3.5179	3.8030	4.4355	5.1598	9
10	2.8394	3.1058	3.3946	3.7072	4.0456	4.4114	5.2338	6.1917	10
11	3.1518	3.4785	3.8359	4.2262	4.6524	5.1173	6.1759	7.4301	11
12	3.4985	3.8960	4.3345	4.8179	5.3503	5.9360	7.2876	8.9161	12
13	3.8833	4.3635	4.8980	5.4924	6.1528	6.8858	8.5994	10.6993	13
14	4.3104	4.8871	5.5348	6.2613	7.0757	7.9875	10.1472	12.8392	14
15	4.7846	5.4736	6.2543	7.1379	8.1371	9.2655	11.9737	15.4070	15
16	5.3109	6.1304	7.0673	8.1372	9.3576	10.7480	14.1290	18.4884	16
17	5.8951	6.8660	7.9861	9.2765	10.7613	12.4677	16.6722	22.1861	17
18	6.5436	7.6900	9.0243	10.5752	12.3755	14.4625	19.6733	26.6233	18
19	7.2633	8.6128	10.1974	12.0557	14.2318	16.7765	23.2144	31.9480	19
20	8.0623	9.6463	11.5231	13.7435	16.3665	19.4608	27.3930	38.3376	20
22	9.9336	12.1003	14.7138	17.8610	21.6447	26.1864	38.1421	55.2061	22
24	12.2392	15.1786	18.7881	23.2122	28.6252	35.2364	53.1090	79.4968	24
26	15.0799	19.0401	23.9905	30.1666	37.8568	47.4141	73.9490	114.476	26
28	18.5799	23.8839	30.6335	39.2045	50.0656	63.8004	102.967	164.845	28
30	22.8923	29.9599	39.1159	50.9502	66.2118	85.8499	143.371	237.376	30
35	38.5749	52.7996	72.0685	98.1002	133.176	180.314	327.997	590.668	35
40	65.0009	93.0510	132.782	188.884	267.864	378.721	750.378	1,469.77	40
45	109.530	163.988	244.641	363.679	538.769	795.444	1,716.68	3,657.26	45
50	184.565	289.002	450.736	700.233	1,083.66	1,670.70	3,927.36	9,100.44	50
55	311.003	509.321	830.452	1,348.24	2,179.62	3,509.05	8,984.84	22,644.8	55
60	524.057	897.597	1,530.05	2,595.92	4,384.00	7,370.20	20,555.1	56,347.5	60
65	883.067	1,581.87	2,819.02	4,998.22	8,817.79	15,479.9	47,025.2	140,211	65
70	1,488.02	2,787.80	5,193.87	9,623.65	17,735.7	32,513.2	107,582	348,889	70
75	2,507.40	4,913.06	9,569.37	18,529.5	35,672.9	68,288.8	246,122	868,147	75
80	4,225.11	8,658.48	17,630.9	35,677.0	71,750.9	143,430	563,068	2,160,228	80

Table A.2 Present value (PV) of unit sum spent after n periods $PV = \frac{1}{(1+i)^n}$

Compound interest rate i from 0.03 to 0.10 (3% to 10%)

n	Interest rate i								n
	0.03 (3%)	*0.04 (4%)*	*0.05 (5%0*	*0.06 (6%)*	*0.07) (7%)*	*0.08 (8%)*	*0.09 (9%)*	*0.10 (10%)*	
1	0.9709	0.9615	0.9524	0.9434	0.9346	0.9259	0.9174	0.9091	1
2	0.9426	0.9246	0.9070	0.8900	0.8734	0.8573	0.8417	0.8264	2
3	0.9151	0.8890	0.8638	0.8396	0.8163	0.7938	0.7722	0.7513	3
4	0.8885	0.8548	0.8227	0.7921	0.7629	0.7350	0.7084	0.6830	4
5	0.8626	0.8219	0.7835	0.7473	0.7130	0.6806	0.6499	0.6209	5
6	0.8375	0.7903	0.7462	0.7050	0.6663	0.6302	0.5963	0.5645	6
7	0.8131	0.7599	0.7107	0.6651	0.6227	0.5835	0.5470	0.5132	7
8	0.7894	0.7307	0.6768	0.6274	0.5820	0.5403	0.5019	0.4665	8
9	0.7664	0.7026	0.6446	0.5919	0.5439	0.5002	0.4604	0.4241	9
10	0.7441	0.6756	0.6139	0.5584	0.5083	0.4632	0.4224	0.3855	10
11	0.7224	0.6496	0.5847	0.5268	0.4751	0.4289	0.3875	0.3505	11
12	0.7014	0.6246	0.5568	0.4970	0.4440	0.3971	0.3555	0.3186	12
13	0.6810	0.6006	0.5303	0.4688	0.4150	0.3677	0.3262	0.2897	13
14	0.6611	0.5775	0.5051	0.4423	0.3878	0.3405	0.2992	0.2633	14
15	0.6419	0.5553	0.4810	0.4173	0.3624	0.3152	0.2745	0.2394	15
16	0.6232	0.5339	0.4581	0.3936	0.3387	0.2919	0.2519	0.2176	16
17	0.6050	0.5134	0.4363	0.3714	0.3166	0.2703	0.2311	0.1978	17
18	0.5874	0.4936	0.4155	0.3503	0.2959	0.2502	0.2120	0.1799	18
19	0.5703	0.4746	0.3957	0.3305	0.2765	0.2317	0.1945	0.1635	19
20	0.5537	0.4564	0.3769	0.3118	0.2584	0.2145	0.1784	0.1486	20
22	0.5219	0.4220	0.3418	0.2775	0.2257	0.1839	0.1502	0.1228	22
24	0.4919	0.3901	0.3101	0.2470	0.1971	0.1577	0.1264	0.1015	24
26	0.4637	0.3607	0.2812	0.2198	0.1722	0.1352	0.1064	0.0839	26
28	0.4371	0.3335	0.2551	0.1956	0.1504	0.1159	0.0895	0.0693	28
30	0.4120	0.3083	0.2314	0.1741	0.1314	0.0994	0.0754	0.0573	30
35	0.3554	0.2534	0.1813	0.1301	0.0937	0.0676	0.0490	0.0356	35
40	0.3066	0.2083	0.1420	0.0972	0.0668	0.0460	0.0318	0.0221	40
45	0.2644	0.1712	0.1113	0.0727	0.0476	0.0313	0.0207	0.0137	45
50	0.2281	0.1407	0.0872	0.0543	0.0339	0.0213	0.0134	0.00852	50
55	0.1968	0.1157	0.0683	0.0406	0.0242	0.0145	0.00874	0.00529	55
60	0.1697	0.0951	0.0535	0.0303	0.0173	0.00988	0.00568	0.00328	60
65	0.1464	0.0781	0.0419	0.0227	0.0123	0.00672	0.00369	0.00204	65
70	0.1263	0.0642	0.0329	0.0169	0.00877	0.00457	0.00240	0.00127	70
75	0.1089	0.0528	0.0258	0.0126	0.00625	0.00311	0.00156	0.00079	75
80	0.0940	0.0434	0.0202	0.00945	0.00446	0.00212	0.00101	0.00049	80
85	0.0811	0.0357	0.0158	0.00706	0.00318	0.00144	0.00066	0.00030	85
90	0.0699	0.0293	0.0124	0.00528	0.00227	0.00098	0.00043	0.00019	90
95	0.0603	0.0241	0.00971	0.00394	0.00162	0.00067	0.00028	0.00012	95
100	0.0520	0.0198	0.00760	0.00295	0.00115	0.00045	0.00018	0.00007	100

Table A.2 (Contd)

Compound interest rate i from 0.11 to 0.20 (11% to 20%)

n	*Interest rate i*								*n*
	0.11 (11%)	*0.12 (12%)*	*0.13 (13%)*	*0.14 (14%)*	*0.15 (15%)*	*0.16 (16%)*	*0.18 (18%)*	*0.20 (20%)*	
1	0.9009	0.8929	0.8850	0.8872	0.8696	0.8621	0.8475	0.8333	1
2	0.8116	0.7972	0.7831	0.7695	0.7561	0.7432	0.7182	0.6944	2
3	0.7312	0.7118	0.6931	0.6750	0.6575	0.6407	0.6086	0.5787	3
4	0.6587	0.6355	0.6133	0.5921	0.5718	0.5523	0.5158	0.4823	4
5	0.5935	0.5674	0.5428	0.5194	0.4972	0.4761	0.4371	0.4019	5
6	0.5346	0.5066	0.4803	0.4556	0.4323	0.4104	0.3704	0.3349	6
7	0.4817	0.4523	0.4251	0.3996	0.3759	0.3538	0.3139	0.2791	7
8	0.4339	0.4039	0.3762	0.3506	0.3269	0.3050	0.2660	0.2326	8
9	0.3909	0.3606	0.3329	0.3075	0.2843	0.2630	0.2255	0.1938	9
10	0.3522	0.3220	0.2946	0.2697	0.2472	0.2267	0.1911	0.1615	10
11	0.3173	0.2875	0.2607	0.2366	0.2149	0.1954	0.1619	0.1346	11
12	0.2858	0.2567	0.2307	0.2076	0.1869	0.1685	0.1372	0.1122	12
13	0.2575	0.2292	0.2042	0.1821	0.1625	0.1452	0.1163	0.0935	13
14	0.2320	0.2046	0.1807	0.1597	0.1413	0.1252	0.0985	0.0779	14
15	0.2090	0.1827	0.1599	0.1401	0.1229	0.1079	0.0835	0.0649	15
16	0.1883	0.1631	0.1415	0.1229	0.1069	0.0930	0.0708	0.0541	16
17	0.1696	0.1456	0.1252	0.1078	0.0929	0.0802	0.0600	0.0451	17
18	0.1528	0.1300	0.1108	0.0946	0.0808	0.0691	0.0508	0.0376	18
19	0.1377	0.1161	0.0981	0.0829	0.0703	0.0596	0.0431	0.0313	19
20	0.1240	0.1037	0.0868	0.0728	0.0611	0.0514	0.0365	0.0261	20
22	0.1007	0.0826	0.0680	0.0560	0.0462	0.0382	0.0262	0.0181	22
24	0.0817	0.0659	0.0532	0.0431	0.0349	0.0284	0.0188	0.0126	24
26	0.0663	0.0525	0.0417	0.0331	0.0264	0.0211	0.0135	0.00874	26
28	0.0538	0.0419	0.0326	0.0255	0.0200	0.0157	0.00971	0.00607	28
30	0.0437	0.0334	0.0256	0.0196	0.0151	0.0116	0.00697	0.00421	30
35	0.0259	0.0189	0.0139	0.0102	0.00751	0.00555	0.00305	0.00169	35
40	0.0154	0.0107	0.00753	0.00529	0.00373	0.00264	0.00133	0.00068	40
45	0.00913	0.00610	0.00409	0.00275	0.00186	0.00126	0.00058	0.00027	45
50	0.00542	0.00346	0.00222	0.00143	0.00092	0.00060	0.00025	0.00011	50
55	0.00322	0.00196	0.00120	0.00074	0.00046	0.00028	0.00011		55
60	0.00191	0.00111	0.00065	0.00039	0.00023	0.00014			60
65	0.00113	0.00063	0.00035	0.00020	0.00011				65
70	0.00067	0.00036	0.00019	0.00010					70
75	0.00040	0.00020	0.00010		Less than 0.0001				75
80	0.00024	0.00012							

Table A.3 Amount of unit sum invested each period for n periods $A = \frac{(1+i)^n - 1}{i}$

Compound interest rate i from 0.03 to 0.10 (3% to 10%)

n	Interest rate i								n
	0.03 (3%)	0.04 (4%)	0.05 (5%0	0.06 (6%)	0.07) (7%)	0.08 (8%)	0.09 (9%)	0.10 (10%)	
1	1.0000	1.0000	1.0000	1.0000	1.0000	1.0000	1.0000	1.0000	1
2	2.0300	2.0400	2.0500	2.0600	2.0700	2.0800	2.0900	2.1000	2
3	3.0909	3.1216	3.1525	3.1836	3.2149	3.2464	3.2781	3.3100	3
4	4.1836	4.2465	4.3101	4.3746	4.4388	4.5061	4.5731	4.6410	4
5	5.3091	5.4163	5.5256	5.6371	5.7507	5.8666	5.9847	6.1051	5
6	6.4684	6.6330	6.8019	6.9753	7.1533	7.3359	7.5233	7.7156	6
7	7.6625	7.8983	8.1420	8.3938	8.6540	8.9228	9.2004	9.4872	7
8	8.8923	9.2142	9.5491	9.8975	10.2598	10.6366	11.0285	11.4359	8
9	10.1591	10.5828	11.0266	11.4913	11.9780	12.4876	13.0210	13.5795	9
10	11.4639	12.0061	12.5779	13.1808	13.8164	14.4866	15.1929	15.9374	10
11	12.8078	13.4864	14.2068	14.9716	15.7836	16.6455	17.5603	18.5312	11
12	14.1920	15.0258	15.9171	16.8699	17.8885	18.9771	20.1407	21.3843	12
13	15.1678	16.6268	17.7130	18.8821	20.1406	21.4953	22.9534	24.5227	13
14	17.0863	18.2919	19.5986	21.0151	22.5504	24.2149	26.0192	27.9750	14
15	18.5989	20.0236	21.5786	23.2760	25.1290	27.1521	29.3609	31.7725	15
16	20.1569	21.8245	23.6575	25.6725	27.8881	30.3243	33.0034	35.9497	16
17	21.7616	23.6975	24.8404	28.2129	30.8402	33.7502	36.9737	40.5447	17
18	23.4144	25.6454	28.1324	30.9057	33.9990	37.4503	41.3013	45.5992	18
19	25.1169	27.6712	30.5390	33.7600	37.3790	41.4463	46.0185	51.1591	19
20	26.8704	29.7781	33.0660	36.7860	40.9955	45.7620	51.1601	57.2750	20
22	30.5368	34.2480	38.5052	43.3923	49.0057	55.4568	62.8733	71.4028	22
24	34.4265	39.0826	44.5020	50.8156	58.1767	66.7648	76.7898	88.4973	24
26	38.5530	44.3117	51.1135	59.1564	68.6767	79.9544	93.3240	109.182	26
28	42.9309	49.9676	58.4026	68.5281	80.6977	95.3388	112.968	123.210	28
30	47.5754	56.0849	66.4388	79.0582	94.4608	113.283	136.308	164.494	30
35	60.4621	73.6522	90.3203	111.435	138.237	172.317	215.711	271.024	35
40	75.4013	95.0255	120.800	154.762	199.635	259.057	337.882	442.593	40
45	92.7199	121.029	159.700	212.744	285.749	386.506	525.859	718.905	45
50	112.797	152.667	209.348	290.336	406.529	573.770	815.084	1,163.91	50
55	136.072	191.159	272.712	394.172	575.929	848.923	1,260.09	1,880.59	55
60	163.053	237.991	353.584	533.128	813.520	1,253.21	1,944.79	3,034.82	60
65	194.333	294.968	456.798	719.083	1,146.76	1,847.25	2,998.29	4,893.71	65
70	230.594	364.290	588.529	967.932	1,614.13	2,270.08	4,619.22	7,887.47	70
75	272.631	448.631	756.654	1,300.95	2,269.66	4,002.56	7,113.23	12,709.0	75
80	321.363	551.245	971.229	1,746.60	3,189.06	5,886.94	10,950.6	20,474.0	80
85	377.857	676.090	1,245.09	2,342.98	4,478.58	8,655.71	16,854.8	32,979.7	85
90	443.349	827.983	1,594.61	3,141.08	6,287.19	12,723.9	25,939.2	53,120.2	90
95	519.272	1,012.78	2,040.69	4,209.10	8,823.85	18,701.5	39,916.6	85,556.8	95
100	607.288	1,237.62	2,610.03	5,638.37	12,381.7	27,484.5	61,422.7	137,796	100

Table A.3 (Contd)

Compound interest rate i from 0.11 to 0.20 (11% to 20%)

n	*Interest rate i*								*n*
	0.11 (11%)	*0.12 (12%)*	*0.13 (13%)*	*0.14 (14%)*	*0.15 (15%)*	*0.16 (16%)*	*0.18 (18%)*	*0.20 (20%)*	
1	1.0000	1.0000	1.0000	1.0000	1.0000	1.0000	1.0000	1.0000	1
2	2.1100	2.1200	2.1300	2.1400	2.1500	2.1600	2.1800	2.2000	2
3	3.3421	3.3744	3.4069	3.4396	3.4725	3.5056	3.5724	3.6400	3
4	4.7097	4.7793	4.8498	4.9211	4.9934	5.0665	5.2154	5.3680	4
5	6.2278	6.3528	6.4803	6.6101	6.7424	6.8771	7.1542	7.4416	5
6	7.9129	8.1152	8.3227	8.5355	8.7537	8.9775	9.4420	9.9299	6
7	9.7833	10.0890	10.4047	10.7305	11.0668	11.4139	12.1415	12.9159	7
8	11.8594	12.2997	12.7523	13.2328	13.7268	14.2401	15.3270	16.4991	8
9	14.1640	14.7757	15.4157	16.0853	16.7858	17.5185	19.0859	20.7989	9
10	16.7220	17.5487	18.4197	19.3373	20.3037	21.3215	23.5213	25.9587	10
11	19.5614	20.6546	21.8143	23.0445	24.3493	25.7329	28.7551	32.1504	11
12	22.7132	24.1331	25.6502	27.2707	29.0017	30.8502	34.9311	39.5805	12
13	26.2116	28.0291	29.9847	32.0887	34.3519	36.7862	42.2187	48.4966	13
14	30.0949	32.3926	34.8827	37.5811	40.5047	43.6720	50.8180	59.1959	14
15	34.4054	37.2797	40.4175	43.8424	47.5804	51.6595	60.9653	72.0351	15
16	39.1899	42.7533	46.6717	50.9804	55.7175	60.9250	72.9390	87.4421	16
17	44.5008	48.8837	53.7381	59.1176	65.0751	71.6730	87.0680	105.931	17
18	50.3959	55.7497	61.7251	68.3941	75.8364	84.1407	103.740	128.117	18
19	56.9395	63.4397	70.7494	78.9692	88.2118	98.6032	123.414	154.740	19
20	64.2028	72.0524	80.9468	91.0249	102.444	115.380	146.628	186.688	20
22	81.2143	92.5026	105.491	120.436	137.632	157.415	206.345	271.031	22
24	102.174	118.155	136.831	158.659	184.168	213.978	289.494	342.484	24
26	127.999	150.334	176.850	208.333	245.712	290.088	405.272	567.377	26
28	159.817	190.699	227.950	272.889	327.104	392.503	566.481	819.223	28
30	199.021	241.333	293.199	356.787	434.745	530.312	790.948	1,181.88	30
35	341.590	431.663	546.681	693.573	881.170	1,120.71	1,816.65	2,948.34	35
40	581.826	767.091	1,013.70	1,342.03	1,779.09	2,360.76	4,163.21	7,343.86	40
45	986.639	1,358.23	1,874.16	2,590.56	3,585.13	4,965.27	9,531.58	18,281.3	45
50	1,668.77	2,400.02	3,459.51	4,994.52	7,217.72	10,435.6	21,813.1	45,497.2	50
55	2,818.20	4,236.01	6,380.40	9,623.13	14,524.1	21,925.3	49,910.2	113,219	55
60	4,755.07	7,471.64	11,761.9	18,535.1	29,220.0	46,057.5	114,190	281,733	60
65	8,018.79	13,173.9	21,677.1	35,694.4	58,778.6	96,743.5	261,245	701,048	65
70	13,518.4	23,223.3	39,945.2	68,733.2	118,231	203,201	597,673	1,744,440	70
75	22,785.4	40,933.8	73,602.8	132,346	237,812	426,798	1,367,339	4,340,732	75
80	38,401.0	72,145.7	135,615	254,828	478,332	896,429	3,128,148	10,801,138	80

Table A.4 Present value of unit sum per period for *n* periods $PV = \frac{1}{i}\left(1 - \frac{1}{(1+i)^n}\right)$

Compound interest rate i from 0.03 to 0.10 (3% to 10%)

n	*Interest rate i*								*n*
	0.03 (3%)	*0.04 (4%)*	*0.05 (5%0*	*0.06 (6%)*	*0.07) (7%)*	*0.08 (8%)*	*0.09 (9%)*	*0.10 (10%)*	
1	0.971	0.962	0.952	0.943	0.935	0.926	0.917	0.909	1
2	1.913	1.886	1.859	1.833	1.808	1.783	1.759	1.736	2
3	2.829	2.775	2.723	2.673	2.624	2.577	2.531	2.487	3
4	3.717	3.630	3.546	3.465	3.387	3.312	3.240	3.170	4
5	4.580	4.452	4.329	4.212	4.100	3.993	3.890	3.791	5
6	5.417	5.242	5.076	4.917	4.767	4.623	4.486	4.355	6
7	6.230	6.002	5.786	5.582	5.389	5.206	5.033	4.868	7
8	7.020	6.733	6.463	6.210	5.971	5.747	5.535	5.335	8
9	7.786	7.435	7.108	6.802	6.515	6.247	5.995	5.759	9
10	8.530	8.111	7.722	7.360	7.024	6.710	6.418	6.145	10
11	9.253	8.760	8.306	7.887	7.499	7.139	6.805	6.495	11
12	9.954	9.385	8.863	8.384	7.943	7.536	7.161	6.814	12
13	10.635	9.986	9.394	8.853	8.358	7.904	7.487	7.103	13
14	11.296	10.563	9.899	9.295	8.745	8.244	7.786	7.367	14
15	11.938	11.118	10.380	9.712	9.108	8.559	8.061	7.606	15
16	12.561	11.652	10.838	10.106	9.447	8.851	8.313	7.824	16
17	13.166	12.166	11.274	10.477	9.763	9.122	8.544	8.022	17
18	13.754	12.659	11.690	10.828	10.059	9.372	8.756	8.201	18
19	14.324	13.134	12.085	11.158	10.336	9.604	8.950	8.365	19
20	14.877	13.590	12.462	11.470	10.594	9.819	9.129	8.514	20
22	15.937	14.451	13.163	12.042	11.061	10.201	9.442	8.772	22
24	16.936	15.247	13.799	12.550	11.469	10.529	9.707	8.985	24
26	17.877	15.983	14.375	13.003	11.826	10.810	9.929	9.161	26
28	18.764	16.663	14.898	13.406	12.137	11.051	10.116	9.307	28
30	19.600	17.292	15.372	13.765	12.409	11.258	10.274	9.427	30
35	21.487	18.665	16.374	14.498	12.948	11.655	10.567	9.644	35
40	23.115	19.793	17.159	15.046	13.332	11.925	10.757	9.779	40
45	24.519	20.720	17.774	15.456	13.606	12.108	10.881	9.863	45
50	25.730	21.482	18.256	15.762	13.801	12.233	10.962	9.915	50
55	26.774	22.109	18.633	15.991	13.940	12.319	11.014	9.947	55
60	27.676	22.623	18.929	16.161	14.039	12.377	11.048	9.967	60
65	28.453	23.047	19.161	16.289	14.110	12.416	11.070	9.980	65
70	29.123	23.395	19.343	16.385	14.160	12.443	11.084	9.987	70
75	29.702	23.680	19.485	16.456	14.196	12.461	11.094	9.992	75
80	30.201	23.915	19.596	16.509	14.222	12.474	11.100	9.995	80
85	30.631	24.109	19.684	16.549	14.240	12.482	11.104	9.997	85
90	31.002	24.267	19.752	16.579	14.253	12.488	11.106	9.998	90
95	31.323	24.398	19.806	16.601	14.263	12.492	11.108	9.999	95
100	31.599	24.505	19.848	16.618	14.269	12.494	11.109	9.999	100
∞	33.333	25.000	20.000	16.667	14.286	12.500	11.111	10.000	∞

Table A.4 (Contd)

Compound interest rate i from 0.11 to 0.20 (11% to 20%)

n	*Interest rate i*								*n*
	0.11 (11%)	*0.12 (12%)*	*0.13 (13%)*	*0.14 (14%)*	*0.15 (15%)*	*0.16 (16%)*	*0.18 (18%)*	*0.20 (20%)*	
1	0.901	0.893	0.885	0.877	0.870	0.862	0.847	0.833	1
2	1.713	1.690	1.668	1.647	1.626	1.605	1.566	1.528	2
3	2.444	2.402	2.361	2.322	2.283	2.246	2.174	2.106	3
4	3.102	3.037	2.974	2.914	2.855	2.798	2.690	2.589	4
5	3.696	3.605	3.517	3.433	3.352	3.274	3.127	2.991	5
6	4.231	4.111	3.998	3.889	3.784	3.685	3.498	3.326	6
7	4.712	4.564	4.423	4.288	4.160	4.039	3.812	3.605	7
8	5.146	4.968	4.799	4.639	4.487	4.344	4.078	3.837	8
9	5.537	5.328	5.132	4.946	4.772	4.607	4.303	4.031	9
10	5.889	5.650	5.426	5.216	5.019	4.833	4.494	4.192	10
11	6.207	5.938	5.687	5.453	5.234	5.029	4.656	4.327	11
12	6.492	6.194	5.918	5.660	5.421	5.197	4.793	4.439	12
13	6.750	6.424	6.122	5.842	5.583	5.342	4.910	4.533	13
14	6.982	6.628	6.302	6.002	5.724	5.468	5.008	4.611	14
15	7.191	6.811	6.462	6.142	5.847	5.575	5.092	4.675	15
16	7.379	6.974	6.604	6.265	5.954	5.668	5.162	4.730	16
17	7.549	7.120	6.729	6.373	6.047	5.749	5.222	4.775	17
18	7.702	7.250	6.840	6.467	6.128	5.818	5.273	4.812	18
19	7.839	7.366	6.938	6.550	6.198	5.877	5.316	4.843	19
20	7.963	7.469	7.025	6.623	6.259	5.929	5.353	4.870	20
22	8.176	7.645	7.170	6.743	6.359	6.011	5.410	4.909	22
24	8.348	7.784	7.283	6.835	6.434	6.073	5.451	4.937	24
26	8.488	7.896	7.372	6.906	6.491	6.118	5.480	4.956	26
28	8.602	7.984	7.441	6.961	6.534	6.152	5.502	4.970	28
30	8.694	8.055	7.496	7.003	6.566	6.177	5.517	4.979	30
35	8.855	8.176	7.586	7.070	6.617	6.215	5.539	4.992	35
40	8.951	8.244	7.634	7.105	6.642	6.233	5.548	4.997	40
45	9.008	8.283	7.661	7.123	6.654	6.242	5.552	4.999	45
50	9.042	8.304	7.675	7.133	6.661	6.246	5.554	4.999	50
55	9.062	8.317	7.683	7.138	6.664	6.248	5.555	5.000	55
60	9.074	8.324	7.687	7.140	6.665	6.249	5.555	5.000	60
65	9.081	8.328	7.690	7.141	6.666	6.250	5.555		65
70	9.085	8.330	7.691	7.142	6.666	6.250	5.556		70
75	9.087	8.332	7.692	7.142	6.667	6.250	5.556		75
80	9.089	8.332	7.692	7.143	6.667	6.250	5.556		80
∞	9.091	8.333	7.692	7.143	6.667	6.250	5.556	5.000	∞

Table A.5 Annual sinking fund (ASF) amount to be invested each year to accumulate to unit sum at end of period $\text{ASF} = \dfrac{i}{(1+i)^n - 1}$

Compound interest rate i from 0.03 to 0.10 (3% to 10%)

n	*Interest rate i*								*n*
	0.03 (3%)	*0.04 (4%)*	*0.05 (5%0*	*0.06 (6%)*	*0.07) (7%)*	*0.08 (8%)*	*0.09 (9%)*	*0.10 (10%)*	
1	1.0000	1.0000	1.0000	1.0000	1.0000	1.0000	1.0000	1.0000	1
2	0.4926	0.4902	0.4878	0.4854	0.4831	0.4808	0.4785	0.4762	2
3	0.3235	0.3203	0.3172	0.3141	0.3111	0.3080	0.3051	0.3021	3
4	0.2390	0.2355	0.2320	0.2286	0.2252	0.2219	0.2187	0.2155	4
5	0.1884	0.1846	0.1810	0.1774	0.1739	0.1705	0.1671	0.1638	5
6	0.1546	0.1508	0.1470	0.1434	0.1398	0.1363	0.1329	0.1296	6
7	0.1305	0.1266	0.1228	0.1191	0.1156	0.1121	0.1087	0.1054	7
8	0.1125	0.1085	0.1047	0.1010	0.0975	0.0940	0.0963	0.0874	8
9	0.0984	0.0945	0.0907	0.0870	0.0835	0.0801	0.0768	0.0736	9
10	0.0872	0.0833	0.0795	0.0759	0.0724	0.0690	0.0658	0.0627	10
11	0.0781	0.0741	0.0704	0.0668	0.0634	0.0601	0.0569	0.0540	11
12	0.0705	0.0666	0.0628	0.0593	0.0559	0.0527	0.0497	0.0468	12
13	0.0640	0.0601	0.0565	0.0530	0.0497	0.0465	0.0436	0.0408	13
14	0.0585	0.0547	0.0510	0.0476	0.0443	0.0413	0.0384	0.0357	14
15	0.0538	0.0499	0.0463	0.0430	0.0398	0.0368	0.0341	0.0315	15
16	0.0496	0.0458	0.0423	0.0390	0.0359	0.0330	0.0303	0.0278	16
17	0.0460	0.0422	0.0387	0.0354	0.0324	0.0296	0.0270	0.0247	17
18	0.0427	0.0390	0.0355	0.0324	0.0294	0.0267	0.0242	0.0219	18
19	0.0398	0.0361	0.0327	0.0296	0.0268	0.0241	0.0127	0.0195	19
20	0.0372	0.0336	0.0302	0.0272	0.0244	0.0219	0.0195	0.0175	20
22	0.0327	0.0292	0.0260	0.0230	0.0204	0.0180	0.0159	0.0140	22
24	0.0290	0.0256	0.0225	0.0197	0.0172	0.0150	0.0130	0.0113	24
26	0.0259	0.0226	0.0196	0.0169	0.0146	0.0125	0.0107	0.00916	26
28	0.0233	0.0200	0.0171	0.0146	0.0124	0.0105	0.00885	0.00745	28
30	0.0210	0.0178	0.0151	0.0126	0.0106	0.00883	0.00734	0.00608	30
35	0.0165	0.0136	0.0111	0.00879	0.00723	0.00580	0.00464	0.00369	35
40	0.0133	0.0105	0.00828	0.00646	0.00501	0.00386	0.00300	0.00226	40
45	0.0108	0.00826	0.00626	0.00470	0.00350	0.00259	0.00190	0.00139	45
50	0.00865	0.00655	0.00478	0.00344	0.00246	0.00174	0.00123	0.00086	50
55	0.00735	0.00523	0.00367	0.00254	0.00174	0.00118	0.00079	0.00053	55
60	0.00613	0.00420	0.00283	0.00188	0.00123	0.00079	0.00051	0.00033	60
65	0.00515	0.00390	0.00219	0.00139	0.00087	0.00054	0.00033	0.00020	65
70	0.00434	0.00275	0.00170	0.00103	0.00062	0.00036	0.00021	0.00013	70
75	0.00367	0.00223	0.00132	0.00077	0.00044	0.00025	0.00014		75
80	0.00311	0.00181	0.00103	0.00057	0.00031	0.00017			80
85	0.00265	0.00148	0.00080	0.00043	0.00022	0.00011			85
90	0.00226	0.00121	0.00063	0.00032	0.00016				90
95	0.00193	0.00099	0.00049	0.00024	0.00011	Less than 0.0001			95
100	0.00165	0.00081	0.00038	0.00018					100

Table A.5 (Contd)

Compound interest rate i from 0.11 to 0.20 (11% to 20%)

n	Interest rate *i*								*n*
	0.11 (11%)	*0.12 (12%)*	*0.13 (13%)*	*0.14 (14%)*	*0.15 (15%)*	*0.16 (16%)*	*0.18 (18%)*	*0.20 (20%)*	
1	1.0000	1.0000	1.0000	1.0000	1.0000	1.0000	1.0000	1.0000	1
2	0.4739	0.4717	0.4695	0.4673	0.4651	0.4630	0.4587	0.4545	2
3	0.2992	0.2963	0.2935	0.2907	0.2880	0.2853	0.2799	0.2747	3
4	0.2123	0.2092	0.2062	0.2032	0.2003	0.1974	0.1917	0.1863	4
5	0.1606	0.1574	0.1543	0.1513	0.1483	0.1454	0.1400	0.1344	5
6	0.1264	0.1232	0.1202	0.1172	0.1142	0.1114	0.1059	0.1007	6
7	0.1022	0.0991	0.0961	0.0932	0.0904	0.0876	0.0824	0.0774	7
8	0.0843	0.0813	0.0784	0.0756	0.0729	0.0702	0.0652	0.0606	8
9	0.0706	0.0677	0.0649	0.0622	0.0596	0.0571	0.0524	0.0481	9
10	0.0598	0.0570	0.0543	0.0517	0.0493	0.0469	0.0425	0.0385	10
11	0.0511	0.0484	0.0458	0.0434	0.0411	0.0389	0.0348	0.0311	11
12	0.0440	0.0414	0.0390	0.0367	0.0345	0.0324	0.0286	0.0253	12
13	0.0382	0.0357	0.0334	0.0312	0.0291	0.0272	0.0237	0.0206	13
14	0.0332	0.0309	0.0287	0.0266	0.0247	0.0229	0.0197	0.0169	14
15	0.0291	0.0268	0.0247	0.0228	0.0210	0.0194	0.0164	0.0139	15
16	0.0255	0.0234	0.0214	0.0196	0.0179	0.0164	0.0137	0.0114	16
17	0.0225	0.0205	0.0186	0.0169	0.0154	0.0140	0.0115	0.00944	17
18	0.0198	0.0179	0.0162	0.0146	0.0132	0.0119	0.00964	0.00781	18
19	0.0176	0.0158	0.0141	0.0127	0.0113	0.0101	0.00810	0.00646	19
20	0.0156	0.0139	0.0124	0.0110	0.00976	0.00867	0.00682	0.00536	20
22	0.0123	0.0108	0.00948	0.00830	0.00767	0.00635	0.00485	0.00369	22
24	0.00979	0.00846	0.00731	0.00630	0.00543	0.00467	0.00345	0.00255	24
26	0.00781	0.00665	0.00565	0.00480	0.00407	0.00345	0.00247	0.00176	26
28	0.00626	0.00524	0.00439	0.00366	0.00306	0.00255	0.00177	0.00122	28
30	0.00502	0.00414	0.00341	0.00280	0.00230	0.00189	0.00126	0.00085	30
35	0.00293	0.00232	0.00183	0.00144	0.00113	0.00089	0.00055	0.00034	35
40	0.00172	0.00130	0.00099	0.00075	0.00056	0.00042	0.00024	0.00014	40
45	0.00101	0.00074	0.00053	0.00039	0.00028	0.00020	0.00010		45
50	0.00060	0.00042	0.00029	0.00020	0.00014				50
55	0.00035	0.00024	0.00016	0.00010					55
60	0.00021	0.00013							60
65	0.00012								65
70				Less than 0.0001					70
80									80

Appendix B
Properties of the Normal Curve

Table B.1 One-tail areas for the normal curve (see Figure 10.12)

z	*0.00*	*0.01*	*0.02*	*0.03*	*0.04*	*0.05*	*0.06*	*0.07*	*0.08*	*0.09*
0.0	0.5000	0.4960	0.4920	0.4880	0.4840	0.4801	0.4761	0.4721	0.4681	0.4641
0.1	0.4602	0.4562	0.4522	0.4483	0.4443	0.4404	0.4364	0.4325	0.4286	0.4247
0.2	0.4207	0.4168	0.4129	0.4090	0.4052	0.4013	0.3974	0.3936	0.3897	0.3859
0.3	0.3821	0.3783	0.3745	0.3707	0.3669	0.3632	0.3594	0.3557	0.3520	0.3483
0.4	0.3446	0.3409	0.3372	0.3336	0.3300	0.3264	0.3228	0.3192	0.3156	0.3121
0.5	0.3085	0.3050	0.3015	0.2981	0.2946	0.2912	0.2877	0.2843	0.2810	0.2776
0.6	0.2743	0.2709	0.2676	0.2643	0.2611	0.2578	0.2546	0.2514	0.2483	0.2451
0.7	0.2420	0.2389	0.2358	0.2327	0.2296	0.2266	0.2236	0.2206	0.2177	0.2148
0.8	0.2119	0.2090	0.2061	0.2033	0.2005	0.1977	0.1949	0.1922	0.1894	0.1867
0.9	0.1841	0.1814	0.1788	0.1762	0.1736	0.1711	0.1685	0.1660	0.1635	0.1611
1.0	0.1587	0.1562	0.1539	0.1515	0.1492	0.1469	0.1446	0.1423	0.1401	0.1379
1.1	0.1357	0.1335	0.1314	0.1292	0.1271	0.1251	0.1230	0.1210	0.1190	0.1170
1.2	0.1151	0.1131	0.1112	0.1093	0.1075	0.1056	0.1038	0.1020	0.1003	0.0985
1.3	0.09680	0.09510	0.09342	0.09176	0.09012	0.08851	0.08691	0.08534	0.08379	0.08226
1.4	0.08076	0.07927	0.07780	0.07636	0.07493	0.07353	0.07215	0.07078	0.06944	0.06811
1.5	0.06681	0.06552	0.06426	0.06301	0.06178	0.06057	0.05938	0.05821	0.05705	0.05592
1.6	0.05480	0.05370	0.05262	0.05155	0.05050	0.04947	0.04846	0.04746	0.4648	0.04551
1.7	0.04457	0.04363	0.04272	0.04182	0.04093	0.04006	0.03920	0.03836	0.03754	0.03673
1.8	0.03593	0.03515	0.03438	0.03362	0.03288	0.03216	0.03144	0.03074	0.03005	0.02938
1.9	0.02872	0.02807	0.02743	0.02680	0.02619	0.02559	0.02500	0.02442	0.02385	0.02330
2.0	0.02275	0.02222	0.02169	0.02118	0.02068	0.02018	0.01970	0.01923	0.01876	0.01831
2.1	0.01786	0.01743	0.01700	0.01659	0.01618	0.01578	0.01539	0.01500	0.01463	0.01426
2.2	0.01390	0.01355	0.01321	0.01287	0.01255	0.01222	0.01191	0.01160	0.01130	0.01101
2.3	0.01072	0.01044	0.01017	0.00990	0.00964	0.00939	0.00914	0.00889	0.00866	0.00842
2.4	0.00820	0.00798	0.00776	0.00755	0.00734	0.00714	0.00695	0.00676	0.00657	0.00639
2.5	0.00621	0.00604	0.00587	0.00570	0.00554	0.00539	0.00523	0.00508	0.00494	0.00480
2.6	0.00466	0.00453	0.00440	0.00427	0.00415	0.00402	0.00391	0.00379	0.00368	0.00357
2.7	0.00347	0.00336	0.00326	0.00317	0.00307	0.00298	0.00289	0.00280	0.00272	0.00264
2.8	0.00256	0.00248	0.00240	0.00233	0.00226	0.00219	0.00212	0.00205	0.00199	0.00193
2.9	0.00187	0.00181	0.00175	0.00169	0.00164	0.00159	0.00154	0.00149	0.00144	0.00139
3.0	0.00135	0.00131	0.00126	0.00122	0.00118	0.00114	0.00111	0.00107	0.00104	0.00100

Values of *z* for selected one-tail areas

Area	*z*	*Area*	*z*	*Area*	*z*	*Area*	*z*
0.00001	4.265	0.0001	3.719	0.001	3.090	0.01	2.326
0.00002	4.107	0.0002	3.540	0.002	2.878	0.02	2.054
0.00003	4.013	0.0003	3.432	0.003	2.748	0.03	1.881
0.00004	3.944	0.0004	3.353	0.004	2.652	0.04	1.751
0.00005	3.891	0.0005	3.291	0.005	2.576	0.05	1.645
0.00006	3.846	0.0006	3.239	0.006	2.512	0.06	1.555
0.00007	3.808	0.0007	3.195	0.007	2.457	0.07	1.476
0.00008	3.775	0.0008	3.156	0.008	2.409	0.08	1.405
0.00009	3.746	0.0009	3.121	0.009	2.366	0.09	1.341
0.00010	3.719	0.0010	3.090	0.010	2.326	0.10	1.282

Appendix C
Use of Current Cost Accounting

Throughout the late 1960s and the 1970s historic cost accounting (HCA) was increasingly criticized. The high rates of inflation over those years meant that the HCA system produced distorted statements, notably regarding profits and asset values (particularly fixed assets).

Several reports were produced in the late 1970s which discussed various proposals for dealing with inflation in accounts, the primary document being the Sandilands Report of 1975 which led to the introduction of CCA via Exposure Draft 18 (ED 18).

The Statement of Standard Accounting Practice 16 (SSAP 16) required current cost accounts to be produced (for all accounting periods commencing on or after 1 January 1980) by the following companies

(a) all listed companies, and
(b) all unlisted companies which satisfy two or all of the following criteria:
  (i) annual turnover of £5 million, or more,
  (ii) assets of *£21* million, or more,
  (iii) average number of UK employees 250, or more.

Certain organizations were exempted from the above; these included charities, wholly-owned subsidiaries of UK companies, property companies, building societies, investment and unit trusts, pension funds and trade unions.

Current cost accounting, under SSAP 16, relies on two principles:

1. that assets must be shown at their value to the business, and
2. that the profit of a company should be determined after the deduction of sums required to maintain the real value of the business

Clearly, great care is required to ensure the correct figures are presented.

Problems may arise in the calculation of profit, which is generally considered to be the amount of total gains arising in the year which, prudently, may be regarded as distributable. The correct provision for depreciation is more complex under CCA.

Naturally, in a period of transition in accounting bases, it is important to maintain comparability between sets of accounts. This was achieved under SSAP 16 by preparing accounts either on the HCA basis and providing CCA supplements or *vice versa*, the former method being rather more usual in which the HCA trading profit is subject to three operating adjustments (depreciation, cost of sales, and monetary working capital) to arrive at the current cost operating profit. A gearing adjustment (which allows for the proportion of assets financed by borrowing) is then applied to determine the current cost taxable profit. Generally, the operating adjustments are calculated by the use of index numbers from the annually published book *Price Index Numbers for Current Cost Accounting*, which is prepared by the Government Statistical Service.

The use of CCA requires frequent revaluations of fixed assets (at least once every five years), and of stocks. The basis of valuation of assets is 'value to the business', i.e. the amount the company would lose if it were deprived of the asset. Such value may be determined from:

(a) the amount which would be realized by the sale of the asset (e.g. sale of a building), or
(b) the cost of purchasing an identical asset (e.g. second-hand machine of the same age and condition), or
(c) the sum earned over the remaining life of the asset by holding and using the asset (e.g. a machine), expressed as a NPV.

Usually the value of an asset to the business will be its net replacement cost ((b) in the foregoing), unless this value exceeds both the net realizable value ((a)) and the economic value ((c)) of that asset; in such circumstances, the value to the business will be the greater of the asset's net realizable value and its economic value.

Depreciation provisions are determined from the valuations of assets, described above. An asset's depreciation in year n is its value to the business at the end of year $(n - 1)$ less its value to the business at the end of year n plus the asset's HCA depreciation in year n. Alternatively, an asset's accumulated depreciation is the deduction which must be made from its gross replacement cost to yield its net replacement cost so the depreciation provision for year n is the change in the deduction which must be made between the end of year $(n - 1)$ and the end of year n.

Stocks are valued at the year ends. By use of indices, the impact of price changes of stocks on the cost of sales is calculated (most simply, by averaging the indices - suitable where stock levels are reasonably constant).

Monetary working capital (trade debtors, prepayments, trade bills receivable and any stocks excluded from cost of sales adjustment, e.g. 'land banks' - *less* trade creditors, accruals and trade bills payable) is adjusted by index numbers to allow for changes in the finance required for monetary working capital due to alterations in the prices of goods and services used and financed by the company. The indices applicable are those for the finished goods the company produces.

Further details and discussion of CCA is beyond the scope of this text. The interested reader is referred to the references following.

References and bibliography

Accounting Standards Steering Committee (1976) *Current Cost Accounting*, Exposure Draft 18, November.

Accounting Standards Steering Committee (1980) *Current Cost Accounting*, Statement of Standard Accounting Practice 16, March.

Gilbert, D. (editor) (1976) *Guidance Manual on Current Cost Accounting*, Tolley.

Holmes, G. and Sugden, S. (1982) *Interpreting Company Reports and Accounts* (Second Edition), Woodhead-Faulkner.

Sandilands, F.E.P. (Chairperson) (1975) *Inflation Accounting. Report of the Inflation Accounting Committee*, HMSO.

Sizer, J. (1989) *An Insight into Management Accounting* (Third Edition), London, Penguin Books.

Appendix D
Recording Transactions

Shareholders

	15,000 (1)

Materials

(2) 4,500	4,000 (5, P&L) 400 (6, P&L) 100 Bal. c/d
4,500	4,500
Bal. b/d 100	

Small tools

(3) 500	50 (7) 450 Bal. c/d
500	500
Bal. b/d 450	

Buildings

(4) 21,000	420 (7) 20,580 Bal. c/d
21,000	21,000
Bal. b/d 20,580	

Wages

(5) 5,000 (6) 200	5,200 (P&L)
5,200	5,200

Cash

(1) 15,000 (4) 14,000 (5) 15,000	4,500 (2′) 500 (3) 7,000 (4) 14,000 (4) 5,000 (5) 1,000 (5) 1,000 (5) 100 (6) 200 (6) 1,000 (8) 9,700 Bal. c/d
44,000	44,000
Bal. b/d 9,700	

Creditors

(2′) 4,500	4,500 (2)

Debentures

	14,000 (4′)

Salaries

(5) 1,000 (6) 100	1,100 (P&L)
1,100	1,100

Plant hire

(5) 1,000	1,000 (P&L)

Work in progress

1,000	1,000 Bal. c/d
Bal. b/d 1,000	

Debenture interest

(4″) 1,400	1,400 (P&L) Bal. c/d
Bal. b/d 1,400	

Retained earnings

Bal. c/d 1,430	1,430 (P&L)

Sales

(P&L) 16,000	15,000 (5) 1,000 (6)
16,000	16,000

Depreciation

(7) 50 (7) 420	470 (P&L)
470	470

Dividends

(8) 1,000	1,000 (P&L)

Profit and loss account

Materials	4,400	16,000	Sales
Wages	5,200		
Salaries	1,100		
Plant hire	1,000		
Depreciation	470		
Interest	1,400		
Dividends	1,000		
Retained earnings	1,430		
	16,000	16,000	

Balance sheet

Assets employed	*(£)*	*(£)*
Current assets:		
Cash	9,700	
Materials stock	100	
Work in progress	1,000	10,800
Current liabilities:		
Debenture interest:		1,400
Net current assets		9,400
Small tools		450
Buildings		20,580
		£30,430

Sources of capital:	*(£)*
Authorized and issued ordinary shares	15,000
Debenture	14,000
Reserves:	
Retained earnings	1,430
	£30,430

Index